Fairy-Tales for Adults

Theories of the Earth in Disarray

By: Ben Tripp M. A. Sc., P. Eng.
Illustrations by; Carolyn Tripp B. A.
Cover by Carolyn Tripp B. A.

Fairy-Tales for Adults
Theories of the Earth in Disarray

By: Ben Tripp M. A. Sc., P. Eng.
Illustrations & cover by Carolyn; Tripp B. A.

Other works by the same author:
1. The Window of Life
   A Theory of the Earth based on Asteroid Impact
2. Elements of Providence during the Genesis Flood
   A Discussion of the Hazards that occurred during the great World-Wide Flood
3. Concerning the Birth of Christ
   A Discussion of the Timing of Christ's Birth
4. The Asteroid Theory of the Flood and the Ice Age
   The Necessity and the Sufficiency of an Asteroid Shower to Cause an Ice Age
5. The non-Myths of the Bible
   The co-relation between Scriptural Chronology and Nature
6. The Impossibility of Extraterrestrial Life
7. Too Much CO2
8. Time is Running Out
9. Climate Change and Holy Writ

More info at; http://benatripp.wix.com/window

About the author:

Ben earned both Bachelor's and Master's of Applied Science degrees in Engineering from the University of Waterloo, Ontario, and has worked as a consulting engineer on such projects as controls for large telescopes and test equipment for the CanadArm. He holds patents for innovations involving the recycling of used tires into fence boards and a novel ground coil arrangement for geothermal heat pumps.

Ben's interest in the current topic, and his related background reading and research span several decades and have culminated in what he purports to be a credible and cohesive discussion of eight theories of the Earth, four of which have received wide recognition while they shouldn't and four of which have not received wide recognition while they should. All discussion and all of the conclusions are based on a wide range of scientific evidence.

It is his hope that these observations and opinions will be helpful to many in their own investigations.

.

Tripp, Ben, 2016
Fairy-Tales for Adults
Theories of the Earth in Disarray

A bibliography has been included in the appendix as well as a list of references. A bracketed number following a word indicates a reference. For example, (46) refers to reference number 46. Also an index of key words will enable particular discussions to be found quickly.

Acknowledgements;

Wendy Speziali; formatting
Carolyn Tripp B. A.; illustrations
Carolyn Tripp B. A.; cover

ISBN   978-0-9936349-6-3 Soft cover version
ISBN   978-0-9936349-7-0 Electronic version

This work is dedicated to my family

To my dear wife:
Judith Anne
The love of my life

To my children:
Bryan, Rebecca, Daniel and Carolyn
The great blessing of my life

To my dear grandchildren:
Evelyn, Ayla, Zoe, Izzy and Ben

And to my bonus child:
Andrea

May this humble epistle assist them in their search for truth.

Table of Contents

.

Foreword

In spite of numerous scientific advances of recent time, several widely-accepted ideas have been retained which do not really have any solid scientific basis at all. Various elements of the scientific community promote some of these erroneous ideas and do so with great vigor – so much so that if anyone should dare to disagree, they are immediately pounced on with a vengeance. Atmospheres such as this are not conducive to real scientific advance. Fortunately, there are numerous areas of scientific endeavor which continuously rely of observation and measurement instead of abstract and unsupportable declarations. Also there are many very capable scientific workers doing excellent and advanced research and who expect that all of the others are carrying on in a similar fashion. 'If I'm doing good work here, I expect that Frank is doing good work there' is a reasonable expectation. When observations and measurements are being made in the present, this is appropriate and indeed a necessary assumption. However many areas of interest to scientists involve the ancient past or the distant universe. In these areas scientific models, or as an outsider might suggest, elaborate mathematical frameworks, have been constructed and present efforts are only addressing very small portions of the overall picture. This approach might even be a basic necessity because everyone cannot start from scratch and hope to make any progress at all. However if one particular overview of the universe becomes well entrenched, an individual worker would have a difficult time proceeding without trying to fit into it. Otherwise that person would likely be ostracized to the detriment of his/her career. Such developments only reinforce the prevailing overview whether or not it is valid.

However problems will certainly arise if the basic framework was faulty in the first place but it might require an outsider to recognize that this was the case. If, for example, someone is in the forest examining a particular species of tree, there isn't much point in telling him that the forest is quite small and that most of the land is covered by grass. What is grass? I'm a tree-man so the world must consist of trees. Why can't you see that? A scenario like this is far from unusual and extends into every sphere of scientific thinking. Specialists have the hardest time grasping that other valid disciplines exist even to the point that a medical specialist in concussion, might completely overlook that the patient had actually suffered a broken neck. In fact, exactly this type of error occurred quite recently when a hockey player was treated for concussion alone only to discover that a vertebra further down was fractured!

The answer, at least in part, might lie in the basic approach. In the following work the systems approach is used wherein all possible aspects of the situation are simultaneously recognized. Ironically, many scientists rely on others to look after the rest and just address one particular area. This could work as long as the systems approach is still recognized as valid and necessary but if there is something basically wrong with the overall systems model then something will certainly be wrong with the details.

Unfortunately, there must be something seriously wrong because major contradictions within several of the present theories of the Earth can be quite readily identified. For example, a currently-popular notion is that the Earth is very old and that it must have gone through a cooling-down phase where it changed from being entirely fluid to only partly fluid. However science tells us that if the Earth had ever been totally fluid, all heavy metals would have sunk to the very center of the Earth. Why then do we find them right up here on the very top as far from the center as it is possible to get?

Historically another situation involved Aristotle, a teacher of the previous era, who promoted a very particular view of the solar system and he would not entertain any other approach. His ideas held sway for almost two thousand years until Nicolaus Copernicus, Galileo Galilei, Johann Kepler and Issac Newton completely overturned his way of thinking. One wonders what progress might have been made earlier if he had listened to Aristarchus, his contemporary, who explained to him that the Sun was at the center of the solar system and that the Earth revolved around the Sun. At that time an estimate was made regarding the distance to the Sun as well as its diameter. These conclusions were reached as a result of careful observation whereas Aristotle's worldview was arrived at with a greater helping of reasoning than observation.

Similarly, the ideas of Galen, a Greek physician of the second century, while much more related to observation, held sway in the medical area for about a thousand years whereas some of them should have been replaced earlier. The same type of thing happened more recently during the time of Ignaz Semmelweiss, a Hungarian physician (working at the same time as Louis Pasteur) of the mid-eighteen hundreds, who was concerned that infection was being transmitted from patient to patient by the doctors themselves with the result that many patients died unnecessarily. In this case, years passed before the medical community took serious notice of his claims and observations.

These types of situations appear to us to be completely absurd so it will come as an even greater shock that the same thing is still happening. In many areas of modern endeavor, original independent thinking is not encouraged because it upsets the status quo. A very recent example of this relates to the Moon. It was long held in a very dogmatic manner that the Moon is very old and therefore must be cold and inactive. If you tried to report observations of the Moon that contradicted these assumptions, your papers would not be accepted for publishing and you would be ostracized from certain segments of the scientific community. In fact, this particular dogmatism prevailed right up until the astronauts went to the Moon and made direct temperature measurements. They found that the Moon is hot. It is also quite active with frequent lights and emissions of gas showing up. Thereafter, it became acceptable to report lunar transients.

Some currently-popular theories have been developed using computer programs! Using a computer to speed up or otherwise assist in making complex calculations could certainly be a powerful tool for research but one should be cautious about giving undue recognition to the results to the point of elevating them to the status of rigid dogma. The period of the Late Heavy Bombardment is in this category and is discussed as having actually happened throughout the solar system. Times are given even though the entire approach has been based on the notion that the solar system formed from a collapsing cloud of space dust - possibly one light-year across. Recent observation shows that this could not have happened because there are now numerous extra-solar systems known that have large planets very close to their host stars. This means that the idea of a collapsing gas cloud should have been set aside, thereby invalidating the idea of the Late Heavy Bombardment period, but instead both ideas are still firmly entrenched with their various conclusions still firmly in place. But just to emphasize the point, the Late Heavy Bombardment is only a computer program so the results should be treated with some degree of caution.

There is no doubt that when a basic long-held-to-be-true idea is shown to be false, it is very difficult to set it aside. If this was done with some ideas, the entire framework could come crashing down. How many post-graduate degrees have been earned that would consequently have no scientific validity? Mountain formation is included in this category. If all of the mountains of the world formed within a very short time (i.e. less than one year) all of the currently-popular ideas related to mountain formation would not be valid. Most schools of geology would be devastated by any such development and if you, as a student, tried such an approach you would never get your advanced degree. The status quo supports the status quo. Do not even bring up such wild ideas or you will not be allowed to proceed at all. However, the evidence shows that all of the mountains of the world did form within a one-year time period. Only an outsider who is not the least bit concerned about getting a degree could ever make such a claim. However the scientific evidence supports this conclusion which therefore cannot be honestly avoided. (see discussion in 'The Window of Life')

Another popular idea is that the Earth became chilly and an ice age developed. In fact, the chill is declared to have lasted for thousands of years. It even seems to be the maxim that 'a chilly Earth is an Earth having an ice age' is valid. However, it is not the least bit valid. Any number of books on climate will clarify that if the Earth ever became seriously chilled on a prolonged world-wide basis, it would not recover. It would become ice-bound instead. If such a development ever happened, it would be irreversible. Why then, would the 'chilly Earth is an Earth having an ice age' idea persist in the light of modern climatology? This reality from science also makes it hard to understand how an ancient Earth could have escaped the deep freeze in the first place because if the Earth is actually 4.5 billion years old, it would have existed when the Sun was much dimmer. The Earth would have been frozen solid from that factor alone because the Earth would not have been in the habitable zone of that ancient Sun. It is inexplicable how it could ever have thawed out because it is well understood that it is the combination of heat from the Sun AND the greenhouse gases that keeps the surface temperature in the habitable range at the present time, even though the Earth is well within the habitable zone of the Sun. What would the temperature have been if the Earth was frozen and water vapour, the primary greenhouse gas, was not available? Science is therefore telling us that there is something seriously wrong with the whole notion of an ancient Earth.

Then, there is the problem of the receding Moon. It is the Moon that holds our angle of inclination just right and does so because of its very particular orbit and distance. Without those features being in place, the Earth would not be habitable – even if it was exactly in the middle of the habitable zone of the Sun – where it most fortuitously seems to be.

The primary hope for any idea that runs against the prevailing dogma, is for it to be advanced by someone who is independent of the need to conform. Independent journalists are in this category as well as scientists who do not require the approval of their colleagues. Immanuel Velikovsky is included in that particular set of writers but even then there was fierce opposition to his writings to the extent that publishers were threatened with not being provided with any more manuscripts if they dared to publish his work. Of course this would upset certain publishers but not all and a publisher was found who wasn't the least bit concerned about such threats. In fact it turned out to be good publicity in the long run.

It could also be the case that an idea doesn't really have any scientific credibility so it probably shouldn't be published anyway. However, in the light of so many works that are offered as science but really aren't, this seems like the lesser of two evils. Unfortunately, the Theory of Evolution is in this latter category but it has become so well entrenched that to suggest that it really isn't a scientific theory at all, is treated with derision and contempt. Unfortunately for its proponents, DNA has been discovered and clearly reveals that the massive genetic gaps from one creature to another make any supposed step from one to the other a leap of faith rather than a scientific reality. Further, there is no suggestion at all concerning how enough complexity developed to get life started in the first place.

It really seems that we have an over-load of contradictions in place. The conclusions reached herein may therefore not be seen as so much of a surprise when they completely contradict so many commonly-held beliefs. In any event there will be no apology. An honest truth-seeker must follow the evidence no matter where it leads him and not really worry too much about what might be popular - particularly when popularity today might not be popularity tomorrow.

Several ideas are discussed herein and in certain cases the same information and argument is relevant to more than one idea. In some of these cases the discussion has been repeated to avoid the necessity of returning to some other section to recover the ideas.

1.0   Introduction

The discussion to follow includes eight ideas. Four of them are widely thought of as being in the fairy-tale category but they aren't and four of them are considered valid scientific ideas but they aren't. One wonders how such a development could ever have happened as all eight of the ideas presented are so diametrically-contrary to currently-popular opinion. However, based on observation and sound scientific principles, all eight of these ideas are shown to be the exact opposite of what is widely-held to be the currently-proper view.

All discussion that is offered herein in support of this upheaval is based on observation and a proper understanding of basic physics and virtually every point made is supported by a reference. In other words, all discussion herein recognizes a great number of commentators working in several different fields of scientific endeavour. Therefore, from a scientific perspective, all conclusions reached herein have been properly referenced and supported.

Since the conclusions reached about these eight ideas are the exact opposite of currently-popular belief, it might be tempting to set them aside. However, it is the author's hope that the plain evidence from nature will be given an honest hearing before this is done.

The first idea discussed is entitled 'The 360 Day Year'. It is well understood that the current length of the year is about 365 ¼ days and not 360 days. After-all this is a repeatable measurement of excellent scientific quality and should only be challenged after considerable investigation. However, it isn't being challenged herein for the present time – only for the ancient past. As is always the problem when dealing with the ancient past, direct measurements cannot be made. Other types of evidence must be used instead. In this case it is recognized that numerous ancient writings have explicitly stated that the year was 360 days long. It can always be suggested that proper measurements were not actually made during those times but this can never be verified - only hypothesized. If a measurement cannot actually be made, then trying hard to challenge it by sounding authoritative is the only option. However, the ancient reports exist so one is left with the option of accepting them as declared or rejecting them outright. However once rejection begins, where does it stop?

If the length of the year was actually 360 days at some time in the past it means that the Earth was orbiting a little closer to the Sun than it is now. So the question then shifts to the possibility that the Earth once actually had a different orbit. Being a little closer to the Sun would mean that the Earth received a little more heat and being 1% closer (which would have been the case with a 360 day year) would mean that about 2% more heat was received. This would still place the Earth within the habitable zone of the Sun which is commonly given as plus or minus 5% but in reality probably isn't this much. It is quite likely that it is closer to plus or minus 2%. However 1% closer would have been quite tolerable because the Earth, on the average, would have been a little warmer. So temperature would not have been a problem but trying to move the Earth into a different orbit certainly would have been a problem. This is probably the main reason that the idea is usually dismissed outright even before it is seriously examined.

However, the Earth has been pummelled with numerous large asteroids - every one of which would have changed its orbit by some degree. Therefore 'how much' is the question not whether or not it happened. This reality shifts the idea from being in the 'outright fairy-tale' to the 'serious-possibility' category. The mechanism to achieve orbital change is examined herein along with the likelihood of this having happened.

Next, the great antiquity of the Earth is discussed. It is commonly and widely held that the Earth is 4.5 billion years old. Of course ideas like this cannot really be called scientific because we cannot return to the ancient past. Instead, ideas are put forward and theories developed to support the viewpoint so the question shifts from being directly addressable to only being indirectly addressable. In other words, either a theory or at least a well-developed idea must be invoked. Unfortunately, if that idea is faulty then the original hypothesis falters. This is exactly the case with the ancient-Earth idea.

The theory that is currently being relied upon in this case is that the solar system formed from a cloud of space dust that collapsed. As it neared the final stages of collapse the planets were formed and the Sun became hot enough to ignite. This wasn't ignition in the usual sense but rather nuclear ignition where the material became so hot that hydrogen atoms were able to fuse together to make helium. Needless to say that would have been very hot and difficult to achieve. Unfortunately, it is not natural for a cloud of gas to collapse and even more difficult to get it hot enough to ignite. Whenever anything gets the least bit warm, it will immediately lose heat. Consequently the cloud of gas had to get hot in such a way that it didn't lose too much heat or the entire scheme would never have worked.

The idea that the Earth was entirely fluid at one time is even more questionable. If it was, continental crust, which is slightly lighter than the rest of the Earth, would have simply spread out evenly over the whole Earth. No explanation has been offered as to why it would clump up and stick out of the water. It should simply have spread out! Even more detrimental to the whole idea is the fact that several very heavy metals are found right up here on the surface. How could this be? If everything was molten, these heavy parts should have sunk right to the center of the Earth. How is it that they are found here on the top? Perhaps the crust never was molten in which case the billion years required to cool down wouldn't be needed.

The Sun also gets into the act because an ancient Sun would have been much cooler than the Sun is now. It is only the present heat from the Sun together with the greenhouse gas effect that barely keeps the Earth in the comfort zone at the present time. A much cooler Sun would have meant that the Earth would have been frozen solid! In fact it would still be frozen solid because the greenhouse gas system would never have been able to get established and contribute to our currently-acceptable temperature. If one is trying to establish the idea that the Earth is very old, they cannot pick and choose the ideas that will be supportive without recognizing all other factors that are also relevant. This has not been done, so the theory that the Earth is very ancient should be set aside.

The Astronomical Theory of the Ice Age is in this same category because it ignores valid scientific factors as well as basic physics. The Earth could never have been subjected to a prolonged world-wide chill because all of the water would have become crusted over with ice and the greenhouse gas effect provided by water vapour would have been lost! There cannot be any exceptions to this. These are simply basic physical realities and any theory of the Earth that is offered must acknowledge them.

On the other hand, The Asteroid Theory of the Ice Age does acknowledge these factors and declares that the Great Ice Age was only able to happen because the necessary amounts of heat and cold simultaneously became available. At the same time, the presence of both heat and cold meant that the greenhouse effect would not be lost. In fact, any theory of the Earth that is offered absolutely must recognize this necessity.

Beringia, an Ice Age Serengeti, is shown outright to be completely in the fairy-tale category. How thousands of animals of any kind could have thrived while an ice age was in progress truly strains the imagination. It would have been dark throughout the entire build-up phase of the ice age and it would have been cold. The snow was piling up until there was enough of it to form ice, thousands of feet thick. An animal has never been born that could have survived this type of trauma. They not only had to survive but when it was all over they had to bury themselves deep in the ground (which must have only then became frozen) because that is where they are presently found. This would have been a most amazing achievement! The basic conclusion is that there never was an Ice Age Serengeti.

Life in outer space has been an intriguing idea for a long time and it seems to have gotten even more attention during the last few years. The argument usually goes something like 'there are so many millions of stars in the universe that surely some of them must have intelligent beings like ourselves'. However, as soon as the matter is investigated at even the simplest level, it is apparent that there really isn't much chance that life exists elsewhere. In order for life to exist numerous life-enabling factors are required and they must all appear in one appropriate place at one particular time. Not the least of these factors is that the Earth absolutely must have an appropriate Moon at the right distance. However, even this criteria can only be met for a relatively brief period of time because tidal action gradually moves the Moon further away from the Earth and slows the Earth's rotation so that it will not be able to provide it's absolutely-necessary Earth-stabilizing function for very much longer. (i.e. 20,000 to 50,000 years) When that factor is lost, the Earth will no longer be habitable, even though it will still be in the habitable zone of the Sun!

The last two ideas discussed herein will be even less familiar to most people. It is widely held that asteroids have hit the Earth from time to time over extreme periods of time. It is pointed out however, that this could not have happened but that most of them had to have arrived as a shower instead. The basic reasoning for this is that a single major asteroid hit would have sent the Earth into the deep freeze - recovery from which would not have been possible. Whenever a major asteroid hits the Earth, as is explained by any number of knowledgeable commentators, the Earth suffers from an asteroid winter. This is because the impact produces a globe-enveloping cloud and the resulting chill would kill most plant life and freeze most bodies of water. Hoar-frost, snow and ice would cover the Earth. The cloud would block life-enabling sunlight and then when the cloud dissipated after a year or two, the surface of the Earth, being covered by reflective material (i.e. ice and snow) would turn away the heat from the Sun. The greenhouse gas, water vapor, would have become seriously diminished and with it the warming effect which is absolutely necessary for the Earth to be habitable. Perpetual winter would result.

Instead, the asteroids must have come as a shower, as they did on Mars and the Moon because in that case a balance between heat and cold could have been maintained. Consequently the greenhouse effect would have been retained and with it, the habitability of the Earth.

The eighth and final idea discussed is Instant Creation. This certainly sounds like a religious idea and it might very well be, but nature provides us with evidence that it actually happened. There is no doubt that this will be unwelcome news for anybody who does not like ideas that sound religious. Science supports the notion however so perhaps another look should be taken before the matter is closed. Since the evidence from nature indicates that it is not a fairy-tale a serious truth-seeker might decide to pause and reconsider the most prevalent view - that the idea is a religious fantasy - whereas the evidence says that it is a scientific reality.

2.0 The 360 Day Year

The current length of an Earth year is very close to 365 ¼ days. There is a possibility that this was not always the case and that in the distant past it was 360 days instead. Changing the length of the year is a very tall order and would be difficult to achieve. Neither do we particularly want or need a different length for our year because of the upsetting effect that it would have on the temperature of the Earth. Having the correct temperature as well as the proper temperature control systems in place is paramount to our existence and it would be hazardous to modify any control factor by even the smallest amount. In order for life to thrive, the average surface temperature of the Earth must be held to within a very narrow temperature range and deviation either way from where it is at the present time would be disasterous. This is, in fact, the reason for the current concern that the temperature might be increasing. While it is not increasing dramatically, the idea that it might be increasing at all is perceived as a matter of grave concern. On any temperature scale one or two degrees is not very much but it is clear to the scientists that study these factors that there would be widespread disruptions in weather patterns as well as the ability of the Earth to produce food for the several billion people that live here if the temperature changed by even such a small amount. There is even considerable optimism that if an increase can be held to two degrees the disruption will be manageable.

A change in the length of the year from 360 days to 365 ¼ days might be seen as a tempest in a teacup because such a small amount doesn't really seem to be significant. However a change in the length of the year would require a change in the location of the Earth's orbit. If the orbit of the Earth changed, there would necessarily be a change in the amount of heat that the Earth received from the Sun. A change in temperature would follow and the concern therefore focuses on whether this change would be tolerable. This type of reasoning follows immediately from the recognition that the Earth currently has a 'Goldilock's' orbit. (186) It is 'just right' so any thought of a change is upsetting. The orbit of the Earth is actually slightly elliptical which means that during the course of a year the Earth moves slightly further from the Sun as well as slightly closer to it. However even this variation works in our favour because of the tilt of the Earth's axis. The tilt gives the Earth the seasons which are not only favourable but necessary to make most of the Earth habitable. The tilt together with the slight ellipticity of the orbit results in an optimal arrangement for heat distribution so we do not want to upset this arrangement with any change in the length of the year. However there is a serious possibility that this has happened, which means that other factors involved in heat distribution and temperature control must also have been modified. If the Earth had ever been a little closer to the Sun, it would expectedly have also been warmer.

2.1 Ancient People

Numerous groups of ancient people recognized the year as having 360 days. This includes the Hindus of India who had the year divided into 12 months of 30 days each. Further they 'describe the Moon as crescent for 15 days and waning for another 15 days; they also say that the Sun moved for 6 months or 180 days to the north and for the same number of days to the south.' This was not just an error because there are other comments that corroborate these times. '... that the Sun remains 13 ½ days in each of the 27 Naksatras, and thus the actual year was calculated as 360 days long.' (8) Certain commentators concluded that these people had a 'wholly confused notion of the true length of the year'. (8) Unfortunately for this type of conclusion, other groups in the ancient world recognized the 360 day year as well. Were all of them also confused? Quoting from other ancient works Velikovsky included the following; '... a passage from the Aryabhatiga, an old Indian work on mathematics and astronomy: "A year consists of twelve months. A month consists of 30 days."' While the above comments were recovered from ancient Indian works other peoples in other regions had the same idea. For example the Persian year 'was composed of 360 days or 12 months of 30 days each'. Similarly the Babylonians understood the length of the year to be 360 days long with comments such as ' ... the walls of Babylon were 360 furlongs in compass, "as many as there are days in the year". The Assyrians had similar convictions. 'The Assyrian year consisted of 360 days; a decade was called a sarus; a sarus consisted of 3600 days.' The Hebrews observed lunar months and a month was always 30 days long with no suggestion at all that it should be a little longer. The Egyptians recognized the year as being 360 days long and various documents discovered in Egypt note that an extra 5 days were only added at a much later time. The Chinese had a 360 day year, 'divided into 12 months of 30 days each.' (8)

It is apparent that many ancient groups of people recognized the year as being 360 days long. This has perplexed certain commentators of more recent time but their perplexity only reinforces the general recognition concerning what the ancients believed. While those ancient people did not have telescopes or any other modern instruments they did have various measuring devices and no shortage of intellectual ability. We recognize them as very careful observers so it seems improbable that all of the peoples mentioned above were in error. In fact there would be serious consequences involved in setting up a year that was too short. Before very long the months would shift so far that a winter month would occur during the summer and so on. In fact, a calendar that wasn't accurate would be useless. This too was recognized by all of the ancients who were forced to add 5 ¼ days to the year at a more recent time. In some cases this was so upsetting that they saw the extra time as not even being part of a proper year so they just did nothing for a few days until a new year started. The extra days are not easily divided into anything because even the length of a month must be adjusted so when the reality of the modified length of the year set in, calendars were adjusted but the new arrangements were not really welcome. (8)

2.2 Energy Requirement

In order to change the orbit of the Earth a great amount of energy would be required. This is simply because the Earth has a lot of mass and is therefore very difficult to move. Alternately, since the Earth is already moving it would be difficult to change that movement. This is a good thing because we really do not want the movement of the Earth to be easily changed as that could readily lead to our peril. On the other hand, the Earth is not tied down so if enough energy became available the movement of the Earth would change. In this case the change of interest relates to the difference in energy that the Earth would have if it had a different orbit.

The energy of any object in orbit around the Sun is directly determined by its mass and its speed. The speed is set by the distance from the Sun. While the Earth orbits at a certain distance from the Sun, any object, large or small, could also orbit at the same distance and its speed around the Sun would be the same as the present speed of the Earth. Even a baseball could orbit the Sun indefinitely at the same distance as the Earth and its speed would be the same as the Earth's speed. In order to change to any other orbit, the speed of an orbiting object must be changed but rather than think in terms of changing the speed, we must think in terms of changing the energy. This is necessary because if an object is to be placed in a higher orbit, while its speed would actually be reduced, its energy would be increased.

In particular, if the Earth once had an orbit with a year that was 360 days long it must have been in a lower orbit. In order to get it into an orbit that was 365 ¼ days long, it would have to be pushed out farther from the Sun. Energy would have been required to do this and any impacting object, either asteroid or comet, that impacted the Earth from behind, would increase its orbital energy and nudge it into a higher orbit.

The total amount of energy that would have been required has been calculated and is included in the Energy Table. (see Appendix) Therein the total energy of the Earth with both a 360 day year and a 365 ¼ day year has been shown. The difference in these two entries is the amount of energy required to change the orbit and hence the length of the year. It goes without saying that this difference really is a lot of energy. Both asteroids and comets have a lot of energy so the question boils down to the matter of size and speed. Both of these can be conjectured of course but a more direct approach recognizes the energy that an asteroid (or group of asteroids) would have if it (they) was (were) orbiting the Sun out in the main asteroid belt. This belt occurs between Mars and Jupiter where the vast bulk of asteroids orbit. There are also numerous wanderers in both directions. Some cross Earth's orbit (Apollos). At least one even shares Earth's orbit as a Trojan (13) Another orbits the Sun with an orbit that synchronizes with Earth's orbit with an 8:5 ratio and simultaneously with Venus with a 13:5 ratio. (13) Similarly there are several asteroids that share Jupiter's orbit and a few have orbits that take them beyond Jupiter. A reasonable approach for the present discussion would be to consider one that was orbiting the Sun about half-way between Mars and Jupiter. The Energy Table includes an entry for such an object, the mass of which has been conjectured. The total energy follows immediately and has also been included in the table. If such an object (or aggregate group of objects) could have been caused to move inwards closer to the Sun, which either an explosion or a collision could have effected, some of their potential energy (energy due to their distance from the Sun) would change to kinetic energy (energy of motion) as they got closer to the Sun. That is, they would speed up. By the time they got to Earth's orbit, they would still have some potential energy but the rest would be kinetic energy. Their speed would have increased and by then they would be moving very fast. The asteroid entry in the Energy Table has been chosen because an object (or group of objects) with that amount of total mass would have enough energy of motion when it reached Earth's orbit to nudge the Earth into a higher orbit. In fact there would be enough to raise the Earth's orbit from a 360 day year to a 365 ¼ day year. The total mass of all of the objects that would have been required to do this is about 0.52% of the mass of the Earth. If we could assume for a moment that the object's average density was similar to the density of the Earth then its size can be estimated. The size that such an object would have would require that a single object would be about 800 miles (1280 km) in diameter. (i.e. about 10% of Earth's diameter) This would be equivalent to eight asteroids about 400 miles (640 km) in diameter. These would certainly be monsters by any comparison and there would understandably be great reluctance to accept such a scenario. However before jumping to a conclusion either way it would be instructive to review the evidence.

There are several ways to cross-check this result. One way is by comparison with another planet. A second way is to review the impacts that have happened on the Earth. A third way is to recognize the remaining inventory of asteroids. The connecting link among these three approaches is the relationship between asteroid size and the specific size of the crater that it would produce on the surface of the Earth. Also the different amount of incoming solar energy that would accompany a different Earth orbit is most relevant to our discussion.

2.3   Asteroid Formation

There are currently three ideas being recognized as the possible origin or the asteroids. None of these three ideas are universally accepted as might be expected but there seems to be considerable support for the location of the formation event which is between Mars and Jupiter. The reason is simply that this is the location where most of the remaining asteroids are still located. Formation has variously been suggested as being caused by an explosion of a small planet or by the impact of some object with a small planet. The other occasionally-suggested explanation is the lack of formation of a planet in the first place with the asteroids being the material that might have formed that missing planet. This third alternative loses credibility when we notice that many of the remaining asteroids both large and (really quite) small are perfectly spherical. If an object is found in space in a perfectly spherical state it must have been cast into space in a liquid state. If a spherical shape is to develop from a group of irregularly-shaped rocks the entire mass of such material must form a sphere almost as large as the Moon. Otherwise the self-gravity of such a mass would not be great enough to bring about a spherical shape. Some of the spherically-shaped asteroids are only a few hundred meters across which means that they do not have enough material to have become spherical by their own gravity unless they were liquid and obtained a spherical shape before hardening. In fact such small objects have hardly any gravity. We also notice that many asteroids are just chunks of irregularly-shaped material suggestive of being possibly the crust of the original planet. Finally, it has also been noticed that many appear as blobs of material which suggests that they were either molten or semi-molten when they were sent out on their own. Then they cooled down before their self-gravity could make them spherical so they retained their irregular semi-molten appearance. It must also be mentioned that the two moons of Mars are irregularly shaped and have this same appearance. They also have numerous small impact sites which could not have happened if they had been solid when the small impactors arrived because the gravity of these moons is too small to cause anything to approach at more than a very slow speed. While the asteroids do move at considerable speed, most of them are moving in the same general direction which would have been the case after either an explosion or an impact. These observations effectively rule out the idea that the asteroids are objects which never formed into a planet. On the other hand they are all supportive of the suggestion that they are the result of a small planet being annihilated.

2.4   Asteroid Size/Crater Size

Strictly speaking, only small asteroids form craters – a bowl-shaped excavation in the ground. Such excavations are usually found in unconsolidated material like soil or sand or gravel but they can also be found in solid rock. For example, the Arizona Crater is a bowl-shaped excavation about 4000 feet across and about 600 feet deep in unconsolidated material. (133) An example in solid rock is the Brent Crater of Southern Ontario which is about

10,000 feet across and about 3000 feet deep. (3) In this case the remaining 'crater' is only a shallow portion of the actual crater which is filled in with broken rock and sedimentary rock - or was possibly never excavated in the first place because most of the fill-in material is broken pieces of the underlying rock. It is quite possible that the rock all the way down to the bottom of the 'crater' was basically broken in-situ by the shock-wave that spread out from the ground-zero location. While the rim is clearly and obviously visible the bowl shape in the rock only became evident when several boreholes were drilled into the surface. When this was done there was no mistake that a bowl-shaped opening in the rock had been formed during the impact. A further example of the classical bowl shape has been identified with the Manson Crater in western Iowa. As with the first two mentioned, a bowl-shaped formation has been identified (311) (which is about 29 miles across and about 3 miles deep) and similar to the Brent Crater, it too has been filled in. The further distinction in the Manson case is that the fill-in material is mostly sedimentary rock. Sedimentary rock is rock, the material for which has been placed by water. In order to form a crater this large, compared to the two mentioned above, would have required a much larger object. The excessive energy would have blasted material completely away from the site. Then later the region must have been over-run by great water flows which deposited the material for both the sedimentary rock and the loose unconsolidated material on top. Now the area is perfectly level and the existence of a crater would never be suspected from any surface feature.

With these types of craters, since there is a definite shape involved, there is a reasonable possibility of identifying how much energy would have been required for their formation. Speculation would still be involved but at least there is something measureable involved. For craters larger than this, the possibility of identifying the classical bowl shape disappears, and with it any measurable rationale or relationship between the crater diameter and the size of the impacting object. Speculation must therefore take over and there seems to be no shortage. For example, for the Chicxulub Crater in Mexico it has been declared that it was 120 miles across and 30 miles deep. (136,5) While the diameter is measurable the depth is not and therefore to declare that it is '30 miles deep' is simply speculation. It is speculation based on using the comparative dimensions from smaller craters. Unfortunately there isn't any way to measure 30 miles deep. There has never been a borehole that deep. While we cannot readily drill that deep neither can we reliably tell what is down there by any other method. Other speculation has adamantly concluded that the Earth consists of various layers but the Kola borehole put an end to that idea because the drillers did not discover any layers. (6) The further difficulty with such a great depth is that we are at, or approaching, the underside of the Earth's crust. When that happens, all speculation concerning depth must be set aside. If the Earth consisted of solid rock, extrapolating from small impacts to large impacts would have some credibility. However the Earth is not solid rock but rather it is a large fluid sphere enclosed in a thin solid crust. While it isn't really possible to specifically determine the thickness of the crust there are several indicators that tell us that it is not very thick. One of these is the borehole just mentioned. The temperature at the bottom was well on its way to incandescence which means that within another few miles the rock would have been glowing dull red and its viscosity would be dropping.

Another indicator is volcanic activity which clearly indicates that the interior consists of red-hot low-viscosity molten rock. Every now and then some of this material comes to the surface and can be seen to flow quite readily. It has variously been suggested that the crust under the ocean might only be 3 km. thick (1.9 miles) and that the crust under the continents might be ten times this thick. When such dimensions are compared to the size of the Earth they are really quite small. More importantly, they are also quite small when compared to the size of the 'craters' at many impact sites. For example, the Vredeforte Crater in South Africa is given as being 300 km. (188 miles) across. Even if the crust was 20 miles thick at that location it would only be 10% of the crater diameter and if it was 40 miles thick it would only be 20% of the crater diameter. Conventional speculation suggests that the size of the asteroid that caused this crater would be in the region of 10 to 15 miles in diameter. However this would be a significant fraction of the crust thickness. The obvious question must therefore be asked. Would a crust 20 miles thick be able to stop a speeding asteroid that was 10 miles in diameter, from punching right through into the interior? The answer is equally obvious – no it would not. It is therefore submitted that once the size of an asteroid becomes a significant fraction of the thickness of the crust of the Earth, the possibility of relating the asteroid diameter to the crater diameter, in the conventional manner, is lost. It is further offered that the situation would be more comparable to a hammer and punch than it would be to a splash type of landing. When a hammer drives a punch into a piece of material, it is the shear strength of the material that resists the entry of the punch. As soon as the impacting force overcomes the shear strength, a hole will simply be punched out. When the hammer is brought down hard enough, a hole is formed which is simply the diameter of the punch. The punched-out-material is pushed ahead of the punch into the under-lying material or space and a tubular opening is formed where there was once solid material. Something similar would happen on the Earth if a large asteroid hit it. The asteroid would punch a hole in the crust and most of the punched-out material would be pushed into the interior. This would happen so fast that a temporary tubular opening would be formed through the crust and into the interior and the size of the 'crater' would be about the size of the asteroid that formed it. Asteroid energies are so very high that once the asteroid size becomes a significant fraction of the crust thickness, the crust would not be strong enough in either shear or bending to stop it so it would punch right through into the interior and thereby, because of the shockwave that would form in front of it, distribute its energy throughout the entire world.

We understand that as a sphere gets larger its volume also gets larger but at a much faster rate than its diameter. Suppose that we have an incoming asteroid with a diameter that is 25% of the thickness of the Earth's crust. An object like this would be very hard to stop and we can imagine a terrible crashing and shattering of both asteroid material and crustal material. Now we further suppose that this asteroid had just enough momentum to shatter the crust all the way through so that a large chunk of material was pushed ahead of the remainder of the body of the asteroid and on into the interior. Now, suppose that we have a second asteroid with a diameter that is 50% of the crust's thickness. The perimeter of crustal material that would be punched out is now twice as extensive as before which would supposedly require twice as much force to shear/shatter it. However because of the cube law of volume, the mass of this larger object would be eight times as great as the first one which means we have proportionately eight times as much displacement force, energy and momentum available as in the first case. If the smaller asteroid could just barely get through, this larger one will go through without any trouble and the entire mass of asteroid and punched-out crustal plug would dive into the interior with considerable speed. This argument can readily be extended. If an asteroid was as large in diameter as the thickness of the crust could anyone expect that the crust would stop it? What if it was twice as great in diameter as the crust thickness? It would be completely reasonable to expect that such a monster would quite readily punch through into the interior and leave a punch-hole in the crust. However the opening/crater would only have a diameter that was twice as large as the thickness of the crust whereas the Earth has craters that are much larger than this. If fact, some of them appear to be more than ten times as great (in diameter) as crust thickness. What are we to make of this? The only logical conclusion is that many of the asteroids that have hit the Earth were much larger than is popularly believed.

The other way to approach this problem is from the time perspective. It takes time for material to accelerate unless it is mechanically forced in which case it must go along with the mover. This phenomenon can be appreciated when we consider the effect that a speeding asteroid would have on the atmosphere. An asteroid plunging through the atmosphere encounters air. It comes upon the air so fast that the air simply piles up in front of the asteroid. The pressure of this air accumulation becomes very high and since air is gas, the temperature will also become very high. Various commentators have recognized this effect and declare that the air would become so hot that it would produce a 'blinding flash'. (7) It is well understood that air is much easier to move than rock. It may therefore be concluded that if air cannot get out of the way of a speeding asteroid, how would the rocky crustal material of the Earth get out of the way? If the crust of the Earth was not able to reduce the speed of the asteroid, the material in front of it, whether it is air or rock would not be able to get out of the way either and so would precede the asteroid until it slowed down enough so that all accumulated material could move aside. For a large asteroid this would not happen until the asteroid was several diameters of itself into the interior of the Earth. In the meantime a temporary tubular-shaped opening would develop in the Earth and extend down into the molten region for several asteroid diameters. The opening would soon refill with molten material which would be expected to crash into the opening and surge upwards above the surface of the Earth. Repeated surging might create multiple rings or apparent crater boundaries. Repeated surging could therefore explain the multiple rings of the giant Vredeforte Crater of South Africa. This type of feature could only develop if the crust was very thin in comparison to the crater diameter.

A similar situation has been observed on Mercury. There the giant Caloris Basin has definite features indicating that the crust of Mercury is very thin and that the molten interior is not very far below the surface because the rim consists of concentric rings that froze in place after the impact. (152) This means that the assembly of material that would form the rings, had to be on the very verge of freezing even as it was being put in place. Only a mixture of solid material and molten material could have this feature enabling these formations to develop. In consideration of the thinness of the crust of Mercury in comparison to the diameter of the crater, it is apparent that the object that formed it would have had a diameter that was a significant fraction of the crater diameter.

One commentator has declared that an asteroid would be moving so fast that it would penetrate to a depth twice its diameter within 1/100 of a second. (546) While this commentator might not have been thinking about very large asteroids, his comment never-the-less illustrates the fact that very little time is available to stop an asteroid from punching right through into the interior.

Molten rock has a completely different structural characteristic than solid rock. While solid rock has shear strength, compressive strength and bending strength, molten rock has none of these characteristics but would simply deform in response to gravity or to allow entry of a solid object and it would not seriously resist the entry of a large asteroid.

The deepest borehole into the crust of the Earth was made in Russia on the Kola Peninsula near Finland. While the plan was to drill down to a depth of 14 kilometers the drilling operation was forced to stop at about 12 kilometers because the bit was getting too hot at 180C. (547)(6) While the bit would have been able to withstand temperatures well above boiling there would have been a limit, beyond which, the structural integrity of the steel in the bit would be compromised. The rock was hot at that level and getting hotter as the depth increased. The drilling crew did not quit in haste. Since they had been drilling for about 20 years it is more than likely that they would have continued for another couple of years without flinching. (6)

While different types of rock will become red hot at slightly different temperatures it is well known that clay begins to glow red at about 700F. At 1200F it will be glowing a bright cherry red. Within another few hundred degrees structural integrity will be completely lost. Something similar would happen with the rock comprising the crust of the Earth. At some particular depth – which presumably varies from place to place – the structural integrity of the rock would diminish and it would no longer have shear strength, bending strength or compressive strength. It would become malleable and would not be able to hold any particular shape. At such a depth, rock in such a condition would not be able to offer any real resistance to the entry of a large asteroid. It would be like throwing a stone into a bowl of syrup. The stone would just sink.

When these realities are applied to the crust of the Earth with respect to the entry of a speeding asteroid, it is clear that probably beyond the 10 mile depth the rock would be too hot to offer any significant shear resistance to an asteroid. Even if this critical depth was 15 miles it is readily seen that we have only barely reached the depth that is comparable to the speculated diameter of the incoming body. A crust with a thickness that is comparable to the diameter of an incoming asteroid would not be able to stop it before it dove deep into the interior. Someplace in the interior it would slow to a halt but would probably have gone several diameters of itself into the interior before this would happen. The asteroid would eventually come to a stop but the shock-wave that formed ahead of it would not come to a stop but would carry on into the interior at very high speed. The relationship of the crater size to the asteroid size therefore is not nearly as well defined as it would have been with an asteroid that was only a small fraction of crust thickness in diameter. Therefore, for large asteroids the possibility that the crater size and the asteroid size are more closely matched is much more likely.

A gradation in the crater size/asteroid size relationship from the classic bowl shape to the punched-hole shape is a reasonable expectation and this is what is observed. As asteroid size increases, craters first get deeper, then get progressively shallower and then disappear altogether leaving a shallow depression and a tell-tale rim but sometimes hardly any sign of either. These relationships will be dependent on the specific crust thickness at the impact site with thicker crusts being able to support larger asteroids before allowing the crater size and the asteroid size to be more closely the same.

If it is valid that a large asteroid would punch through the crust and dive deep into the interior most of its energy would be embedded in the shockwave that would form in front of it. While the remains of the asteroid and the fractured plug of the crust would come to a stop someplace in the interior, the shock wave would not stop but continue propagating through the interior until it reached the other side of the Earth. For a straight-in approach the shock-wave would go straight through to the antipode. If the approach was not straight in, the shockwave would be somewhat off the side but still come up under the far side. This appears to be what has happened on the Moon, on Mars and on Mercury. Also, we have at least one such region on the Earth, located in Canada and called the Laurentian Plateau that tells a similar tale. Over this 3,000,000 square mile area we find all manner of broken crust, dislocated blocks, fissures and dykes. Dykes are narrow sections of rock which seem to have been formed by material oozing up from underneath just as we would expect if the crust had been fractured all the way through. It further follows that if a large area of chaotic terrain is found it might indicate that an asteroid hit the Earth some place on the far side. Of course, if impact sites can be masked by other material, so can regions of chaotic terrain but at least it provides one more tool for discovery.

With respect to the Martian situation at least one commentator has recognized that crater size and asteroid size are closely related for large asteroids and offered that the size of the asteroids would be 90% of the size of their respective craters. (548) This would necessitate that the interior of Mars be molten at the time of impact and this does appear to have been the case because all of the large impact sites have regained the spherical shape of the planet. On Earth this might be written off as having been filled in by some other type of activity but on Mars the fill-in options are much more restricted.

2.5 Mars Craters

Further understanding can be gained from noticing what might have happened on other planets. The most obvious one to consider is Mars. Numerous landers have been successfully sent to Mars and the surface has been repeatedly observed by orbiters as well. Chaos is everywhere. On one side, which we will refer to as the impact side, there are numerous impact sites. In fact, there are more than three thousand craters with diameters of more than twenty miles on the impact side. (76) This is however only the first factor to be recognized. The second is the size of the largest of these craters. The following list identifies them.

Number	Name	Diameter (miles)	Diameter (km)
1.	Hellas	999	2300 (496)
2.	Isidis	684	1094
3.	Argyre	481	770
4.	Huygens	291	466
5.	Schiaperelli	282	451
6.	Cassini	241	386
7.	Antoniadi	222	355
8.	Schroeter	185	296

Number	Name	Diameter (miles)	Diameter (km)
9.	Unnamed	175	280
10.	Herschel	158	253
11.	Kepler	150	240
12.	Newcombe	144	230
13.	Secchi	139	222
14.	Schmidt	133	213
15.	Flaugergues	132	211

These are very large impact craters by any comparison and with only a few exceptions are larger than anything that has been recognized on the Earth up to the present time. Why was Mars so favoured with large impacts when Earth is so much larger? Why wouldn't a similar situation have happened on the Earth?

With so much impact material involved, is it reasonable to suspect that the orbit of Mars would have changed as a result of all of these impacts? The energy required to change the orbit of Mars from a circular orbit at Mars' present perihelion to its current elliptical orbit has been calculated and the energy estimate as been included in the Energy Table. (see appendix) While this represents a lot of material it is less than 1% of the mass of Mars. If asteroid material is the remainder of a previously-existing planet that orbited between Mars and Jupiter, this small amount does not seem like very much. The arrival of all of these asteroids at the surface of Mars was an overwhelming disaster. Widely time-based impacts would have been bad enough but since it was a shower (i.e. more than 90% of the impacts are on one side) the planet would have shuddered and vibrated and sustained major structural failure. The numerous monster volcanoes, bulges, trenches and chaotic terrain all testify to this being the case. If a swarm of asteroids hit the Earth a similar pattern of destruction would be expected.

2.6 Mercury

The planet Mercury is the closest planet to the Sun in the solar system. As mentioned, it has an impact feature called the Caloris Basin which is very large when compared to the size of the planet and even very large when compared to other planets. At approximately 800 miles (1280 km) (139) across, this feature is a significant fraction of the planet's diameter of 3000 miles (4800 km) which means that the impact would have modified Mercury's orbit by a very measurable amount. The orbit of Mercury is highly elliptical further suggesting that it might have been hit by something significant. If an orbit is perfectly circular it is more difficult to imagine that the circularity would have been caused by an impact because the impact would have had to occur in a very precise way to result in a perfect circle. This by no way means that it could not have happened – just that it would seem to be much less probable.

The question of more immediate interest relates to the energy requirement to achieve an orbital change. The approximate total energy of Mercury has been calculated and included in the Energy Table. Also the energy that would have been required to change from a circular orbit at the same distance as Mercury's current closet approach to the Sun, has been calculated. ('Closest approach' is called 'perihelion' and is about 28.5 x 10(6) miles (45.6 x 10(6) km).) The difference in the total energy level between these two orbits would have had to have been supplied by some external source in order for the orbit to have changed by such a large amount. An asteroid could have done this but it must have been very large or else there would have had to have been a swarm of them. However the surface of Mercury does not show a number of large impact sites. While the mass of the hypothetical asteroid has been calculated and is less than the mass of the asteroid swarm that could have changed Earth's orbit, Earth has a large number of impact sites to be supportive of such a consideration. Even if the impactor that created the Caloris Basin were almost as large as the crater, it would not have been large enough to effect an orbital change from circular to the current ellipse. A smaller change would have happened instead. On the other hand, if Mercury originally had a circular orbit that was in between the current orbital extremes, an impactor with a diameter similar to the diameter of the Caloris Basin would have had enough energy to effect the current elliptical orbit!

2.7   Earth

The Earth has a large number of impact sites and more are being identified on a regular basis. The following is a list of the currently-recognized large impact sites.

Number	Name	Location	Diameter (km)	Diameter (miles)
1	Sudbury	Canada	250	156
2	Manicouagan	Canada	100	62
3	Chesapeake Bay	USA	90	56
4	Chicxulub	Mexico	170	106
5	Vredefort	Africa	300	187
6	Acraman	Australia	90	56
7	Popigai	Russia	100	62
8	Kara	Russia	65	41
9	Morokweng	Africa	70	44
10	Puchezh-Katunki	Russia	80	50
11	Tookoonooka	Australia	55	34
12	Yarrabubba	Australia	50	31

Other sites have been noticed as being possible impact sites. The following list includes some of them.

Number	Name	Location	Diameter (km)	Diameter (miles)
1	Czech Rep.	Czech Rep.	450	281
2	Takla Makan	China	1600	1000
3	Congo	Africa	1000	625
4	Tenitz	Kazakhstan	550	344
5	Shiva	India	500	312
6	Impact structure	Australia	600	325

Number	Name	Location	Diameter (km)	Diameter (miles)
7	Nastapoka arc	Canada	500	312
8	Wilkes land	Antarctica	480	300
9	Bedout	Australia	250	156
10	East Warburton	Australia	200	125
11	Ullapool	England	150	94
12	Maniitsoq	Greenland	100	62
13	Rubielos de la Cerida	Spain	80x40	50x25
14	Vichada	Columbia	50	31

The fact that more than one list exists underscores the question of recognition of any particular site as well as the problem of getting all interested parties to agree. In addition several very large impact sites have only very recently been recognized with the reason being that they have been masked from our view.

A great number of asteroids have hit the Earth with only the largest 'confirmed' being listed above. Small asteroids continue to hit the Earth and the total amount of material that is added to the mass of the Earth every year has been declared to be millions of tons. (17) This material would be made up of chunks from as large as a garage down to very tiny particles. The atmosphere protects us from this type of bombardment resulting in a very small number of them getting through to the surface of the Earth. However it is still a significant number. The relatively large percentage of open space on the Earth's surface works in our favor but every now and then a chunk of material will hit a house or other structure. Open space was in our favor when a meteorite recently landed on the ice of Antarctica. Only a very tiny fraction of Antarctica is inhabited (by research teams) so the chances of somebody being hit there are quite remote.

While the ongoing arrival of small pieces of material is of some interest what really concerns us for the present discussion is the amount of asteroid material that hit the Earth possibly causing the orbit of the Earth to change. The mass required is shown in the Energy Table in the Appendix by dividing the mass of the Earth by the assumed aggregate mass of the impacting asteroids. This is 0.0052 or 0.52% of the mass of the Earth. While this represents a lot of material it is still only ½ of 1% of the mass of the Earth and can be compared to several other masses of material both measured and hypothesized.

What was the volume of the material that seems to have hit the Earth? How can it be roughly identified? The first clue is the number of impact sites. The second is the size of these sites. In order to arrive at a possible volume, an assumption has to be made concerning the relationship of asteroid size to 'crater' size. It is hereby recognized that there will never be any agreement on either of these points. This is acknowledged with the recognition that in many cases disagreement today leads to greater understanding tomorrow. It will be assumed that for large 'craters' the asteroid size was 100% of the 'crater' size. This compares with the opinion of at least one commentator who placed the percentage at 90%. (548) The rationale in both cases is that the asteroid would simply 'punch' through any surface which has a thickness that was only a small multiple of the asteroid diameter. This assumption is simply based on the incredibly high speed and energy of an impacting asteroid. If air cannot even get out of the way, how would solid material like the granite of the Earth's crust get out of the way? It will further be assumed that as crater size diminishes the asteroid that made it will become (comparatively) increasingly smaller. The final result would be a small crater in unconsolidated material. In this case the asteroid size might only be $1/100^{th}$ of the crater size. Also, the size relationship would be expected to vary with crust thickness wherein 'the thicker the crust the smaller the asteroid' in comparison to its crater diameter. In any event any object that was only 50%, or less, of the size of the largest would not have contributed to the overall aggregate mass substantially. With these various assumptions in mind it is estimated that the aggregate amount of asteroid mass that hit the Earth was about 23.5 x 10(30) joules from the currently-recognized possibilities. (see appendix 5)

2.8 Crater Masking

The major problem in identifying impact sites on Earth is the presence of water and vegetation. Water covers approximately 70% of the Earth's surface so it would be reasonable to expect more sites under water than there are above. In fact by simple extrapolation there should be twice as many under water as there are above water. If this was the case, the above lists could be multiplied by three.

While water could simply cover a site there is a second way that water could have masked an impact site. If an impacting asteroid caused the Earth's water to move on a global scale, it would have washed material across the location where the asteroid landed thereby making the impact site much less obvious. Water moving on a large scale would carry with it a load of material - some of which would be appropriate for forming sedimentary rock. The massive Chicxulub Crater in Mexico is one example of this type of crater masking. The Chicxulub Crater is hard to study because it is buried under several kilometers of sedimentary rock. (16) At the Manson Crater in Iowa the situation is similar except in this case the sedimentary rock fills the Mid-Continental Rift in which the crater is situated. This rift (i.e. massive split in the crust of the Earth) runs down through Iowa and actually takes up about 50% of the width of the state. In the Manson Crater case, the crater was only discovered because the asteroid had impacted the layer of sedimentary rock within the rift area forming an opening in it. This caused confusion for a well driller who was expecting a layer of sedimentary rock. (4) Both the crater and the rift also include a layer of unconsolidated material which could also have been placed by water. (Conventional wisdom states that a glacier brought the material to fill both the rift and the crater but this would have necessitated the glacier pushing a pile of material ahead of itself that was several hundred miles high, which is not credible.)

Moving water on a massive scale is exactly what would be expected when an asteroid hits the Earth. Even a relatively small one would generate massive waves or tsunamis that were miles high and capable of moving completely across continents. In this manner an asteroid would be able to cover its own crater opening and thereby wipe out evidence that it actually came. Such massive water movements would come complete with thousands of tons of entrained material. Depending on the speed of the flow, the entrained material would either keep moving or settle out. Either way it would make it difficult to notice a crater site even if it was quite large. Either currently-standing water or ancient moving water makes the discovery of impact sites more difficult but water that surged around the Earth when the asteroids arrived would make crater identification difficult on land as well as in water.

Also there is the question of vegetation. When an area is covered by rain forest or some other form of camouflage, a large impact site could easily be overlooked. This might seem dubious until we recall that it has only been very recently that both the Chesapeake Bay site and the Chicxulub site have been identified. This is not because they were not big enough. They were simply masked by a large amount of other material and so were not noticeable.

Silt is the fourth agent that could cover an impact site - particularly underwater. In some areas of the ocean the layer of silt is hundreds of feet deep. What lies hidden beneath all of this silt?

Craters would be expected to occur with a gradation in size which means that the above list could probably be continued all the way down to much smaller sizes. Further, there were another 23 on the source list that were not listed above because they are less than 50 km across. However a crater that is 50 km across is still a large crater and the uncertainty related to the relationship between crater size and asteroid size only generates more ambiguity. If crater size is actually only marginally larger than asteroid size, the Earth would certainly have taken a terrible pounding when the asteroids arrived and the overall turmoil would be the main reason that so much crater masking has happened.

If all of these masking factors could be stripped away the Earth might appear more like the Moon or Mars where the impact sites are almost too numerous to count. As mentioned above more than 3000 greater than 20 miles in diameter have been counted on Mars and several of them are very large. Also the Moon has more than 200,000 craters with diameters greater than one kilometer. Since the cross-section of the Earth is much greater than either the Moon or Mars and the Earth is much more massive than either one, why wouldn't hundreds of thousands be expected here as well? Also from the above lists we notice that Australia has six large sites. Why is Australia so favored? Could it simply be that since Australia has such a large area of desert, impact sites are easier to notice?

In recognition of all of the current uncertainties one is tempted to wonder what the situation will appear like in another hundred years. If the current rate of discovery continues, the list of recognized craters will be much longer at that future time than it is at the present time.

2.9   Remaining Asteroids

There are still numerous asteroids in orbit around the Sun. While it is suspected that there are more than one million altogether (9) most of them are quite small. The following list identifies the largest that are currently known. (322)

Number	Name	Diameter (km)	Diameter (miles)
1	Ceres	998	624
2	Pallas	605	378
3	Vesta	534	334
4	Hygeia	448	280
5	Euphrosyne	347	217
6	Interamnia	347	217
7	Davida	321	201
8	Cybele	307	192
9	Europa	283	180
10	Patienta	275	172
11	Eunomia	270	161
12	Juno	248	155
13	Psyche	248	155
14	Doris	248	155
15	Undina	248	155

While these are the largest, some commentators suggest that there are 2000 more that are greater than one kilometer in diameter. (11) The further problem being the fact that all of the remaining asteroids have not been discovered and that asteroid discoveries continue at a rate of 100,000 per year. This only exasperates the problem further. (12) This is not very comforting and basically means that there is more material still in orbit around the Sun than anyone can imagine. It is also likely that more material hit the Earth than anyone would like to imagine. The aggregate mass of the remaining has been estimated with respect to the mass of the Earth as being approximately 0.05% of the Earth's mass. (544) It has also been suggested that Ceres, the largest, embodies most of the remaining asteroid mass. If we allow that it actually embodies 50% of the remaining asteroid mass, then the total mass of all of the remaining asteroids would be about 1/2000th of the mass of the Earth. (i.e. 0.05%) (545)

2.10   Asteroid Mass Comparison

Mass Comparison Table	
1. Calculated mass to change Earth's orbit	0.52% of Earth's mass (see App. 5)
2. Mass of remaining asteroids	0.05% of Earth's mass (494, 495)
3. Estimated mass of asteroids into Mars, the Moon& Mercury	0.36% (see App. 5)
4. Estimated mass of asteroids that have struck Earth	0.58% (see App. 5)

With these possibilities in mind as well as the uncertainty in the actual number of asteroids that have hit the Earth it can be seen that there is a serious possibility that there was enough impacting asteroid mass (i.e. 0.52% of the mass of the Earth) to have met the criteria for changing the orbit of the Earth from a 360 day year to a

365 1/4 day year. (This type of approach shows plausibility (i.e. reasonable possibility). In consideration of the many unknowns involved this is as far as reasoning can go.)

2.11   A Warmer Earth

If the Earth on the average had been a little warmer in the distant past, it would be supportive of the idea that the Earth was slightly closer to the Sun at that time with a slightly shorter year. The existence of a water vapour canopy enveloping the Earth would also have caused the Earth to be warmer and certain evidence is supportive of the idea. The existence of a vapor canopy would have enabled the Earth to accommodate a little more incoming energy because it would have caused it to be distributed more evenly over much of the Earth's surface. By intercepting some of the incoming heat well above the surface, as would have happened with a vapour canopy in place, the surface would have been prevented from over-heating so being a little closer to the Sun could have been more easily accommodated. However before exploring this possibility the volatility and restrictiveness of the Earth's habitable zone must be reviewed.

2.11.1   The Earth's Habitable Zone

If the length of the Earth-year changed from 360 days to 365 ¼ days it means that the Earth was pushed out a little further from the Sun where it would receive a little less heat. It would still have to be in the thermally-habitable zone but out closer to the outer boundary of it. This criteria fits with the understanding that the Earth (from its present orbit) could still be in its habitable zone if it was up to 5% closer to the Sun but would be on the very outer fringe of that zone if it was only 1% further out. (100) Being closer to the Sun in a 360 days/year orbit would mean that the Earth would have had a slightly warmer climate and would have been closer to the mid-point of the thermally-habitable zone. The heat impinging on the Earth from the Sun changes with the square of the distance change which in this case means that the Earth with a 360 days/year orbit would have been about 1% closer to the Sun and would have been receiving about 2% more solar energy. (Please refer to the Energy Table in the Appendix) With the Earth's present heat-distribution system, a 2% increase would have had a very negative effect at the equator. This is understood with respect to the very small increases that are currently happening due to global warming. If these changes continue, some equatorial regions will become uninhabitable within another hundred years or so which in turn means that the general habitability of the Earth will have been reduced.

A 2% increase in incoming energy would expectedly result in a 2% increase in temperature but 2% of what. A reasonable approach to this question would consider what the temperature of the surface of the Earth would be if there wasn't any solar energy hitting it. The surface temperature would not simply be absolute zero because the interior of the Earth is warm. Even if no heat at all was impinging on the surface, it would still be well above the coldest temperature imaginable – absolute zero. The temperature on the dark side of Mars would be indicative – to a first degree – of the temperature that could be used as a starting point to determine how cold the surface of the Earth would be if there was no sunshine. Mars has neither a large ocean nor a significant atmosphere to either retain or distribute heat. (It does appear to have a molten interior which should help to keep the surface above absolute zero.) Therefore when the Sun goes down the surface cools down significantly. In fact the cold side of Mars has been measured to be about -75 degrees F. Using this as a starting point and allowing that the average surface temperature over land at the equator on Earth is 75F, a 2% increase would be 3 degrees F. Three degrees is not very much but the resulting increase in air temperature would enable the air to hold more water vapor which, because of its greenhouse effect, would cause a further rise in temperature. What if it rose another 3 degrees? If it did, some of the equatorial regions would be above the body temperature of animals as well as the temperature at which seeds can germinate. If this was the case, some regions near the equator would become uninhabitable resulting in a general reduction in the habitability of the Earth. (All of this tells us that temperature control and stability at the surface of the Earth is a delicate matter not to be assumed or taken lightly.)

However, if the Earth's heat distribution system was different at that time and provided means to distribute incoming heat more evenly around the entire surface, 2% more energy would have been quite manageable. It might have meant that the equatorial regions were a little warmer but more appropriately that the rest of the Earth - in particular the extreme latitudes - were considerably warmer. This arrangement would have been quite tolerable because all areas further from the equator than the 45th parallel could easily handle more heat at the present time without causing any discomfort at all. It therefore appears that two factors need to be addressed; a. Is there any evidence that the entire Earth was warmer in the distant past? and b. Is there a possible mechanism that could have enabled this to have happened?

2.11.2 Positive Feedback Loops

In recognition that the thermally-habitable zone has occasionally been declared by some commentators as having considerable latitude, the volatility of changing it must be noted. This volatility is due, in part, to positive-feedback loops. A positive-feedback loop is the same as a viscous cycle. Once it gets going it feeds itself and gets going all the more. A fire is like this. If you start a very small fire at the bottom of a large haystack, that very small fire will soon engulf the entire stack – even if it was one hundred feet high. The very small fire is hot and the heat causes a small column of air to rise. As it rises, more air is drawn in and the fire burns a little brighter. This causes more air to be drawn in and the fire builds up. The greater heat of this larger fire brings in even more air until the entire stack of hay will be burning. This also happens with a forest fire. The heat rising due to a burning tree will draw in more air until a huge fire builds up and the inrushing air becomes of hurricane force even capable of sweeping fire-fighters right into itself. Wherever stability is required, positive-feedback loops are to be avoided.

The incoming heat from the Sun together with the complement of greenhouse gases keeps the surface temperature of the Earth at its present value. While the heat from the Sun is not easily changed, the greenhouse gases can be changed and changing them is to be avoided. The two most important ones in Earth's atmosphere are water vapour and carbon dioxide.(165) Increased emissions of CO2 during the last few centuries have resulted in an increase in atmospheric CO_2 from 280 ppm to almost 400 ppm. This is not good news but at least it is happening at a fairly slow pace so there is some time to recognize what is happening. Water vapor, on the other hand, is more easily changed and, in fact, can change itself due to the positive-feedback mechanism. A warm atmosphere can hold more water vapor. This will cause more warming and hence more water vapor. Then more warming will cause even more water vapor. Since water vapor is a greenhouse gas, this increase leads to a bit more warming still, and so on, and this process is known as positive feedback. (166) There are other processes that would contribute to temperature increase in a similar manner and one of them is the characteristic of water whereby the amount of CO_2 that it can retain is dependent on its temperature. Cold water can hold more CO_2 than warm water. (168) The oceans of the world are basically very cold - in fact just a little above freezing - even at the equator only a few hundred meters below the surface. (167) It follows immediately that if the ocean warmed up it would release some of its CO_2 burden and thereby cause more warming. A similar phenomenon can develop in the other direction if too much cooling happened on a world-wide basis. In this case snow and ice would reflect more of the Sun's heat and result in more cooling. More cooling reduces the release of water vapor and so on. '... the reduced greenhouse effect from water vapour, causing a further reduction of temperature ... the temperature would drop so much that Earth would become completely frozen over forming a "snowball Earth".' (169) (It must be noted that an ice-covered Earth is not an Earth having an ice age. It is simply an Earth covered by ice and snow and an Earth that would not be the least bit habitable.)

These various processes make one wonder how the system works at all and they draw significant attention to the reality that the habitable zone is very narrow and that temperature control and regulation of the Earth cannot be modified without serious consequences. There must be built-in stability and control mechanisms or the system will collapse completely. (169) This makes one wonder how this could be achieved on a far-away planet when it's achievement here on Earth has such a demanding criteria even when so many control and regulation mechanisms are in place!

2.12   A Warmer-Earth Mechanism

There is a possibility that prior to any initial impact of an asteroid upon the Earth, the atmosphere of the Earth was structured differently than it is now. If there had originally been a region in the upper atmosphere, which, instead of containing atmospheric oxygen and nitrogen, contained mostly water molecules, (Please refer to the diagram, 'The Water Vapor Layer' on page 46) a hot torrential downpour would have resulted from the expanding pressure wave generated by an incoming asteroid. Consequently, such a region of atmosphere would have been totally destroyed. This hypothetical atmospheric structure has been referred to as a vapor canopy because it would have been like a blanket enveloping the entire Earth – including our present atmosphere - and it would have had a significant moderating effect on the climate of the whole world. Would it have been possible to have had a region of atmosphere like this and if so, what characteristics would it have had?

If the Earth had been receiving 2% more heat there would have been a very negative effect on the habitability of the Earth throughout the equatorial regions which the presence of a vapor canopy would have offset in two ways. Consisting of water vapor the canopy layer would have absorbed most of the incoming infrared energy thereby preventing it from getting to the surface of the Earth where it would have caused over-warming with the negative consequences already mentioned. If the incoming infrared energy had been intercepted and trapped in a water vapour layer, the surface a few miles below would not have over-heated. The warming effect within the vapour layer would have enabled it to transfer heat away from the equatorial zone altogether. Secondly, being a greenhouse gas the water vapour in the vapour canopy would have absorbed the incoming infrared energy and re-radiated some of it right back into space just as the greenhouse gases do at the present time as they return some of the infrared heat coming up from the Earth, right back to the Earth. The result from these two characteristics of water vapour - arranged in an enveloping blanket enclosing the Earth - would have been an Earth that was not over-heated at all – even though it was slightly closer to the Sun.

2.12.1   The Basic Physics

In order to speculate that a vapor canopy could have existed above the entire present atmosphere of the Earth, first, the basic physics which would be necessary for this to be possible, must be recognized.

While it will probably never be possible to identify just how much water there might have been in this hypothesized upper layer of the atmosphere, the problem can be approached by speculating on the amount and then identifying the accompanying physical factors associated with such speculation. Suppose, for example, that there was a layer of water vapor above the present atmosphere and that it was equivalent in weight to about one-third of our present atmosphere. The present atmosphere weighs 15 pounds for every square inch of the Earth's surface. This is referred to as atmospheric pressure and what it means is that for every square inch of the surface of the Earth, there are 15 pounds of air pressing down. If we could put this small column of air on a scales, the scales would read 15 pounds. Then, if we could somehow add more air to the Earth, it too would be pressing down and the little column of air would have become heavier. From this we can speculate that if (instead of having more air on top of our present air) there had been a layer of water molecules on top of the atmosphere instead, we can see that our little column of air would have been a little heavier than it is now. Another way of saying this is if the Earth's atmosphere had more air in it, or if there was some other gas floating on top of it, it would be heavier. Atmospheric pressure would be higher. As suggested, if the additional amount was about one-third of the present atmosphere, there would be another 5 pounds so the total weight of the square inch of atmosphere would have been 20 pounds and atmospheric pressure would have been 20 pounds per square inch.

If such a layer of water vapor had existed and it was then squeezed completely out of existence, a layer of liquid water about ten feet deep would have rained down on the entire Earth. As rain, locally, this represents a storm, which is comparable to a very large hurricane. A hurricane has been observed to produce nearly four feet of rain. (355) Therefore ten feet of rain would be similar to the rainfall from two and one-half such hurricanes. A hurricane however, is a local event covering at most a few hundred square miles. The atmosphere, on the other hand, covers the entire Earth so the comparison would be to several hurricanes raining down on the entire Earth at the same time. If a monster hurricane produced ten feet of rain, serious local flooding would result. However, the flood water would soon run into rivers and lakes and find its way to the ocean. On the other hand, if ten feet of rain fell on the entire world, including the oceans, this much more extensive flood would alter the topography of the whole Earth.

2.12.2   Temperature Considerations

In order to have a water vapour layer above our present atmosphere, it must remain warmer than a very particular temperature which is referred to as the critical temperature. This temperature is called critical because if it was any lower, the hypothesized water vapour layer would condense into rain. Above this critical temperature, the water would exist as vapour. Below the critical temperature, the water would exist as a liquid. Since a liquid cannot remain floating in the air, but would simply fall to the Earth, the entire idea would fall with it. Below this so-called critical temperature, any water above the atmosphere would not have been able to remain as vapour (or gas) but would have changed to a liquid. This is the way nature behaves in the present atmosphere. If, after a warm summer day when a lot of moisture has evaporated into the air, evening comes and it cools down, dew forms on everything. This happens because it becomes too cool to enable the water vapour in the air to stay as an invisible vapour, so it changes into water, which condenses on all of the cool surfaces. This is also similar to the way moisture forms on the surface of a glass of cool water. In the vicinity of the glass, the air is cool and it is too cool to enable the water in this region of air to remain as an invisible vapour so it appears on the glass as a liquid. If the glass was subsequently warmed up, the water on it would change back to a vapour and disappear back into the air. Therefore, in order to hypothesize that a layer of water vapour existed above our present atmosphere, it must be recognized that it had to have been above a certain very particular critical temperature. If it was too cool, the individual water molecules would not have enough energy (speed) to bounce into each other fast enough to bounce away again. Water molecules have a certain amount of attraction towards each other. When this attractive force is greater than the tendency to bounce away, the water molecules will start to stick together. They are now at the 'critical' temperature and exist partly as individual molecules (vapour) and partly coalesced or as groups of molecules (liquid). For any particular pressure, there will be a temperature where theory predicts that the individual molecules of water would be expected to start coalescing. While the particular theory involved is quite valid, the expected results will only be obtained if the vapour includes dust. Particles of dust provide landing places for the atoms of water and make it easier for them to come together and form droplets. When tiny droplets form, fog is visible. The situation is dramatically different if there is no dust present. Also, in this case, only a very particular type of dust, called hydroscopic dust, will do. (Hydroscopic simply means that it attracts water.) Some types of dust are not hydroscopic including the dust from meteorites. If the appropriate type of dust is not present, it becomes very difficult to form droplets of water even if the temperature drops well below the expected critical level. As long as the water molecules stayed separate and just kept banging into each other, the water vapour layer would have remained invisible and would have just kept floating on the oxygen-nitrogen layer.

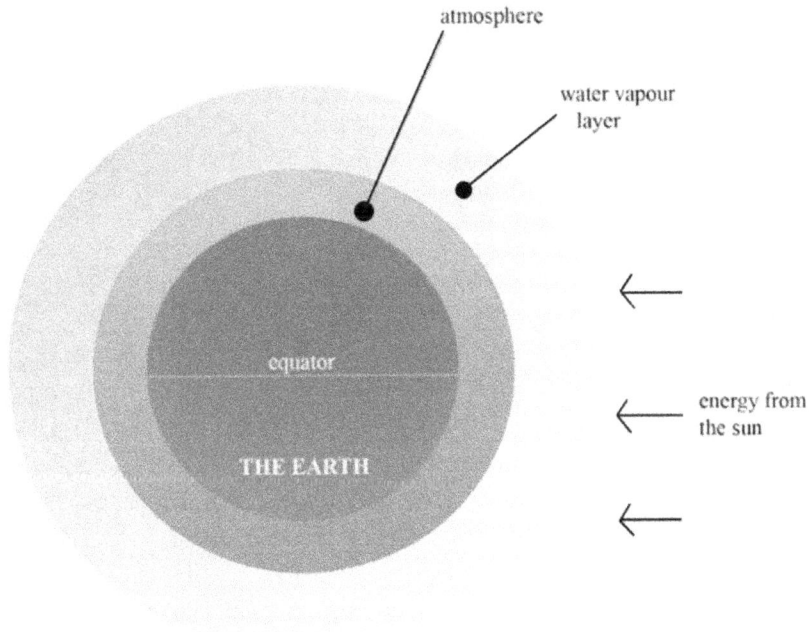

The Water Vapour Layer

2.12.3   Canopy Temperature Chart

If water vapour is to remain in vapour form, this is equivalent to saying that it must be above its boiling point. However, due to the above-mentioned anomaly, it may actually be able to cool down much farther than expected and still remain as vapour. The following chart gives several examples of the theoretically expected, fog-forming temperatures as well as the lower anomalous temperatures for eleven examples which involve different amounts of water.

Example Number	Portion of atmosphere	Liquid water equivalent	Pressure Psi	atm	Expected temp. to form fog		Anomalous temp to form fog	
1.	10%	3.3 ft.	1.47	0.1	115 F	46 C	59 F	15 C
2.	20%	6.6 ft.	2.94	0.2	141 F	61 C	79 F	26 C
3.	25%	8.25 ft	3.67	0.25	149 F	65 C	86 F	30 C
4.	30%	9.9 ft.	4.41	0.3	157 F	69 C	91 F	33 C

Example Number	Portion of atmosphere	Liquid water equivalent	Pressure Psi	atm	Expected temp. to form fog		Anomalous temp to form fog	
5.	40%	13.2 ft.	5.88	0.4	169 F	76 C	101 F	38 C
6.	50%	16.5 ft.	7.35	0.5	179 F	82 C	108 F	42 C
7.	60%	19.8 ft.	8.82	0.6	187 F	86 C	115 F	46 C
8.	70%	23.1 ft.	10.28	0.7	194 F	90 C	120 F	49 C
9.	80%	36.4 ft.	11.76	0.8	201 F	94 C	125 F	52 C
10.	90%	29.7 ft.	13.23	0.9	204 F	96 C	130 F	54 C
11.	100%	33.0 ft.	14.7	1.0	212 F	100 C	134 F	57 C

In this chart, example number 6 indicates the minimum temperature which the base of a water vapour layer above the atmosphere must have in order to remain as invisible vapour and not turn into fog. (If the amount of water involved was 50% as heavy as the present atmosphere.) In this example, the equivalent weight of one-half of our atmosphere would have been floating on top of our present atmosphere in the form of water vapour. If all it precipitated out, 16.5 feet of liquid water would have fallen down onto the entire Earth. With this extra material above our present atmosphere, the atmosphere would be heavier and atmospheric pressure would be 1.5 times 15 pounds per square inch or 22.5 pounds per square inch. The theoretical minimum temperature to keep this much water in vapour form is about 179 degrees F. However, if there wasn't any hydroscopic dust mixed in with the water molecules, the temperature might have to drop down as low as 108 degrees F or even lower before any fog would form. Similarly, example number 2 shows that 79F might be needed to form fog if there was about 6.6 feet of water in the vapour layer. This temperature is within the comfort range for people meaning that such a circumstance would not present any threat to human survival. If the vapour layer contained any amounts of water greater than this, the minimum temperature at the bottom of the water vapour layer would probably have been above the comfort zone. (This would not necessarily have been any threat to human survival either.) Of course for any elevation above the bottom of the vapour layer, the required minimum temperatures would be even lower and would continue to drop off as elevation increased. It does mean however that for any amount above 20%, the temperature at the bottom of the vapour layer would probably have been above the comfort zone for people on the surface of the Earth. Therefore, for all of these cases, having a water vapour layer on top of the atmosphere would have been like having a warm blanket around the entire Earth, isolating it from the bitter cold of space.

2.12.4 Atmospheric Pressure Profiles

Atmospheric pressure decreases as the height (or distance) above the surface of the ground is increased. (Please refer to the diagram Atmospheric Pressure Profile) The pressure drops fairly rapidly through the lower elevations and then the decrease rate reduces at higher elevations. The pressure variation follows the familiar half-life curve or time-constant curve. Mathematically, this type of relationship is called an exponential and it has a very particular mathematical formulation. Atmospheric pressure very closely follows this type of relationship. In the diagram, pressure variation is shown as curve number 1. Three other curves are also included to show how atmospheric pressure would be increased if there was a layer of water vapour above the present atmosphere. From these curves the height of the base of the postulated vapour layer may be determined and this information has been identified on the diagram. For example, curve number 2 (the 25% of present atmosphere added-on-top case) shows that the height of the base of the water vapour layer would be about 7.5 miles. (Note the checkmark on the diagram.) From the Atmospheric Temperature Chart mentioned above, the theoretical temperature to ensure the water at the base of this particular vapour layer remains as a vapour would be 149F. The anomalous or lower possible temperature would be 86F. The other anomalous temperatures for the examples shown on the diagram are not very much above temperatures, which are commonly reached on the surface of the Earth.

Atmospheric Pressure Profile

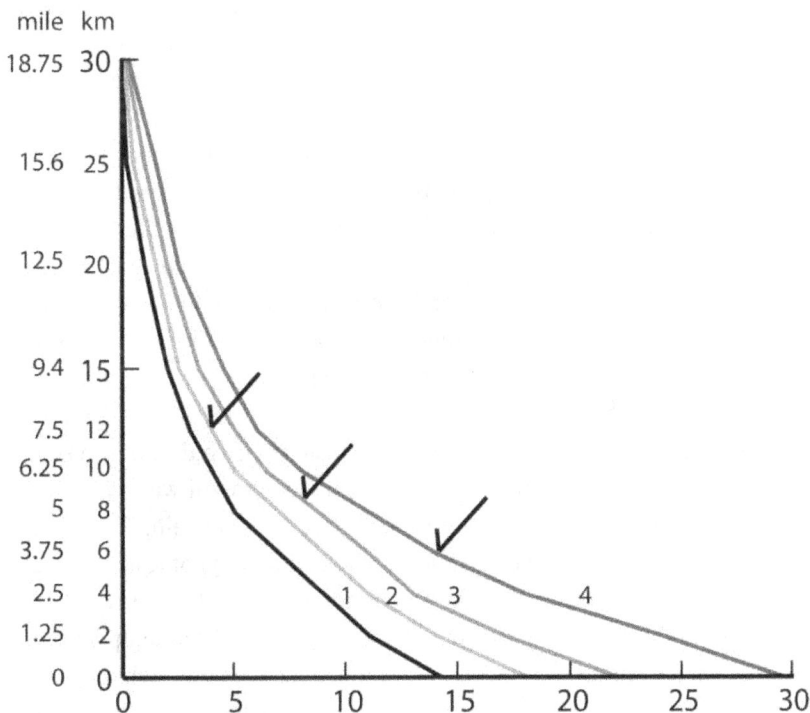

Line No. 1 Atmospheric pressure at the present time
Line No. 2 Atmospheric pressure with a 25% vapour layer addition
Line No. 3 Atmospheric pressure with a 50% vapour layer addition
Line No. 4 Atmospheric pressure with a 100% vapour layer addition

Arrows indicate the top of the atmosphere and the bottom of
the vapour layer. Please refer to the Appendix for further
discussion.

2.12.5   Stability of Atmosphere with a Canopy

Usually, air which is high up in our atmosphere is cooler than air at the surface of the Earth. This is understood from the way in which all gases behave as their pressures are varied. If air, which is near the surface of the Earth is raised to a higher level it will cool down. It cools because its pressure will drop as the elevation increases. The atmosphere of the Earth always behaves like this and means that most of the time it is stable. However, occasionally these stable conditions are upset and storms result. If either the temperature at the surface of the Earth becomes hotter than usual or the temperature high above the Earth becomes colder than usual, the atmosphere is recognized as being unstable and either thunderstorms or tornadoes are expected. This may be understood by considering the results of raising a quantity of air upwards from the surface of the Earth. As air is raised, it cools by expansion. It cools simply because as it is raised, it expands which directly results in cooling. Usually, as air rises up from the surface of the Earth, it will cool and reach a temperature which is very close to the temperature of the surrounding air at that height. If this does not happen and the air cools but is still hotter than the surrounding air, this rising air will rise up even further. It will, of course cool some more. Again, if the surrounding air is still colder, the rising air will now be rising at an increased rate and it will form an updraft, which will draw replacement air in at the bottom. If this rising air includes moisture, it might reach an elevation where the entrained moisture chills to its condensation point whereupon it will then fall out of the rising air column as rain. This, however, only compounds the problem because now the heat of condensation of the rain-forming moisture heats the resulting drier air so it rises even faster and a thunderstorm develops. This is also the way in which a hurricane develops, because a hurricane starts off like several thunderstorms linked together. Tornadoes often accompany thunderstorms and are usually formed at the perimeter of the rising air column. As the air rises, it also rotates. The rotation might cause a tornado to form. All of these stormy results begin with air which is too warm near the surface or too cold higher up.

On the other hand, if the air high above the Earth was warm instead of cold, the atmosphere would be very stable vertically and there would not be any storms. In this case, if a column of warm air started to rise, it would encounter air, which was still warmer and so it would just settle back down again. Therefore, if the atmosphere included an upper layer of vapour and the temperature at the base of this hypothesized vapour layer was at the anomalous temperatures or higher as discussed above, the atmosphere would be very stable. Any vertical air movement would be restricted to the lower regions and could only occur if the temperature at some particular elevation was a little cooler than a quantity of air would be if it had just been brought up from the surface of the Earth to this same elevation. (Several examples of the temperature-elevation relationship are included and discussed later in this chapter.)

2.12.6   Stability Factors

2.12.6.1   The Influence of Dust

If water molecules in the atmosphere are forced together, droplets will form and rain will result. This is not the usual procedure in nature however, where the coming together of the water molecules is greatly facilitated by dust. The dust in the air, (which in this case is understood to be hydroscopic dust), provides a location for the water molecules to start condensing or clinging together. This process is occasionally encouraged by manually seeding the clouds. If rain is badly needed, and a likely-looking cloud appears, an airplane might be dispatched to drop small dust-equivalent particles into the cloud to facilitate the water molecules to aggregate. If there is enough water in the cloud, rain could result.

The opposite situation also holds. If there is no dust in the air, it is very difficult to get a water droplet to form. It will not form easily and as long as the air remains above the dew-point (or the theoretical condensation temperature discussed above), rain will not be expected. Even at temperatures much lower than the dewpoint (i.e. the anomalous temperature discussed above), rain still might not happen.

2.12.6.2 The Buoyancy of Water Vapour

When moisture is added to air, the air becomes lighter. At first this would appear to be incorrect. If two things are added together, the resulting mass should be as heavy as the two original materials added together. However, this is not the situation with gases such as air. If you add one gas to another, there will still be the same number of molecules in any particular space as there was with either gas by itself. More space would certainly be required for these gases, but in any original volume of this space, the original number of molecules would still be present. Therefore, when water vapour is added to air, because a water molecule is lighter than either of the two types of air molecules, the mixture of air and water vapour will result in a lighter gas. Therefore, when moisture is added to dry air, the resulting damp air will be lighter than any similar volume of dry air and the damp air will rise up and float on top of the dry air. In the atmosphere, any air which includes moisture will be found on top of air which does not contain moisture. A portion of atmosphere, which was entirely water vapour, would be lighter than either atmospheric oxygen (O2) or atmospheric nitrogen (N2) and so it would float on top of our present atmosphere with no tendency to sink into it. The demarcation between the two regions would be quite abrupt.

2.12.6.3 Noctilucent Clouds

Noctilucent clouds, or night-shining clouds, occur about 50 miles above the Earth. They are quite faint and may only be seen for a short period of time right after sundown. Later, when the Sun is further below the horizon it will not shine on these clouds. Apparently, they were first observed shortly after the great explosion at Krakatoa in 1883 and have been up there ever since. (356) Krakatoa is recognized as the most powerful explosion that the world has ever witnessed and it did drive moisture up to extremely high altitudes. Moisture at such a high elevation cannot come down. If it should drift lower hypothetically, it would be warmed, boil and rise back up again. (It would boil because of the extremely low pressure.) Exactly the same situation would exist with the hypothesized vapour canopy. The existence of noctilucent clouds therefore provides direct evidence that a vapour canopy layer would be quite stable in the upper atmosphere and would just remain there for years until something significant occurred to disturb it and bring it down. An incoming asteroid could provide such a disturbance.

2.12.7   Basic Criteria for a Canopy

There are two basic criteria, which must have been in place, if the atmosphere of the Earth at one time included a water vapour layer. First, the water vapour layer must have been above a particular minimum temperature (i.e. the critical temperature) just to remain as water vapour. Secondly, the temperature at the surface of the Earth must have been in the comfort zone or animal life could not have existed and the "window of life" would already have closed.

The temperature of the Earth is determined by the characteristics of the atmosphere and by the incoming solar energy. The energy which approaches the Earth is virtually always the same but the way in which the atmosphere deals with this energy will determine the actual temperature which results at the surface of the Earth.

2.12.8   Energy Distribution & Transport

While virtually all of the energy that heats the Earth comes from the Sun, the way in which it is either absorbed or reflected determines the temperature at the Earth's surface. Solar energy consists of infra-red energy (53%) and visible light energy (38%). There is also a small amount of ultraviolet energy. (About 9% of the total) (65). Due to the presence of water vapour in the air, on the way through the atmosphere, some of the infrared energy is absorbed and some of the visible light energy is scattered. A little less heat energy (infrared is also called heat) is therefore available to heat the Earth and the Sun is not quite as bright as it is above the atmosphere. At the surface of the Earth, some energy is reflected and some is absorbed. The reflected portion of the visible light enables things to be seen. If no light reflects from an object, or if there isn't any light to even shine on it (i.e. after sundown) that object cannot be seen.

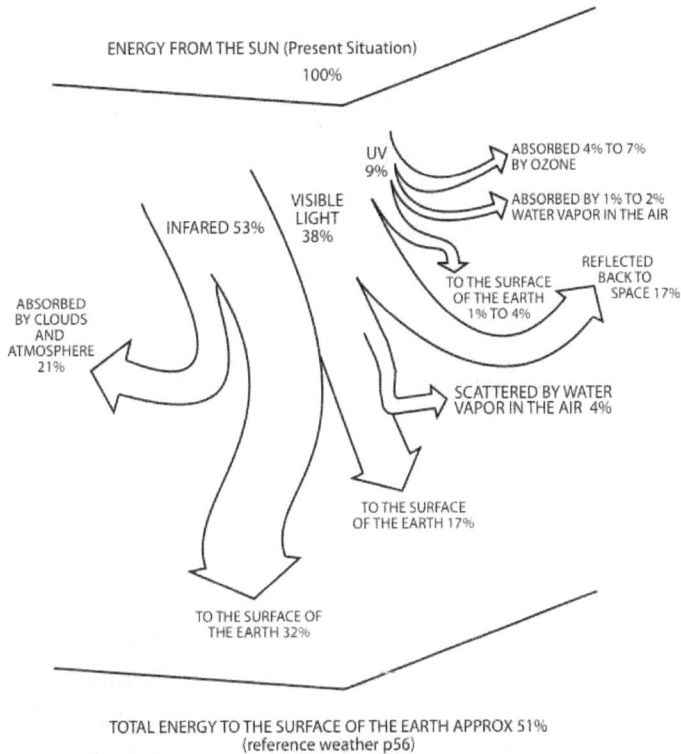

ENERGY FROM THE SUN (Present Situation)
100%

UV 9%

ABSORBED 4% TO 7%
BY OZONE

ABSORBED BY 1% TO 2%
WATER VAPOR IN THE AIR

VISIBLE LIGHT 38%

INFARED 53%

TO THE SURFACE
OF THE EARTH
1% TO 4%

REFLECTED
BACK TO
SPACE 17%

ABSORBED
BY CLOUDS
AND
ATMOSPHERE
21%

SCATTERED BY WATER
VAPOR IN THE AIR 4%

TO THE SURFACE
OF THE EARTH 17%

TO THE SURFACE OF
THE EARTH 32%

TOTAL ENERGY TO THE SURFACE OF THE EARTH APPROX 51%
(reference weather p56)

SOLAR ENERGY DISTRIBUTION
WITHOUT A VAPOUR LAYER

None of the quantities shown are either exact or consistent. They
all shift slightly with location, time of year, and atmospheric content.

Reflected visible light energy is referred to as albedo. If the albedo is 1, everything is being reflected. If it is 0.5, only one-half of the visible light impinging on it is being reflected. Forests have an albedo of 0.1, oceans 0.2, deserts 0.45 and clouds possibly 0.5 depending on their dust content. This means that forests are absorbing 90% of the incoming visible light energy because their albedo is only 0.1 or 10%. (64) Similarly oceans absorb 80% because their albedo is only 0.2 or 20%. The average for the whole Earth from the perspective of space is 0.44 (357) because clouds are brighter than either forest or ocean and reflect more energy back to space (and there is always a certain amount of cloud). This means that 44% of the visible energy from the Sun will be reflected back to space. Also the atmosphere will absorb some of the infrared energy and some of the ultraviolet energy. The net result is that approximately 51% of incoming solar energy reaches the surface of the Earth. Some of this energy will be used for plant growth but most of it will heat the Earth. Without a vapor canopy the temperature of those areas, which are directly under the Sun during the day (i.e. the tropics), commonly rises above the comfort zone while the polar regions are commonly below the comfort zone. Temperate regions are in the comfort zone much of the time.

If there was a layer of water vapour above the atmosphere enclosing the entire Earth, this situation would be dramatically changed and all areas of the Earth would be in the comfort zone. Water absorbs infrared energy and it scatters visible light. Absorption results in warming and scattering results in less light being available to see things. A layer of water vapour enclosing the atmosphere of the Earth would absorb virtually all of the incoming infrared energy. Of the visible light which was being scattered, some of it – possibly 50% - would be directed toward the Earth with the result being a brighter sky and a dimmer Sun. Suppose that all of the infrared energy was absorbed and that 50% of the visible light energy was scattered and that only the remaining half reached the surface. Of course, a water vapour layer would completely absorb all of the ultraviolet energy.

Solar energy available at Earth's surface with a water vapour layer above the atmosphere:
Energy =   100% - (53% (all the infrared energy) + 50% of the visible energy + 9% (all of the UV energy))
    =   100% - (53% + 50% of 38% + 9%)
    =   100% - (53% + 19% + 9%)
    =   100% - 81%
    =   19%

Therefore it may be concluded that with a water vapour layer surrounding the Earth, a little less than one-half as much solar energy would reach the Earth's surface as it does at the present time. (i.e. 19% compared to 51%) The primary difference is that in the vapour layer case the heating component (i.e. the infrared) would be above the surface in a medium, which could distribute the energy more efficiently. The diagram, Solar Energy Distribution with a Vapour Canopy, shows these assumptions and relationships.

At the present time, when we do not have a vapour canopy above our atmosphere, the energy received from the Sun is not well distributed around the Earth. Fortunately, the oceans act as a temperature regulation mechanism and provide a thermal flywheel effect which helps to distribute the Sun's energy over the cycle of day and night as well as from season to season. The Earth is dependent on winds and ocean currents to distribute the energy between the equator and the poles. Unfortunately, air has very little ability to retain heat with dry air having the least. It is the dry air of the Hadley Cell circulation systems that is currently the most widespread mechanism in operation transferring heat from the tropics to areas both north and south. The water in the great ocean currents has much greater ability to transfer heat but ocean currents only operate in a way that influences a few areas near their pathways. In summary, present energy transfer mechanisms are not very efficient, resulting in significant temperature discrepancies between the equator and the poles. The Sun shines right straight down on the equator but it hardly shines at all on the North Pole. This situation is partially remedied by movement of both air and water away from the equator. All of the weather patterns of the world result from this movement, including storms which are concentrated, intensive, energy transfer mechanisms. The greater the energy discrepancy between heated zones and unheated zones, the more intensive both weather patterns and storms become.

The energy, which would be received by the Earth from the Sun, would be the same with a water vapour canopy as it would be without it. However, the greater energy transfer ability of the water vapour in the vapour canopy would have resulted in solar energy being distributed over the entire Earth much more efficiently. The reason this would happen is because of the nature of water. Water is able to absorb and hold heat better than almost anything else. With a layer of water enclosing the Earth, the means to transfer energy and distribute it around the Earth would be much improved. Ocean water can and does transfer energy but ocean water is not nearly as free to move over the entire Earth as a layer of water in the upper atmosphere would have been.

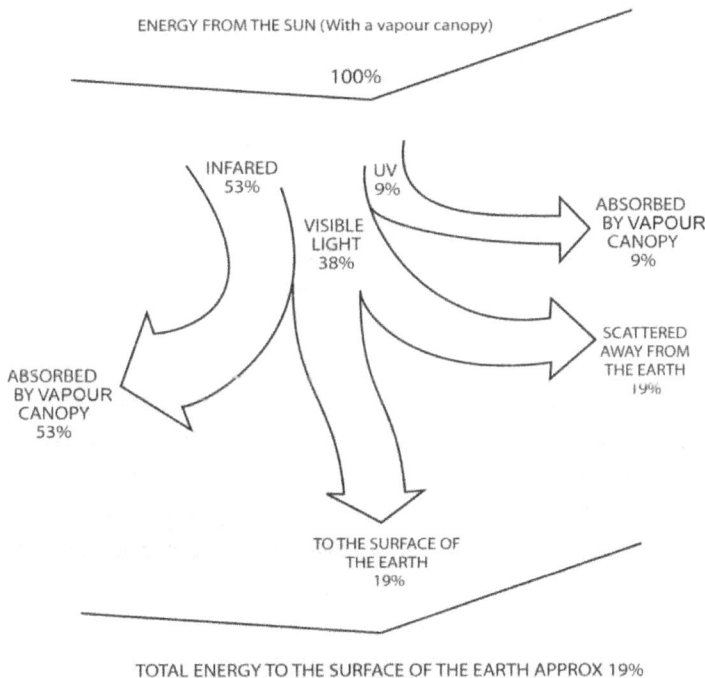

ENERGY FROM THE SUN (With a vapour canopy)

100%

INFARED
53%

UV
9%

VISIBLE
LIGHT
38%

ABSORBED
BY VAPOUR
CANOPY
9%

ABSORBED
BY VAPOUR
CANOPY
53%

SCATTERED
AWAY FROM
THE EARTH
19%

TO THE SURFACE OF
THE EARTH
19%

TOTAL ENERGY TO THE SURFACE OF THE EARTH APPROX 19%

SOLAR ENERGY DISTRIBUTION
WITH A VAPOUR CANOPY

Regardless of the energy output from the sun, the quantities shown would have been
more consistent with the presence of a vapour canopy.

As shown in the 'Atmosphere Pressure Profile' diagram in paragraph 2,12,5, most of the material in the atmosphere, whether it is vapour in the vapour canopy or air in the lower layer, is concentrated near the bottom. Since the bulk of the water vapour would be in the lower part of its layer, the incoming heat energy from the Sun would be mostly absorbed and concentrated in this region. At the same time the upper levels are more exposed to space and due to heat loss would be cooler. Solar energy would therefore warm the lower regions more than the upper regions. When water vapour is warmed, just like air or any other gas, its density will decrease. Further, if a region of vapour were warmed slightly more than a horizontally-adjacent region, there will also be a horizontal density differential. The amount of horizontal temperature differential, which the Sun can produce in one day, is shown on the diagram 'Vapour Canopy Temperature Increase'. There are therefore two types of density gradients conductive to movement, vertical and horizontal.

Whenever the water vapour was warmed by the Sun, its density would have dropped causing it to rise upward while the surrounding cooler and denser regions would have moved horizontally to restore density uniformity. A circulation pattern would therefore have been set up, with the vapour at the upper elevations moving away from the directly heated region to be over the surrounding slightly cooler regions on either side. At the lowest level of the vapour canopy, horizontal movement would have been toward the hottest area directly under the Sun. In the next layer, slightly higher up, the warm vapour would be moving pole-ward. In this manner a global circulation pattern would have developed and become established in the water vapour canopy. Some flow will short circuit and not follow the maximum-distance route all the way. The pattern of circulation would continue as long as there was any difference in density to provide the driving force. Since the Earth is a better radiator of heat than the vapour layer would have been, the circulation pattern would have been augmented by heat loss from the surface of the Earth during night-time. With a vapour canopy, energy from the Sun would be distributed around the entire Earth by a very efficient energy transfer engine. Surface temperatures would therefore remain in the comfort zone with the equatorial regions being cooler than they are now. Polar areas would have been warmer.

As the Earth rotates, the movement would not cease. As long as there was a region of water vapour, which was warmer than adjacent regions, the movement would continue. As mentioned, some movement would be short-circuited and all of the moving vapour would not make it all the way to the extreme northern or southern regions. However the Earth is a sphere, not a flat plate, so as the vapour currents travel away from the area directly under the Sun, they would be converging and the flow would be accelerated. The heat transfer effect would thereby be improved.

Curiously, the basic energy transfer mechanism, which our present atmosphere provides, operates in a similar fashion. The term 'basic' is appropriate because, the mechanism (i.e. solar energy) causing the air near the equator to rise and move towards the poles is continually happening. The Hadley Cell circulation and the Ferrel Cell circulation move energy away from the equator. (358) However, what would have been basic and readily recognized patterns of flow, are significantly modified by the existence of both land and water surfaces on the Earth (The fact that water freezes and turns into ice further disrupts any possibility of regular flow patterns.). If the surface of the Earth was either all water or all land, this would not be the case and patterns of circulation and energy transfer would be very similar in our present atmosphere as they would have been in a vapour canopy layer.

Basic flow patterns would be affected by the rotation of the Earth in either case which effect is referred to as the Coriolis Effect. Atmospheric flow would not occur directly from equator to pole but would proceed along a curved pathway. Northward flow higher up would curve to the west and southward flow in the lowest layer would also curve to the west. In this manner the vapour over any particular area would be constantly changing and would not be simply travelling to the pole and back. With this arrangement, any particular 'parcel' of vapour returning to the equator, would be many miles to the west of where it left the equator as it rose and proceeded northward.

With an atmosphere, which included a vapour canopy, the basic cycle of energy gain and energy loss would be operating far above the surface of the Earth, which would be continually covered by this warm blanket and not exposed to the cold of space as it is now. The temperature over the entire Earth would consequently have been within a fairly narrow band. The blanketing effect relates partly to the ability of the vapour layer to reabsorb heat coming up from the surface of the Earth. This together with the more efficient energy transfer ability of water compared to air would have resulted in the temperature around the Earth being within the comfort zone for humans everywhere.

Vapour Canopy Temperature Increase
(in one day)

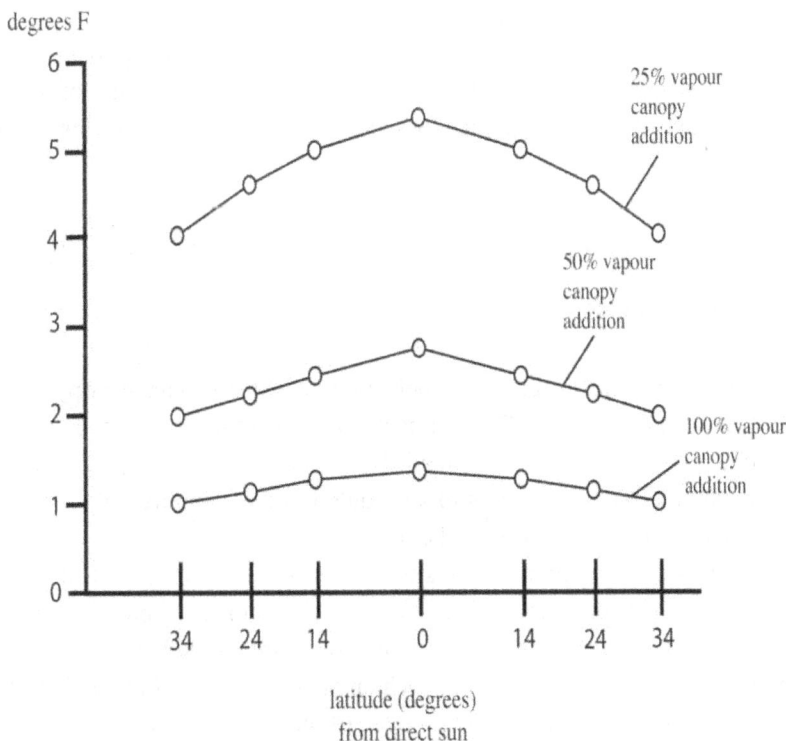

As vapour within the vapour canopy rose, it would expand. At the reduced pressure higher up, the critical temperature (i.e. the temperature at which fog will form) is reduced. Consequently it would be even more difficult for the water molecules to condense and form fog. Therefore if the lowest vapour layer remained above the critical temperature, all higher layers would also. These are the layers that would be involved in the transfer and distribution of heat around the Earth. The vapour layer would not be a good radiator of heat (i.e. it is not black) so the vapour moving toward the poles would not lose all of its heat. Then as it dropped down toward the tropopause and returned to the equator, it would absorb any heat being given off by the Earth. Consequently, the entire circulation pattern would operate within a fairly narrow band of temperature. Much higher up above the circulating vapour currents, heat would be lost to space. This would, of course be necessary to keep the Earth from overheating but it might also result in making the upper regions of vapour visible. It would have become visible if it had condensed into fog or formed ice crystals. Lower down the temperature leveling and blanketing effect of the lowest layers would have kept the entire Earth within a temperature zone that would have been quite comfortable for humans and animals alike. The net result of having a water vapour canopy is that a similar amount of solar energy would be involved in heating the Earth, as it is at present, (i.e. if the Earth was the same distance from the Sun) but it would have been distributed over the entire Earth much more efficiently, thereby making the temperature around the world much more uniform.

It is true that the stratosphere of the present atmosphere does warm up as the ozone absorbs the ultraviolet energy from the Sun. It thereby might have provided a global heat transfer mechanism, except that the temperature within the present stratosphere increases with elevation. The rightward bulge in the temperature curve (as shown in the diagram, 'Atmospheric Temperature Profile') indicates a no-pass-zone where there is no upward movement of air. Therefore, there is no upward movement of air in the stratosphere and no global air circulation pattern in the stratosphere. It therefore does not contribute to transferring solar energy from the tropics to the poles.

2.12.9   Temperature Profiles

2.12.9.1   Atmosphere with Vapour Canopy

As mentioned, the temperature profile of an atmosphere with a water vapour canopy layer is shown in the diagram, Atmospheric Temperature Profile. The leftmost line represents the present atmosphere. The other three, lines 2, 3 and 4 each represent an atmosphere loaded with a water vapour layer. There is also a line in this diagram which shows the temperature that a parcel of air would have if it simply rose and cooled by expansion. This will be referred to as the reference line. It must be noted that it is the slope of the reference line and not its actual location on the chart that is important. This slope tells us how a parcel of air would behave at any particular altitude for any one of the four atmospheres represented. If, for example, the temperature curve of an actual atmosphere sloped more to the left than the reference line, it indicates that the air is colder the higher up we go and that any parcel of air that rose upwards in such an atmosphere would be able to keep on rising. The temperature of the surrounding air would not inhibit its upward movement but would actually enable it because the rising air would still be warmer than the surrounding air. As mentioned previously, the reason that thunderstorms form is because it is colder-than-usual high up in the atmosphere. They form because upward movement of air is enabled by the colder upper temperatures. Any region, where the atmospheric temperature curve sweeps to the left as higher elevations are reached, indicates where upward atmospheric activity could occur. This is the type of temperature profile that would be expected within the vapour canopy (but not from the lower atmosphere into the vapour canopy). Therefore upward movement of vapour within the vapour layer would happen. As a result of upward movement, the temperature would drop due to expansion. Also some heat within the vapour would be lost to outer space. When this vertical activity was coupled with horizontal movement away from the hot equatorial regions, the thermal energy from the Sun would have been distributed much more evenly over the Earth than it is with our present atmosphere.

Atmospheric Temperature Profile
(Illustrative Only)

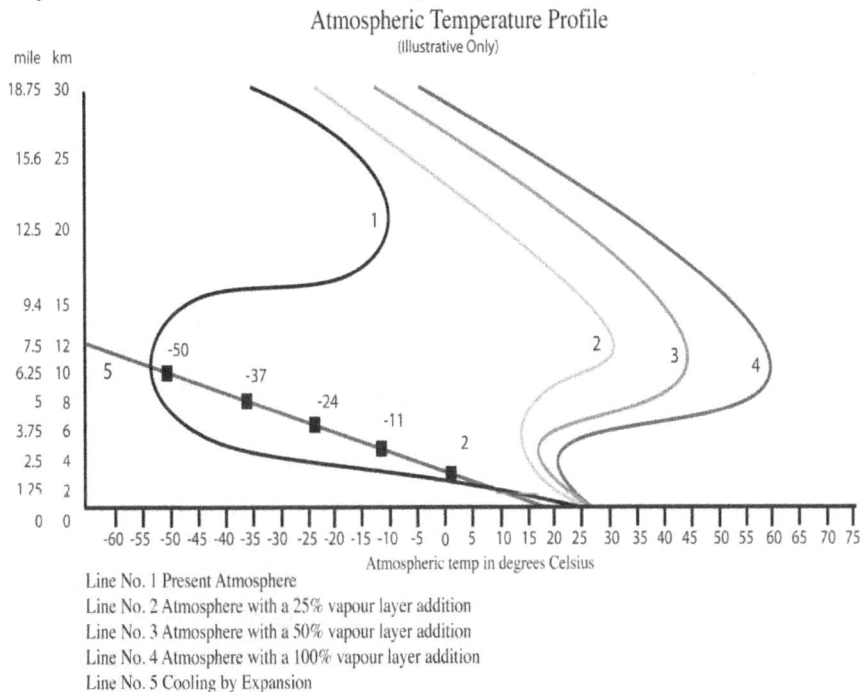

Line No. 1 Present Atmosphere
Line No. 2 Atmosphere with a 25% vapour layer addition
Line No. 3 Atmosphere with a 50% vapour layer addition
Line No. 4 Atmosphere with a 100% vapour layer addition
Line No. 5 Cooling by Expansion

On the other hand, if the slope of the atmospheric temperature curve was a little steeper than the reference line, upward movement of any parcel of air would be inhibited. At any elevation, if the slope is steeper than the reference line, a no-pass zone would exist. In this diagram, any region of atmosphere, where the temperature curve slopes to the right on the way up, emphatically indicates a 'no-pass zone'. Vertical activity cannot occur through such a region because the regions, which are higher up are warmer. A rising air column would therefore be stopped. As shown, the lower layer of the vapour canopy would be quite warm and the temperature curve at the very least would be almost vertical and more expectedly curve to the right. Because of this stability feature, the oxygen-nitrogen region of the lower atmosphere would never mix with the water vapour canopy region above it. The shape of the temperature profile would vary with latitude, time of day and season but it would retain the basic no-pass zone feature because the water vapour in the canopy would preferentially absorb and retain incoming infrared energy from the Sun. In this manner the no-pass zone would be maintained and stable separation of the lower atmosphere and the upper vapour canopy would be guaranteed. An atmosphere which included a water vapour canopy would have a definite 'no-pass zone' at the junction of the oxygen-nitrogen lower layer and the water vapour upper layer. An atmosphere such as this would have a temperature profile very similar to our present atmosphere, which actually includes two 'no-pass zones' – one in the stratosphere, (as shown - elevation between 5 and 15 miles) and one much higher up in the thermosphere, (not shown - which is of no consequence at the surface of the Earth).

2.12.9.2   Present Atmosphere

The present atmosphere of the Earth consists of four layers - so identified because of their particular temperature characteristics. These four layers are: troposphere, stratosphere, mesosphere and thermosphere.

The troposphere is the lowest layer and includes most of the clouds and all of the weather and other activity, which affects all of us on the surface of the Earth. In this layer the temperature normally drops as altitude is increased until an altitude of six or seven miles is reached. At this elevation, the temperature may be -70 F. From basic physics it is understood that when pressure is reduced, temperature will drop. Therefore if a quantity of air is raised to a higher elevation, its temperature will drop as it expands. As mentioned above, storms develop in the troposphere because the surface of the Earth becomes warmer than the temperature required for stability or the temperature higher up becomes colder than the temperature required for stability. This stability relationship is also shown in the diagram, Atmospheric Temperature Profile, and is simply the temperature that any parcel of air would have if it was raised and cooled by expansion. (i.e. the reference line in the diagram) If stability is upset by either a colder-than-usual region above or a hotter-than-usual region at the Earth's surface, there will be upward vertical movement of air, updrafts will be formed and storms will result.

In the troposphere, thunder-clouds form and may rise to great heights. However, they do not usually rise higher than the top of the troposphere. This limitation is readily observed as thunder-clouds rise and reach an elevation where vertical movement stops. Further movement will only be horizontal and the top of the cloud will spread out in an anvil shape. At this elevation, the cloud has reached a 'no-pass zone' or a ceiling, where further vertical movement is inhibited. This boundary, called the tropopause, is usually quite well defined and indicates the top of the troposphere. The layer above is called the stratosphere. There are two exceptions to this basic expectation. If a storm develops a stronger-than-usual updraft, vertical movement will continue into the stratosphere. However, even the most powerful storms will reach an altitude where the surrounding air is the same temperature as the rising air. The updraft will then peter out. For example, a storm may rise to 60,000 feet before the rising air cools by expansion and becomes as cool as the surrounding air at this elevation. Then the storm will not rise any higher but since more air is still coming up from below, the air at the top must move aside. The anvil top will then develop and may be readily seen from the ground. The cloud usually spreads out both ways showing that it has now become too cold to rise any farther.

While the temperature drops as elevation is gained up through the troposphere, the exact opposite happens in the stratosphere. As the elevation through the stratosphere increases, the temperature increases. The tropopause defines a basic boundary for vertical movement because at this level, the temperature no longer decreases with increasing elevation but starts to increase instead. Even if the temperature remained constant with increasing elevation, a boundary would still exist because any quantity of air, which started to rise into such a region, would cool due to being raised and would soon be surrounded by air, which was warmer. Therefore it would stop rising and just settle back down. A 'no-pass zone' would still result. Further, if the air in the stratosphere continued to get cooler with elevation, but not as cool as a quantity of air would, if it was simply raised, there would still be a 'no-pass zone'. As things presently exist, the temperature rises as we progress upward through the stratosphere, so the no-pass zone is quite extensive and therefore well guaranteed.

Ozone is produced and concentrated in the stratosphere. As the ozone absorbs the ultraviolet energy from the Sun, the air is warmed. A major side benefit for those who dwell on the surface of the Earth, is that the harmful ultraviolet energy is intercepted before it reaches the surface causing harm to numerous forms of life, including humans. Due to the absorption of the ultraviolet energy, the air in the stratosphere warms from around -70F at the base of the layer to +40F near the top. (364) As we proceed upward through this layer the temperature curve therefore slopes to the right, which clearly indicates a 'no-pass zone'. Also it is so extensive that no air from the lower troposphere would ever be able to pass up through it.

The ultraviolet energy from the Sun is intercepted and absorbed by the ozone in the stratosphere and the air in the stratosphere is thereby warmed. At night, most of the stratosphere is in the dark and is not being heated. It will therefore cool down a certain amount before the Sun comes up again. However, it will only cool down a few degrees before the Earth rotates far enough to enable the Sun to rewarm it to the temperatures of the previous day. A similar situation would exist within a vapour canopy layer. It would cool down a few degrees at night before rewarming again the next day. A no-pass zone would remain in both cases because the temperature curve will always slope to the right as elevation increased.

One of the great tragedies of losing the ozone layer would be the loss of the rightward bulge in the temperature curve. If this should happen, the lower atmosphere would no longer be isolated from the upper atmosphere and storm clouds, with their entrained updrafts, could rise upwards for many miles. In such a case, the power of the updrafts would be greatly increased along with the destructiveness of the storms. Unfortunately, this appears to be happening as evidenced by the recent reports of storm clouds at 75,000 feet as well as by reports that the ozone is being destroyed.

Above the stratosphere is the mesosphere where the temperature drops to the neighbourhood of -130F before rising dramatically in the uppermost layer, the thermosphere, to over +2500F. Temperatures at this level have no effect on the surface of the Earth because there are only a few widely-separated air molecules at these great elevations.

2.12.9.3 Atmosphere Comparison

The thermal profile of the present atmosphere may be readily compared to the thermal profile of the postulated ancient atmosphere, which may have included a water vapour canopy. Both of these profiles may be considered along with the atmospheric pressure profile discussed earlier. If the ancient atmosphere included the equivalent of a 25% water vapour layer, the 'no pass zone' separating the vapour layer from the oxygen-nitrogen layer would occur at approximately 7.5 miles up. This, coincidentally, is approximately the level of the tropopause in the present atmosphere, which is also a 'no-pass zone' for present weather systems. If the ancient atmosphere included a 50% water vapour layer, the demarcation level would have been lower at approximately 5 miles, which is lower than the present tropopause but not very much.

2.13 Warmer-Earth Evidence

2.13.1 Fossil Evidence

If the above speculation is correct and the atmosphere at one time included a moisture layer on top of an oxygen-nitrogen layer, atmospheric pressure at sea level would have been greater. The fossil record provides evidence that the atmosphere of the Earth was heavier than it is now and that atmospheric pressure at sea level was indeed greater.

Fossil dragonflies have been found which had wingspans which were more than two feet wide. (359) We certainly do not have any insects that large today. In fact it would be impossible because of the way that insects breathe. Insects, like all other creatures, need oxygen throughout their bodies but insects do not actually breathe. 'Insects don't have lungs; they rely on a network of tubes connected to openings called 'spiracles' that run down the side of the abdomen. Insects can pulse their abdomen and flap their wings to ventilate the spiracles, but as the insect gets bigger the proportion of the body that is taken up by spiracles gets rapidly out of hand.' (179) Air is admitted to the insect body through the skin as well as by the spiracles. Air is moved through the skin and along the tubes by atmospheric pressure rather than by the forced flow of air as in mammals which have lungs and a pumping system. Therefore, if atmospheric pressure was greater, the air could move further into the insect, thereby enabling a larger insect to develop. Current atmospheric pressure is not able to force air into insects very far, so they cannot grow to be very big. Apparently this was not the case in the distant past and insects could and did grow to be much larger. What would a dragonfly with a two-foot wingspan sound like?

There was another ancient creature, the pterodactyl, which provides evidence that atmospheric pressure was formerly greater. These were large creatures with very large wings, which appeared to be fully operational and intended for flying. Their bones were hollow and air-filled like the bones of birds. Analysis of their brain cavities indicated that they had massive (several times as large as a bird's) flocculi (the region of the brain that integrates signals from joints, muscles, skin and balance organs. (360)) and this may have been due to their large wings. (361) However, it appears from an analysis of the size of these creatures compared to the size of their wings that they could not have become airborne at present atmospheric pressure. (361) Consequently someone decided that they should be depicted climbing up cliffs so they could glide back down. One of the most interesting pieces of evidence for the previous existence of a vapour canopy is the pteranodon, an enormous flying reptile with wing spans of up to seven meters (twenty-three feet), whose remains are found in the Cretaceous sediments. Experts long debated whether the creature would have had sufficient muscular strength for powered flight. They concluded that it must have lived on cliffs and sailed on the updrafts above the sea in search of fish. … an even larger flying reptile, the pterosaur, with an estimated wing span of fifteen meters (fifty-one feet), was reported … in 1975. …there is no means by which it could have left the ground.

However, if the atmospheric pressure was at that time twice as great … the giant reptile could just have become air-borne. (362)

More recent analysis of the fossils of these great creatures indicates that due to the existence of a small bone near the end of the wing, the lift that the wing could have generated would have been improved by 30%. (363) If this was the case, pterodactyls and pterosaurs could possibly have become airborne with present atmospheric pressure but certainly with a small percentage more.

If these creatures were indeed flyers, their large wings (which were otherwise appropriate for flying) would have been useful rather than just being in the way. Flying is much more desirable than cliff-climbing and much easier on the feet.

2.13.2   Carbon14 Evidence

The vapour canopy would have consisted primarily of water molecules and would have enclosed the atmosphere of the Earth completely. With such an arrangement, certain components of cosmic radiation would have been attenuated before reaching the nitrogen-oxygen layer. In particular, neutrons are readily absorbed by water and this is the reason that ordinary water is not generally used to control a nuclear reactor. If the neutrons in a reactor are absorbed they can no longer affect the reaction, which will therefore slow down too much. (365) Similarly, a layer of moisture above the oxygen-nitrogen layer would have absorbed most of the high speed incoming neutrons from space thereby shielding the nitrogen underneath. As a result very few neutrons would have penetrated down to the nitrogen layer to change nitrogen into carbon14. Therefore, very little carbon14 would have been formed. When the vapour canopy was removed the production of carbon14 would have increased abruptly. If this happened recently it would have caused a discrepancy between the production and decay rates of carbon14 (which there is), and it would make coal appear older (i.e. it has a low C14 count indicating that it is about 50,000 years old) than it is in reality.

2.13.3   The Young Age of Coal

Coal, which is commonly understood to have formed during the Carboniferous Period, is thought of as being very ancient (The carboniferous period spanned 'about 80 million years beginning about 350 million years ago.' (170)) But if this were the case, its carbon14 count would be virtually zero. (Since the half-life of carbon14 is only about 5730 years, (366) after 20 or 30 half-lives the carbon14 count would be negligible and hence not meaningful) As it is, carbon14 commonly indicates that coal has an age of 50,000 years. Unfortunately, this is not old enough to agree with the assumption that the carboniferous period was millions of years ago. An alternate explanation which also recognizes the evidence that human artifacts have been found in coal (367) would be that coal was formed within the last few thousand years and that the plants from which it was formed had not been exposed to very much cosmic radiation (i.e. due to the vapour canopy). The vapour canopy hypothesis therefore accommodates the human artifact evidence simultaneously with the low carbon14 count.

If coal has been formed quite recently the carbon14 count should indicate only a few thousand years instead of 50,000. (368) The low count might therefore be evidence of a vapour canopy having been present while the coal-forming plants were growing.

2.13.4 Carbon14 Rate Discrepancy

Carbon14 is being continually produced in the atmosphere by the action of incoming cosmic radiation on atmospheric nitrogen. (369) It is also, because it is radioactive, continually decaying and changing back to nitrogen. Both the production and decay rates have been determined and the decay rate is considerably less than the production rate. At face value this simply indicates that the carbon14 production process only started a few thousand years ago and that the decay rate has not had enough time to catch up to the production rate. If the carbon14 production process has been going on for a long time, (e.g. for 30 or 40 or 50 half-lives) the two rates would be equal. Since they are not equal, it is appropriate to ask why. It may be that the loss of the vapour canopy allowed the production level of carbon14 to be increased, in which case, the discrepancy is evidence for the canopy's existence.

2.13.5 Evidence from a Universally Warm Earth

There is evidence that the Earth was once universally warm. Since a vapour canopy would have caused the Earth to have been universally warm, evidence of a universally warm Earth supports the vapour canopy hypothesis.

2.13.5.1 From the North

Included in this evidence are the trees of Axel Heiberg, Ellef Ringnes and Amund Ringnes which are islands in the far north of Canada. All of these locations are far above the Arctic Circle and far north of the present treeline. Countless numbers of trees (i.e. logs) are found lying across these islands. (216) Included among the logs are numerous tree stumps which appear in some cases, to have been undisturbed since they were placed. Neither are many of these trees fossilized (i.e. turned into stone) but appear simply dried out. They are not partially decayed either. The wood is both sawable and burnable.

Ellesmere Island - the northern part of which is even farther north than Axel Heiberg Island – was once forested. Arctic wood in a "polar desert" has been discovered that is "so well preserved that the wood can still burn, and even the most delicate tree structures, such as leaves are present". "The dead trees look just like the dried-out dead wood. The place is currently a polar desert where winter temperatures can dip down to 50 below zero and there is not enough light to grow trees (171)

The conclusion, that a warm climate once existed at Axel Heiberg Island, was clearly stated in a report in Science magazine. 'An ancient tropical paradise, complete with turtles and crocodile-like lizards … a hot steamy world … home to huge heat-loving lizards called champsosaurs … daytime temperatures hovered between 25C and 35C', captures the flavour of the report. 'Anywhere a champsosaur is found could not have been very cold.' (371) The bones of champsosaurs were recovered from what appeared to have been the remains of a fresh-water lake. It was the further conclusion of the authors that the warmth was caused by carbon dioxide in the atmosphere and that even the deep oceans were warm at 18C in the Arctic which is just a little bit below room temperature.

As well as trees, there are 'wood-bearing alluvial sediments containing plants of temperate climates'. In this regard the Canadian Arctic has similarities with the Russian Arctic 'where references have been made to the coastal bluffs of the island of New Siberia as the "Wooden Hills" '. (193) Fossil forests are found on the New Siberian Islands. 'In New Siberia (Island) on the declivities facing the south, lie hills 250 or 300 feet high, formed of driftwood ... Other hills of the same island, and of Kotelnoi, which lies further to the west ... (the trunks of trees) lie flung upon one another in the wildest disorder, ... wood hills ... consist of carbonized trunks of trees, with impressions of leaves and fruit ... these desolate islands were covered with great forests, and bore a luxuriant vegetation.'(193)

Finding a woolly mammoth in the far north would not be surprising as they are thought of as being suitable for cool weather. However, finding elephants and rhinoceros would certainly be suggestive that the area was once much warmer. 'The New Siberian Islands ... as well as the islands of Stolbovoi and Belkov ... the soil of these desolate islands is absolutely packed full of the bones of elephants and rhinoceroses (and woolly mammoth) in astonishing numbers.' (194)

More evidence is available at Spitzbergen Island, which is an island north of Norway and about as far north as Axel Heiberg Island. Coal has been found at Spitzbergen. (198, 372) The material which would later form coal would have required a warm climate to grow. Coal is understood to have been formed from the remains of plants. These plants had to have grown (and plant growth is only possible) when the appropriate heat, light and nutrients are available. Since the ice pack, the extreme weather and the lack of enough sunlight currently inhibit plant growth on Spitzbergen Island, it may be concluded that the heat and light conditions of that ancient time were appropriate for the plants to grow and so they did.

Fossil palm trees have been found in Alaska. (373) This type of tree usually grows in much warmer locations. Since there are no palm trees either growing in Alaska or being carried to Alaska at the present time, these trees must have grown there some time in the past or they were carried there some time in the past.

There are really only two options to explain the presence of trees and fossil animals in the frozen north. The trees were either carried to these locations or they grew in these locations. Either way the climate of the Earth must have been a great deal warmer at that time than it is now.

Antarctica also provides evidence of having been warm as 'extensive coal measures are found'- at least seven seams of coal from three to seven feet thick. (27, 40) How this material could have washed in from elsewhere is most difficult to explain. The only reasonable conclusion is that the material grew in-situ and that the weather was warm enough to enable this growth. In addition to the coal, dinosaur fossils and tropical trees are found (72) while Spitzbergen, so far north of Norway can only claim dinosaur footprints to go along with coal as evidence of a warm climate. (198)

While a universally warm Earth is commonly acknowledged, it is occasionally attributed to CO_2. While CO_2 is a heat retention gas, it is certainly not a heat transport mechanism and would not eliminate temperature disparities between the equator and the poles. While there is some evidence that the level of CO_2 was higher during the time that the Earth was universally warm, (56) and that it would have contributed to the retention of the heat that got through the vapour layer, it could not on its own have accounted for the equity of temperature that once existed between the equator and the poles.

2.13.5.2   Submarine Canyons

Submarine canyons are evidence of a universally warm Earth because they are direct evidence that the ocean was warm. Submarine canyons always appear as extensions of large rivers right out into the ocean. Some of these formations are vast canyons that are hundreds of miles long and continue down for thousands of feet right to the bottom of the sea. Like river-cut land canyons, sea canyons are deep and winding with steep wallsand narrow floors. Also they appear quite youthful and may have something to do with the Great Ice Age. (55) Indeed they do relate to the Ice Age and were formed when the Ice Age ended. The cold melt-water from the great ice fields flowed down to the sea and because it was colder (i.e. denser & heavier) than the sea water, it flowed along the bottom of the sea and carved the canyons. If the ocean had been as cold as the melt-water, the submarine canyons would never have been formed. This assumes that the ocean water was a little less salty than it is at the present time but since salt is being continually brought to the ocean, this is a reasonable assumption. However, since the density of water decreases by more than 4% as its temperature is increased from freezing to boiling, even if there had been some salt in it, the cold melt-water would still have been denser than an ocean that was several degrees warmer than it is now. This enabled the cold melt-water to flow straight to the bottom and form the submarine canyons on the way down.

2.13.6   Gigantism Evidence

As well as the fossil evidence discussed above, there is further evidence that at one time the Earth was home to many very large creatures. From the ground, fossils, the evidence of extinct creatures, are continuously being recovered. While a range of size is commonly found, some of them were very large. The dinosaurs, for example, stood several times as high as man and some specimens reached lengths of nearly 100 feet. Tyrannosaurus Rex was about 47 feet long. Triceratops and Stegosaurus were in the range of 20 to 25 feet. (397) However, the dinosaurs were not the only large creatures. Baluchitherium was a very large horse-like mammal, 18 feet at the shoulder. (398) None of these extinct creatures have recognizable modern counterparts but there are many that do. These include: Rhinoceros (the size of a two story house); Raccoon (large and as dangerous as black bears); Guinea Pig (large as rhinos); Giant Teratorn (a flying bird with 25 ft. wingspan, 10 feet from beak to tail, standing 6 feet high when on the ground; (399)), Titanus (flightless bird 10 ft. high and approximately eight hundred pounds); Woolly Mammoth (about twice as massive as present-day elephants and 15 feet tall); Crocodile (51 feet long); Sabre-toothed Tiger (9 feet long); Great Shaggy Bison (7 1/2 feet tall); Man (9 1/2 feet tall (400)); Pterosaur (35 foot wingspan (401)); Megatherium (18 feet long (402)); Beaver (large as black bears); American Lion (25% larger than present day African Lion); Jefferson's Ground Sloth (size of an ox); Short-Faced Bears (tall as a moose); Camels (about one-fifth larger than present camels). (413) We must also include the Dinohyu Hollandi, a giant swine that stood six feet high. (403) From Peru comes evidence of giant penguins. These creatures stood between four and five feet high and according to the report, had cousins which were even larger. It is also curious that they lived where the climate was warm whereas present-day penguins are more at home where the climate is cold. (404) A discussion of gigantism would not be complete without mentioning snakes. The ancient giant snake, named Giant Titanoboa was about 45 feet long, too wide to go through a normal door and may have weighed more than a ton. This discovery in South America included 'turtles and giant crocodile-like dinosaurs'. (60) Another giant creature also deserves mention. There was a giant, rhinoceros-sized creature, possibly related to modern wombats but it was 3 yards long, 6 feet tall at the shoulder, and weighed 2 tons. (414)

There is no doubt whatsoever that many ancient creatures were very large, including humans. Since nothing happens without cause, what caused so many ancient creatures to be so large? The possible link between this gigantism and the hypothesized vapour canopy may be the Hypothalamus Disregulation Theory. (405) The basic premise of this theory is that having a higher level of oxygen available in the hypothalamus would positively affect both gigantism and longevity. Actually the Hypothalamus Disregulation Theory discusses the combined effect of greater CO_2 along with higher oxygen pressure. The CO_2 would dilate the blood vessels allowing more oxygen to reach the hypothalamus. There is, of course, a serious limit concerning how much more atmospheric pressure both animals and man can safely tolerate. Appropriate studies have apparently shown that up to almost two atmospheres may be acceptable. (406) (Higher pressures have been maintained in deep sea diving experiments where the higher pressure had a positive effect on healing. (407) Also high pressure oxygen has been used in the treatment of fractures wherein the healing rate in rats was accelerated by 25%. (408)) If the Hypothalamus Disregulation Theory is correct, slightly higher atmospheric pressure combined with a greater level of CO_2 would have resulted in larger animals. Of course, the required higher atmospheric pressure could have readily been provided by a vapour canopy. Therefore, if the link between gigantism and higher atmospheric pressure is valid, gigantism may be cited as evidence of a vapour canopy.

2.13.7   Plankton Evidence

Evidence of greater oxygen pressure has been obtained from an analysis of the shells of small extinct, ocean creatures. (409) From a study of oxygen isotopes retained within the structure of these little creatures, certain investigators have concluded that oxygen pressure was once 35% compared to the present 20%. From a preliminary viewpoint this represents a 75% increase above present oxygen levels and could have been achieved in several different ways. One way would be to have the present amount of nitrogen and simply have more oxygen in the atmosphere. In this case, with more oxygen, atmospheric pressure would have been higher.

If, for example, there was 50% more oxygen, it might be represented as 30 units instead of 20. Since nitrogen is 80 out of 100 at present, the total with 10 more oxygen would be 30 plus 80 or 110. With this arrangement, the percent of oxygen has only increased to 30 out of 110 or to 27.3%. Similarly, if the oxygen level was doubled to 40 units, the percentage of oxygen would be 40 out of 120 or 33%. 35% oxygen would therefore represent more than twice as much as there is at present and would have resulted in an atmosphere which was approximately 20% heavier than the present one.

Alternatively, the vapour canopy hypothesis provides a consistent and valid explanation for an ancient oxygen level of 35%, which, as mentioned is twice as much as there is at the present time. If there had been a burden on top of the present atmosphere amounting to 75% of the present atmosphere, ancient oxygen levels would have been at the 35% level. If this burden had been provided by a vapour canopy, it would have weighed 75% as much as our present atmosphere and the postulated higher level of oxygen would be explained by the greater load on top of the atmosphere instead of by actually having more oxygen. There would not have been any more oxygen, only a higher pressure of oxygen (as well as the whole atmosphere). It is therefore submitted that the greater amount of oxygen which is thought to have existed in ancient time, is evidence of a water vapour layer above the present atmosphere, which layer was approximately 75% as heavy as our present atmosphere. The problem of explaining what became of the extra oxygen is thereby solved. With the vapour canopy hypothesis, there wasn't any extra oxygen at all but only greater oxygen pressure. This greater pressure was lost when the canopy collapsed.

2.13.8   Illumination in the High Arctic

There is a second way that the trees of Axel Heiberg Island provide evidence for a vapour canopy. The trees found on Axel Heiberg Island in the high Arctic of Canada are mummified. Numerous stumps are found positioned in the ground just like they would be if that is where they had grown. (410) The tree trunks are broken from the stumps. Since there is no sign of decay, the conclusion must be reached that the trees grew in this location and were catastrophically destroyed. The only other suggestion that has been occasionally put forward is that Axel Heiberg Island was further south when the trees grew. Subsequently it drifted north to its present location. (410) However, drifting would have required many years during which time the trees would have rotted. They did not rot. They were simply knocked down, buried and frozen before rotting took place.

Many of the stumps and logs are found in a mummified state. The wood is still wood and has simply dried out and has not become fossilized. There are also fossilized trees on Axel Heiberg. Either case implies that nearly as soon as the trees were knocked over, they were protected from rotting. The preservation of much of the associated material including nuts, bark and leaves, is so good that it is almost indistinguishable from the litter on the floor of a modern coniferous forest. (410) From this evidence as well, it may be concluded that an extensive forest once grew where all of this forest material is found today.

However, trees cannot presently grow where the trees of Axel Heiberg are found. (396) Even allowing that the temperature may have been warm enough, the amount of light available this far north is considered inadequate for tree growth. It is completely dark for several months every winter and during the summer the Sun is low in the southern sky. At the present time it would be impossible for any forest flora to exist in the northern regions (even if it was warm enough) because there isn't enough light. As photosynthetic organisms, problems exist for plants in the Arctic. At Resolute on Cornwallis Island the long winter night begins on November 4 and ends when the Sun again rises again on February 5. (396))

It is certainly true that no trees of any consequence grow this far north today. The bitter cold could be sited as sufficient to prevent tree growth but the lack of light is also considered sufficient, even if the climate was much milder. It is reasonable to conclude that the trees of Axel Heiberg Island and the other islands of the high Canadian Arctic as well as Antarctica, present a great dilemma for many theories of the Earth because their existence cannot be explained. However, there are three types of atmospheric phenomena which collectively point to a solution to this dilemma.

2.13.8.1   Tunguska Explosion

The Tunguska Event involved the explosion of an object in the atmosphere and this had physical effects on the countryside of central Siberia. Trees were flattened (411) but in addition to this immediate and localized result, a great deal of dust was stirred up and spread far and wide throughout the atmosphere. It spread across both Europe and Asia. Usually it would be expected that so much dust would block out the Sun, which may have partially been true, but in this case, it redistributed the Sun. A letter-writer to the London Times reported being able to read a book at night by the sunlight which was reflected from the dust in the atmosphere. (395) The night sky was lit up. London, England is located at approximately the 50th parallel of latitude north. For a location like this to be illuminated at night by reflected sunlight, the light must be transmitted around the Earth from wherever it is shining. At summer solstice, which occurred barely one week earlier than the date of the report, sunlight would have been continuously available at the Arctic Circle which is only 17 degrees of latitude north of London. To reach London, the light from the Sun would have to bend/refract/reflect around the Earth for 17 degrees which is equivalent to saying that a beam of sunlight would be bent 17 degrees from its original path. At summer solstice, when the Sun is sitting on the horizon at midnight at the Arctic Circle, it will be about 182 miles above the Earth at London. There is no atmosphere at this altitude so the sunlight could not have been simply reflected to London. Therefore, it must have been bent and refracted around the Earth at a much lower altitude.

The Tunguska explosion occurred on June 30, 1908. It would be expected that by that date all of the ice and snow from the previous winter would have melted. However the explosion stirred up the surface of the Earth with considerable vigour and the Tunguska River is nearby. In fact, the event is named after the river and its associated valley. It is therefore likely that some water was thrust into the upper atmosphere along with whatever dust and other debris was blown loose. In any event, whatever was projected high above the Earth was sufficient to carry sunlight from at least the Arctic Circle to lower England. (This discussion assumes that the letter writer was reporting from London. If this person was elsewhere, a different conclusion is needed.)

2.13.8.2   Noctilucent Clouds

The second type of atmospheric phenomena suggestive of how sunlight could propagate beyond the horizon is the Noctilucent or night-glowing clouds. These very faint clouds extend for over one million square miles. (393) Their unusual feature is that they exist 80 km. (50 miles) above the Earth. (394) This means that they may be seen long after sunset because at this altitude the Sun can still be seen when it is 9 degrees of latitude below the horizon at ground level. These clouds are therefore reflecting sunlight 9 degrees past the sunset line.

The Noctilucent clouds appeared about two years after the volcanic explosion at Krakatoa in 1883. (393) It has been postulated that the massive energy from that explosion, (which was heard in Australia 3500 km. away and sent both atmospheric pressure waves and water waves completely around the world) projected dust and water vapour into the upper stratosphere. At this altitude, water vapour or ice particles are trapped and will not settle out. Lower down in the stratosphere, it is warmer because ozone traps ultraviolet solar energy and keeps the air relatively warm. Therefore if water vapour settled lower it would boil, (because atmospheric pressure is extremely low at this altitude) and ascend back up again. Similarly, over an equatorial desert, rain might not reach the ground. As it descends into warmer regions, it often evaporates and drifts away unseen.

The other possible source of water vapour for the noctilucent clouds is incoming material from space. It is possible that small comets come into the atmosphere and because they would be mostly water instead of rock, they would break up far above the surface of the Earth. A portion of their material could then remain trapped in the upper atmosphere and contribute to the noctilucent cloud effect. (392)

2.13.8.3   Atmospheric Ice Crystals

The third type of atmospheric condition which is suggestive of how sunlight could have reached Axel Heiberg Island involves ice crystals. Whenever an atmospheric cloud of ice crystals occurs when the Sun is low in the sky, secondary suns may be seen on either side of the Sun. These refractions of sunlight are called sundogs. There are usually two sundogs because the light is being bent horizontally as it passes through the layer of ice crystals. Therefore with Sundogs, sunlight is being bent 22 degrees. This means that if the layer of ice crystals extended higher up (until it was 22 degrees higher than the Sun) a complete semicircle of sunlight would be seen instead of two elongated suns on either side. Also, if for some reason the Sun could not be seen directly due to an intervening obstacle, the sundogs would still be seen. In this effect, the ice is simply refracting the sunlight 22 degrees causing a solar image to appear 22 degrees away from a direct line between the Sun and an observer. Therefore if an observer was 22 degrees above the ice cloud, a single sundog would be expected. Similarly, if the Sun were below the horizon, but still shining into a cloud of ice crystals, a sundog could be formed and bend the light downward 22 degrees to an observer. Alternatively, when the Sun is high in the sky and a layer of ice crystals is in between, a complete halo will be seen around the Sun. It is simply characteristic of a cloud of ice crystals to refract light 22 degrees.

2.13.8.4   Atmospheric Refraction

Fourthly, at both sunrise and sunset, bending (i.e. refracting) of sunlight occurs. In these cases, the light is being refracted by the atmosphere to make the Sun appear red. This happens because red is the component of light which bends the least. The refraction, in this case, is very slight, but it makes the Sun appear above the horizon whereas it might actually be below it. In this manner, sunlight curves slightly around the Earth and extends slightly above the Arctic Circle at winter solstice.

2.13.8.5   Looming

There is a fifth type of light bending that also occurs in the atmosphere. The more familiar phenomenon involves air which becomes increasingly warmer toward the ground. In this case, light from distant objects will bend upward and cause a mirage to form. In particular, the light from a blue sky is often bent upward in such a way that the observer thinks he sees a lake whereas he is actually seeing the sky. Alternately, if the air gets warmer with elevation (as it would with a vapour canopy) an inverted mirage might occur. This is called looming because an image of a distant object will be seen looming over the horizon. (15) In this case light is being bent downward enabling an object which is over the horizon to be seen.

2.13.8.6   Whitehorse, Yukon

Whitehorse is a city and the capitol of The Yukon Territory, Canada. Its latitude is 60.7 degrees north. This is only 6 degrees of latitude south of the Arctic Circle where, at summer solstice (June 21), there is perpetual sunshine and the Sun doesn't quite disappear below the horizon at midnight. Whitehorse, a little further south, doesn't receive direct sunlight at midnight at that time but neither does it get completely dark for several weeks every year around the time of summer solstice. ' ... as the calendar ticked toward summer solstice ... in the land of the midnight Sun ... people go for late-night canoe rides ... in a golden twilight that is measured in hours. Whitehorse has a ...'24 Hours of Light Mountain Bike Festival; held the last weekend of June ... the only 24-hour race in the world "where lights are illegal"'. (243) Regions south of Whitehorse are also partially illuminated through the night at summer solstice - in decreasing amounts the further south we go. This illumination scenario makes it clear that even with our present atmosphere there is illumination around the Earth much further than where the Sun is able to shine directly.

2.13.8.7   Crepucular Rays

One more type of atmospheric illumination should be recognized at this time as it occurs quite frequently. Crepucular rays are a common phenomenon involving the Sun's rays and can appear quite beautiful and awe-inspiring as they streak across the sky. The phenomenon occurs when the Sun's rays are made visible by haze or dust in the atmosphere. (304) If the Sun is hidden below the horizon, occasionally rays of light radiate outward from it like giant headlights. If there was moisture very high above the surface of the Earth these rays could provide illumination for the surface of the Earth well beyond the civil twilight zone.

2.13.8.8   Illumination Summary

There are several ways by which a vapour-canopy-loaded atmosphere could have caused Axel Heiberg Island to have been illuminated to a much higher level than it is now. Temperature inversions would have been expected in such an atmosphere because the vapour canopy may have stayed warmer at night than the surface of the Earth. Whenever this would have happened, looming may have developed and sunlight may have been bent around the Earth and illuminated areas which would otherwise have been dark. Secondly, having water in either vapour or ice form at very high altitudes, would have allowed the Sun to shine by reflection on areas far beyond the reach of its direct rays. This primary reach may have been extended by secondary and even higher order reflections until there was light far beyond any location where the Sun was on the horizon. Thirdly, ice crystals in the vapour canopy may have bent sunlight as in the sundog effect. As with reflection, there may well have been secondary bending and widespread scattering causing sunlight to reach and shine on land which would have otherwise been well into the shadow area.

In order to illuminate Axel Heiberg when the Sun was on the horizon further south at the Arctic Circle, the amount of bending required from any combination of the above possibilities, would have been about 15 degrees of latitude. This is further than either the noctilucent cloud example but similar to the Tunguska cloud effect. It is further than we would expect from looming. However, since the vapour canopy would have extended from about seven miles up to about fifty miles, a much greater amount of reflective and refractive material would have been involved. The possibility of propagating sunlight far beyond the horizon consequently comes much further into the realm of feasibility. Further, it may not have been necessary to provide an abundance of light all year to enable the trees of Axel Heiberg to grow. They may have done quite well with reduced light for several weeks every winter as currently happens with boreal forest much further south than Axel Heiberg. However, with so much reflective material available, it is also possible that the areas in the far north were never really dark due to 'a visible water heaven, scintillating with light' (390). In any case, the trees of Axel Heiberg Island may be explained by the vapour canopy hypothesis and they therefore provide evidence to support it.

In the above discussion about illumination over London, England, 17 degrees of latitude are involved whereas only 15 degrees of latitude are involved in the Axel Heiberg situation. Both of these cases involve amounts of latitude which fall within the three definitions of twilight. These include Civil, Nautical and Astronomical twilight which respectively end when the Sun is 6, 12 and 18 degrees below the horizon. At the end of civil twilight, the brightest stars are visible and the horizon is clearly defined. At the end of Nautical twilight the horizon is no longer visible. At the end of Astronomical twilight the indirect illumination from the Sun is less than the light from the stars. From these definitions we can conclude that with a little help from an upper atmosphere that was loaded with water vapour and ice crystals, light could conceivably have reached Axel Heiberg Island every day of the year. (Please refer to Ice Bound, by Jerri Nielsen, page 147, published by

Hyperion, New York for a brief discussion of this topic.) This notion is reinforced by the reality that when the Sun sets in the high arctic it approaches the horizon at a very shallow angle. This means that it will be just barely over the horizon for some time after it directly disappears and darkness will be delayed.

2.13.9 Noctilucent Clouds Again

Noctilucent clouds not only illustrate that sunlight can be reflected far beyond the horizon where it would otherwise not normally shine, they also illustrate that a layer of vapour in the upper atmosphere can remain in place indefinitely. The noctilucent clouds have been in place for more than one hundred years. If they had been gradually deteriorating, they would probably be gone by now. Their stability provides evidence that even when mixed with air, vapour can remain in the upper atmosphere for a very long time.

2.13.10 Ancient Human Records

There are numerous accounts from ancient mythology that a water heaven existed above the atmosphere. '... numerous mythological accounts of the ancient Earth... tell of a visible water heaven, scintillating with light.' (390) 'In ancient Egypt, the heaven was regarded as an ocean parallel with that on Earth. The Sun god travelled in a barge through this ocean which surrounds the world. This watery heaven was the god Canopus.' (391) In Indian religious literature, 'the idea of an upper or heavenly sea is frequent'. The Greek word for heaven may mean 'there waters' and it was located above the upper air paralleling the Hebrew idea of waters above the expanse and not in it. In the Babylonian creation account, Tiamat was a water ocean half of which formed the sky. Indirect evidence for a vapour canopy comes from a Persian account when it refers to there being neither 'cold wind nor hot wind...enabling both mankind and animals to increase at an alarming rate'. This favourable state was followed by a flood and cold weather. From Polynesia reference is made to the sky being low followed by a situation where the sky retreated to its present position. In a Sumerian account, the waters above were maintained up there by a metal vault, possibly made of tin, as tin was thought of as 'metal of heaven'. (390)

'The ancient peoples of Mexico referred to a world age that came to its end when the sky collapsed and darkness enshrouded the world. (Could this have been the post-impact dust cloud?) The Chinese refer to the collapse of the sky which took place when the mountains fell. (An asteroid impact would have brought down the vapour canopy and it would also have formed and destroyed mountain ranges.) The tribes of Samoa in their legends refer to a catastrophe when "in the days of old the heavens fell down". The Lapps make offerings accompanied by prayer that the sky should not lose its support and fall down. The primitives of Africa ... tell about the collapse of the sky in the past. The tribes of Kanga and Loanga also have a tradition of the collapse of the sky which annihilated the human race.' (389)

The ancient Hebrew account is explicit and correlates the above reports. '... and God proceeded to make the expanse (where birds fly) and caused the division of the waters which were down underneath the expanse from the waters which were up over the top of the expanse, and it came to be so.' (Holy Bible, 388)

The exact meaning of ancient accounts, such as these, can never be firmly established, but never-the-less, they are all more than a little bit suggestive of a water layer above the atmosphere in which case they are supportive of the vapour canopy hypothesis.

2.13.11   Absence of Ancient Deserts

Many of the great deserts of the world have records made by humans that indicate these areas were once anything but desert. With respect to the Sahara we have: 'Drawings on rock of herds of cattle, made by early dwellers in this region, were discovered by Barth in 1850. Since then many more drawings have been found. The animals depicted no longer inhabit these regions, and many are generally extinct. It is asserted that the Sahara once had a large human population that lived in vast green forests and on fat pasture lands. Neolithic implements, vessels and weapons made of polished stone, were found close to the drawings. Such drawings and implements were found in the Eastern as well as the Western Sahara. Men lived in these "densely populated" (flint) regions and cattle pastured where today enormous expanses of sand stretch for thousands of miles. It appears that a large part of the region was occupied by an inland lake or marsh known to the ancients as Lake Triton.' (387) The Arabian deserts have similar information. 'In the southern part of the great Arabian desert, ancient ruins, almost entirely obliterated by time and the elements, and vestiges of cultivation are silent witnesses of the time when the land there was hospitable and fruitful; it was as copiously watered and luxuriously forested as India on the same latitude. Orchards covered Hadhramaut and Aden. It was a land of plenty, paradise on Earth, but following a sudden catastrophe, Arabia Felix turned to a barren land.' (386) Similar evidence comes from farther east. 'Like the Sahara and Arabian deserts, other deserts of the world disclose the fact that they were inhabited and cultivated sometime in the past. On the Tibetan plateau and in the Gobi Desert remains of early prosperous civilizations were found and occasional ruins surviving from those times when the great barren tracts were cultivated. (385)

The fact that deserts currently exist in all of these regions and that they occupy a band at these latitudes that extends all the way around the world is not coincidental. The deserts exist in these areas due to the present pattern of atmospheric circulation. In particular, deserts exist where the descending air column of the Hadley Cells comes to the surface of the Earth and feeds the trade winds.

The warm air at the equator rises as far as the troposphere, drifts north (or South) and then sinks back to Earth's surface about thirty degrees north and south latitudes. Some of the air from these latitudes moves back to the low pressure at the equator and this airflow is known as the trade winds. The circulations that rise in the tropics, sink at 30 degrees and flow back to the equator are known as Hadley Cells. (384) When the warm moist air at the tropics rises, it cools and loses its moisture load. In fact the temperature drops three degrees Celsius with every thousand meters of altitude. Since cool air cannot hold nearly as much water as warm air, the moisture comes out in the form of tropical downpours. High above the equatorial regions, therefore, there is a constantly replenished layer of chilled, recently-dried air – which is then pushed away, to both the north and the south, by more warm, moist air rising from below. This cold dry air comes back down in a very dry state. When it hits the surface it causes the world's deserts. The deserts are not randomly distributed around the planet. Most of them are arranged in two bands girdling the planet north and south of the equator right where the Hadley Cells bring their hot, dry air down to the surface. (383)

The conclusion from all of this evidence is that the deserts of the world exist due to atmospheric circulation patterns. If these patterns did not exist, the deserts would not exist. It follows that if these areas were once luxurious forests and pasture lands, the Hadley Cells, which currently cause all of these deserts, did not exist. If they did not exist, the current global air circulation patterns obviously did not exist. A total lack of atmospheric circulation such as this may be explained by the existence of a vapour canopy. With the canopy in place, there would not have been the great weather-causing air circulations of the present time, but a stable, uniformly warm climate which supported "vast green forests, fat pasture lands, copiously watered and luxuriously forested … a land of plenty, paradise on Earth." (382)

2.13.12   The Existence of Life on Earth

The Earth is teeming with life. There is life everywhere; in the air, in the soil, in the water and all over the land. Included in the results of even one large asteroid impact would be massive worldwide destruction. Everything would be destroyed. Ocean waves would repeatedly overrun the continents and earthquakes would shake the ground until nothing could withstand it. The devastation would be multi-faceted and many "windows of life" would close. These unthinkable events could be sufficient to terminate life on the Earth. Then, when these various catastrophes were finally settling down, a post-impact winter would be in progress. The atmosphere would be so full of dust and smoke that the Sun would not shine at all. A taste of this has been experienced in recent time when volcanic activity produced cloud cover that lowered the temperature of the entire world. It has been estimated that Tambora ejected an incredible 35 cubic miles of rock, dust, and debris. This created a thick layer of dust in the atmosphere that circled the planet and lasted for years and it caused the frigid summer of 1816. (381) A more recent example involves a volcanic eruption on Iceland which produced a cloud that drifted south-east until it was over Europe. The grit in the air was considered harmful to aircraft engines so all planes were grounded until it dissipated. Events such as these are devastating enough but they only involved one eruption. How would life on Earth cope, if there were several thousand eruptions as well as several major asteroid impacts?

If plants are deprived of sunlight for several years and if the temperature remains below freezing for several years, plants will not survive. Without plants, animals will not survive. Without either plants or animals, humans will not survive.

The only way to enable survival would have been to abbreviate the post-impact winter and wash the dust and other particulate matter, out of the atmosphere within months, not years. The atmosphere would have to be cleansed in order for the Sun to shine and life, after the major disruption of both impacts and volcanic eruptions, to become re-established. However, impacts did happen and life really is here. Therefore, it can only be concluded that within a reasonable time, the atmosphere must have been cleansed of the dust and debris to enable life to carry on. The cleansing agent could have been, in part, the hypothesized vapour canopy. If this were the case, the existence of life on the Earth is evidence of a vapour canopy having existed. It would have been terminated by even a single major asteroid impact but as it collapsed, it would have partially cleaned the atmosphere and the Sun would have been allowed to shine much sooner than would otherwise have been possible. This, in turn, would have enabled numerous forms of life to regenerate and become re-established within a reasonable period of time.

While the absence of the Hadley Cell circulation pattern explains the absence of deserts, its development subsequent to loss of the vapour canopy also helps to explain the cleansing of the atmosphere. If the atmosphere had not been cleansed following the impact, a post-impact winter would have persisted reducing the possibility of survival. The atmosphere simply had to be cleansed within a reasonable period of time. As the vapour canopy collapsed, it brought down a lot of dust. As the canopy disappeared, the oxygen-nitrogen layer expanded and in fact became the whole atmosphere. The vapour canopy had also shielded the oxygen from ultra-violet light so that ozone had not yet been produced. (380) There was therefore no increase of temperature up through the stratosphere and consequently no tropopause. With the collapse of the vapour canopy, the remaining atmosphere consisting of oxygen and nitrogen would have expanded and cooled by expansion which would have also prevented tropopause development. Rising warm moist air from the equator could therefore have moved unrestricted to very high elevations before heading north. Therefore, when the Hadley cell circulation pattern first developed it was not restricted to the troposphere as it is now. The air rose much higher before moving north. Consequently the rain that poured down washed dust from much higher altitudes than it could now. Also, when the air was able to continuously circulate to higher elevations, more dust became entrained in it and was brought down to lower elevations where it would be more easily washed out. Therefore the development of Hadley cell circulation helped to clean the air over the tropics earlier than it was cleaned farther north. The post-impact winter in the tropics was quite brief and solar energy reached the surface of the Earth much sooner than it did in the mid and high latitudes. Later, as ozone was produced, a tropopause developed and the Hadley circulation was restricted to lower altitudes. Its original absence explains the lack of ancient deserts and its appearance helps to explain why a post-impact winter didn't persist and terminate life on Earth altogether.

The development of Hadley Cell circulation, with its attendant band of deserts where there had not been any, therefore supports the vapour canopy hypothesis.

2.13.13   The He3 Dilemma

While Ccrbon14 is being produced by cosmic-ray neutrons, a heavy isotope of hydrogen, tritium, is also being produced from deuterium (another isotope of hydrogen i.e a form of hydrogen). Tritium is unstable and decays rapidly by beta decay to an isotope of helium, He3. But there is too much He3 in the atmosphere to be accounted for by this process operating at present rates over geologic time. A factor which would have increased the amount of tritium in the past would be a warmer Earth was warmer with an atmosphere containing much more water vapor, and the process (of generating tritium from deuterium) might have been operating at a much higher rate than at present. (379) Since the vapor canopy hypothesis provides an explanation for the present excess atmospheric He3, the excess He3 is evidence supporting the existence of the ancient vapor canopy.

2.14 Objections to a Vapor Canopy

2.14.1 Greenhouse Gas Objection

In order for the Earth to maintain temperatures well within the habitable range, the greenhouse gases must be within an appropriate range. If the Earth was warm from pole to pole there would have been much more water vapor in the atmosphere. This would suggest that due to the greenhouse gas effect, the average temperature around the world would have been too high for life to exist. However this would not have been the case with a vapor canopy in place because the incoming energy from the Sun would have been intercepted well above the surface of the Earth. A much smaller portion of the incoming energy would have reached the surface and the heat-retention effect of the greenhouse gases near the surface would have been reduced from what it is at the present time. The corollary to this fact is that; IF AT ONE TIME THE EARTH HAD BEEN WARM FROM POLE TO POLE THERE MUST HAVE BEEN A VAPOUR CANOPY IN PLACE TO PREVENT OVERHEATING DUE TO THE GREENHOUSE EFFECT OF THE EXTRA WATER VAPOUR.

2.14.2 Heat of Condensation Objection

One of the basic objections to the idea of a vapor canopy is that when water condenses, it releases a lot of heat. This is a basic fact of nature and this principle is applicable any time that water changes from vapor form to liquid form. However, the conclusion that the heat of condensation released by a collapsing vapor canopy would cause the Earth to overheat, necessitates several assumptions which might not be valid.

The entire canopy had to condense. However, it is more likely that it was already partially condensed and included ice crystals and small water droplets rather than 100% individual water molecules. The ancient accounts of a water canopy suggest that it was visible (378) and the fact that Axel Heiberg Island had enough light to enable tree growth also suggests that the canopy was visible. Visibility indicates that ice crystals and/or small droplets are present. It may therefore have been in a partially-condensed state all the time so a reduced amount of heat would be produced at the time of its collapse.

The basic objection ignores the fact that if an over-riding canopy collapsed, the atmosphere beneath it would have expanded. Expansion always results in cooling. Apart from any other cooling mechanisms, the expansion cooling would have been significant. (377)

The total elapsed time to condense is also a consideration. If the entire mass condensed and fell in one hour, the rate of energy release would have been high. However, if the process took one or two months, the thermal energy released during collapse would have been spread out, resulting in a much lower rate of energy conversion and fewer temperature problems, if there were any.

The collapse of the hypothesized vapor canopy would have partially been the result of vertical movement of atmosphere initiated by the incoming asteroids directly as well as both volcanic action and water wave action. These activities would all have developed vertical air currents causing great masses of air to be driven much higher above the Earth than one would normally expect. However anytime that a shower of asteroids comes to Earth the situation is far from normal. The result of all vertical air movement is cooling due to both expansion as well as exposure to the cold of space.

A condensing vapor canopy can be compared to a thunderstorm. When rain falls from a cloud during a thunderstorm, the water vapor in the air does condense, and the accompanying heat of condensation is released. A thunderstorm may produce several inches of water in an hour and every drop of this condensing water will release heat into the air. In fact, the heat of condensation augments the updraft and is part of the reason that thunderstorms happen. Thunderstorms can produce a lot of rain. 'On Saturday, July 31, 1976 … in Northern Colorado … at Glen Haven some 14 inches of rain fell in three and one-half hours. At the height of the storm the rate of precipitation held steady at five inches per hour for 30 to 45 minutes.' (376)

Another comparison, which involves the release of the heat of condensation, is a hurricane. It has been recorded that Tropical Storm Claudette brought 45 inches of rain to an area near Alvin, Texas in 1979. (376)

However, neither during a thunderstorm nor during a hurricane, does the heat of condensation, which is continually being released, become a problem for people on the surface of the Earth. In fact there doesn't seem to be any effect from this heat at all. It doesn't increase the temperature at the surface. It seems that the heat is dissipated somehow in the clouds, but in any event, there are never complaints about overheating the Earth.

If an atmosphere included a water vapor layer, which could produce several feet of water, and all of this water condensed during a catastrophe, it would be like the rain from several hurricanes coming ashore. If they arrived over a period of a month or more, there would certainly be a lot of rain to deal with but there wouldn't be any heat of condensation to worry about. Similarly, if the collapse of a water vapor canopy happened over a reasonable period of time, there wouldn't be any heat of condensation to worry about either.

2.14.3   Potential Energy Objection

It has occasionally been argued that the potential energy of a collapsing water vapor canopy would overheat the Earth when the water fell and hit the Earth as rain. The potential energy of any object, which is elevated, is converted to kinetic energy when the object starts falling. Much of this second type of energy is dissipated on impact, usually by dislocating the impact surface. If a vapor canopy high above the Earth started to fall as rain, there would be a lot of rain and flooding would certainly result. However, when Tropical Storm Claudette dropped almost four feet of rain in only a matter of hours (376), nobody complained about the Earth overheating due to the impact of the falling rain. A collapsing canopy might have produced two or three times as much rain as Claudette but it would have been spread out over a much longer time period. Consequently any heating effect would have been even less noticeable. Therefore it may be concluded that water hitting the Earth from a collapsing canopy, would not have overheated the Earth.

2.14.4   Exposed Rock Objection

It is understood that when rock is freshly exposed to the air it will absorb carbon dioxide (CO_2). In light of this reality, one of the results of the impact of an asteroid will be a reduction of the amount of CO_2 in the air.

When an asteroid hits the Earth there are several ways by which rock will become exposed. First of all there is simply the impact. An asteroid – even a small one – will blast an opening in the ground exposing rock which was previously hidden. Rock will be broken and ejected from the site. Some of it will be thrown for great distances and in the case of a major event like the Sudbury impact in Northern Ontario, these distances – being commensurate with the size of the crater – could easily be several hundred miles. Erratic boulders can be explained this way. When an asteroid hits the Earth, rocks would be flying for hundreds of miles. Since rocks from Canada are found all across the northern USA, it is readily seen that a great deal of exposed rock can result from an impact.

The second major way that an impact will result in exposed rock will occur at the antipode. Here the shockwave from a large impact (like the one in South Africa) would push up on the underside of the crust on the far side of The Earth and violently elevate it, breaking it into numerous fragments in the process. A great deal of rock can be freshly exposed in this manner. This appears to be what happened on the Moon, on Mars and on Mercury. On Earth we note that the antipode for the large South African impact is the Laurentian Plateau of Canada and the Northern USA. This is an extensive (three million square miles) region of chaotic broken rock and deserving of some explanation. When this region was broken, a lot of CO_2 would have been absorbed from the air.

It seems that prior to the impact of the asteroids, the entire Earth was warm. Both water vapor and CO_2 would have been in abundant supply. In fact with our present understanding of Greenhouse Gases, the Earth would have been too warm for life to exist. Since the current average surface temperature of +15C appears close to optimal, the presence of even more water vapor and CO_2 would have resulted in overheating. Instead of there being abundant life, there would not have been any life. However, this was not the case as life during that time appears to have been abundant.

The explanation lies with the vapor canopy. With a vapor canopy in place, part of the incoming heat from the Sun would have been interrupted high above the Earth. The Greenhouse Gases lower down were therefore dealing with a much reduced energy input. Life on Earth would therefore have been possible.

Then the asteroids came. A great deal of rock was broken and CO_2 started to be absorbed. However, the vapor canopy was destroyed. With the loss of the vapor canopy and in order for it to have been possible for life to carry on, some of the CO2 in the air as well as much of the water vapor in the lower atmosphere, had to be removed or the surface of the Earth would have over-heated. A simple temperature decrease (due to the loss of the vapor canopy) would have removed some of the water vapor and the exposed rock would have removed some of the CO2. More CO2 would have been removed during the coming years as as the ocean cooled and absorbed it. The destruction of the vapour canopy in conjunction with the absorption of CO2 by freshly exposed rock meant that, from the greenhouse perspective, that life on the Earth could carry on.

2.15 A Warmer Earth Conclusion

If a water vapor layer had surrounded the entire Earth in ancient time, the temperature around the Earth would have remained within a fairly narrow range all year similarly to the way that it does now in the Farrow Islands in the North Atlantic. While these islands would be considered to be in the 'north', because they lie between Scotland and Iceland, because they are planted squarely in the path of the Gulf Stream, their temperature remains steady all year. Because they are planted squarely in the path of the Gulf Stream, their weather is moderate with an narrow range of temperature of from 37 to 51 degrees F, all year. (375)

A water vapour layer surrounding the Earth would have been of great benefit. The key to its retention would have been atmospheric stability. As long as the vapor layer was dust free and not seriously disturbed, it would just float year after year providing protection for the Earth. If the atmosphere was seriously disturbed, however, the whole thing would have come crashing down. As well as dealing with this huge quantity of water, the Earth would lose its warm blanket and subsequently chill to some lower temperature range.

If a large asteroid had crashed down through this ancient water-covered atmosphere, there are at least three ways in which the protective water vapor layer could have been destroyed. As has been discussed earlier, first the expanding shock-wave would have caused the water to condense. If the water molecules remained stuck together after the shock-wave passed, a torrential downpour would have resulted. Secondly, the descent of the asteroid would have drawn atmosphere in behind itself. As this material moved horizontally into the region through which the asteroid had just passed, and then downward behind it, a donut-shaped pattern of air movement would have resulted. The associated upward movement would place air high up in the vapor layer. Immediately, the water portion of this new mixture of water and air would precipitate because air simply cannot hold very much moisture. (374) Condensation of the vapor into rain would release the heat of condensation. The upward movement of air would thereby be accelerated. More air would be drawn into the upper atmosphere from the lower atmosphere and more vapor would be mixed in with it. More precipitation would result. (Please refer to the diagram, 'Vapor Capacity of Air', to see just how much vapor air can hold.) Such a region of instability might have expanded outward indefinitely until the whole Earth was involved. In this manner, the entire vapor layer might have been brought down by a single asteroid. The third difficulty would develop from any significant volcanic activity. Upward motion of effluent from a volcano would have driven air up into the water vapor layer and, as before, disaster would result. Stability within the atmosphere would have been the key to retention of the vapor layer and an incoming asteroid would have totally upset stability. It must be noted at this time that, as discussed elsewhere herein, that asteroid formation was a very definitive event. Prior to the asteroid formation event, there were not any asteroids to worry about thereby ensuring the stability of the vapor canopy,

An asteroid would have upset stability in several ways and the vapour canopy would not have survived. In addition to the numerous devastating effects discussed in the previous chapter, the Earth would also have to deal with a rainfall several feet deep over its entire surface. The flooding and erosion from this much flood water would be difficult to visualize. Possible benefits would have included extinction of some of the forest fires (caused by the shock wave) as well as removal of volcanic dust but removal of the warm blanket from above the atmosphere would have caused the Earth to start cooling to some lower temperature range. Polar areas would have chilled first. But, due to the enormous heat retention of a universally warm ocean, its temperature would not drop instantly but would drift down over several hundred years. (242) If the volcanic dust and the smoke from the forest fires caused the air over the land to chill to below freezing before the oceans cooled substantially, an ice age would commence. (This is discussed in a subsequent chapter as well as in The Asteroid Theory of the Ice Age.) The cooling ocean would absorb more CO_2, which would reinforce the chilling effect. Atmospheric pressure would have been reduced by the removal of the vapor canopy and any affect this and the reduced CO_2 levels may have had on aging or gigantism would start to show up during the following years. Reduced CO_2 levels would also result in less vigorous plant growth. All plants would, comparatively speaking, be stunted. The lush forests of the high arctic would never again exist.

2.16   360-Day Year Conclusion

In order to raise the Earth from an orbit which would have given it a 360 day-year, to one with a 365 ¼ day-year, a considerable amount of energy would be required. In fact it would require an asteroid or a group of asteroids with a total mass equivalent to about 0.5% of the mass of the Earth coming in from behind the Earth from about half-way between Mars and Jupiter.

There are two factors, both of which must be valid, for the asteroid-mass requirement to have been met. The first one relates to the relationship of crater diameter to asteroid diameter. Large crater diameters must approximately represent the diameters of the asteroids that created them. This would expectedly be the case when the diameter of the incoming asteroid was a significant fraction of the thickness of the Earth's crust. In cases like this, the Earth's crust wouldn't really be able to stop them before they broke right through into the interior thereby dissipating much of their energy far from the impact site. Consequently, the crater diameter would not reflect the energy of the asteroid and therefore not indicate its size in the conventional manner.

The second factor relates to the number of times that the Earth has been hit by large asteroids. This is unknown. Since large asteroid-impact sites continue to be identified, it is suspected that there are many more that still remain hidden. (We recall that the monster Chicxulub site is buried under 2-3 km of sedimentary rock.(16)) One can only wonder how many more will be recognized in another hundred years. From the above lists of 'known' hits and 'possible' hits (par. 2,7) it is clear that much more asteroid material has hit the Earth than has usually been recognized.

From a review of the size and number of both recognized impact sites and suspected impact sites it is clear that the criteria for orbital change could have been met (i.e. A single asteroid 1280 km. Dia. with a density similar to the average density of the Earth. See par. 2,2 Energy Requirement) if asteroid diameters had been significant fractions of the crater diameters. It would have required a very large object to form the Takla Makan Basin, elevate the Tibetan Plateau and cause the Himalayan Mountains to be concentric to the Basin. The Congo Basin is similarly demanding of an explanation. In fact the Takla Makan impactor could have effected the suspected orbital change on its own. Consequently it can be concluded that changing the orbit of the Earth and with it the length of the year is well within the realm of possibility from the scientific perspective. Taking the ancient records at face value would only correlate such a conclusion.

If the Earth had once been closer to the Sun it would have received more heat. The average surface temperature would consequently have been higher and more water vapour (the most important greenhouse gas) would have been involved. Consequently the Earth would probably have been too warm. However, if there had been a water vapour canopy in place the incoming heat would have been intercepted prior to reaching the surface and it would have been spread more evenly over the entire Earth. Polar regions would have been warmer as well as more illuminated. Fulfilling both of these criteria can explain the presence of mummified forests in these areas. Otherwise they are very difficult to explain and in fact no other explanation is currently available.

The possibility of the ancient year being 360 days long should be retained with the expectation that future scientific observations regarding the number and size of the asteroids that have hit the Earth to be more clearly recognized. This, in turn, would enable the possibility of a 360 day-year to also be more clearly recognized because it is patently obvious that we do not have a complete picture yet. Otherwise an important element of our collective history could be lost. It is clear from the information already known that the notion of a 360 day-year is far from being a 'Fairy-Tale for Adults'.

3.0   The 4.5 Billion Year-Old Solar System

The idea that the Earth has a 4.5 billion year history is widely accepted as the way things are and there are several other ideas that would appear to correlate and support the basic idea. As an example, the universe appears to be very large and since light can only travel at a finite speed – albeit a very fast speed – it would have taken a great amount of time for the light to get to us from all of those distant places. Another idea that is supportive of long time periods is the way that the Sun is understood to be powered. If nuclear reactions power the Sun, since an enormous amount of energy is released when hydrogen atoms fuse together, it can be seen that due to the Sun's great volume there must be enough fuel to power it for billions of years which, by the way, does not say anything about when the process might have started. Support for long time periods also comes from the supposed temporal spacing of the arrival of the asteroids. Some of them are thought to have impacted the Earth more than one billion years ago. (2) Others have arrived more recently but claiming that their arrival has been spread over hundreds of millions of years is commonly done. (2) Therefore declaring that the Earth has been in existence for several billion years fits right in with the overall picture that all of these other ideas support. However things are not always as they first appear and considerable evidence exists which suggests that the Earth is not really very old at all. The following discussion will review some of this contradictory evidence enabling us to conclude that the Earth is actually much younger than is commonly understood to be the case.

3.1   The Moon

The Moon and its interaction with the Earth is directly related to the possible age of the Earth. There are several aspects of this interaction that bear on this discussion and collectively they make it difficult to understand how long time frames of billions of years could be possible.

3.1.1   The Earth's Axial Tilt

The Earth's Axial Tilt) is caused and determined by the Moon. If there wasn't any Moon there would not be a stable and predictable axial tilt and the Earth would not have predictable seasons. Further, if the Axial Tilt was fixed and the distance to the Moon was fixed, the question of age would not come up but since neither of them is fixed they have a bearing on both the time that the Earth could have been habitable as well as on the age of the Moon itself.

As the Earth orbits around the Sun it follows a pathway that is properly called an orbital plane or the 'ecliptic'. (The ecliptic is actually the apparent pathway that the Sun follows against the background of stars but since this is also the pathway that the Earth follows, using the same term is reasonable. This would be more obvious if we could observe the movement of the Earth against the background of stars from the location of the Sun.) Other planets describe similar pathways as they orbit the Sun and these other pathways are very close to the apparent pathway that the Earth follows (against the background of stars). This similarity of movement enables conjunctions to happen. At such times two or three other planets might appear in the sky in relatively close formation. Of course the planets are always a great distance from each other but occasionally they appear in the sky seemingly close together. They very seldom exactly line up but the fact that they appear quite close indicates that they orbit the Sun in almost the same plane as the Earth.

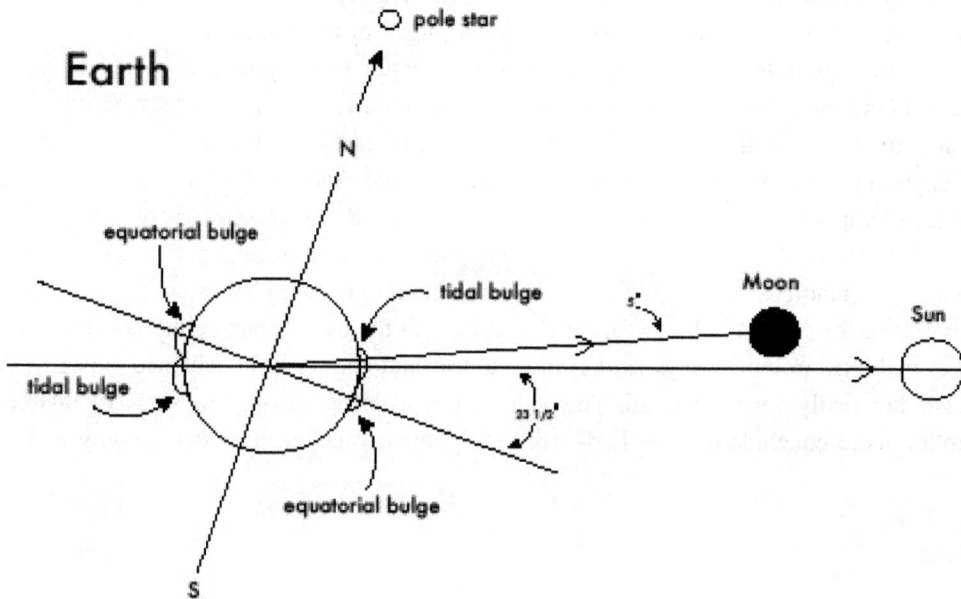

The Axial Tilt 23 1/2°
Illustration shown for summer in northern hemisphere. Bulges shown for illustration purposes only. Actual bulges are more spread out than shown.

As the Earth orbits around the Sun it also rotates on its own axis and this enables us to have day and night. However the Earth's axis of rotation isn't perpendicular to its orbital plane and the difference is called the Axial Tilt. With this arrangement the equator is tipped up by 23 ½ degrees to the Earth's orbital plane and this forms the Axial Tilt. This means that as the seasons progress, the Sun will move up and down in the sky but always stay within 23 ½ degrees of the equator.

Seasons are more than interesting, they are essential. If there wasn't an Axial Tilt there would not be any seasons and the Sun would shine right down on the equator all the time – or would it? It would if the Earth rotated on its axis in exactly the same way year after year but why would we expect this to happen? Why would it just stay the same way all of the time? Why wouldn't drifting be involved? What forces are keeping the Earth rotating on its axis with the North Pole pointing at almost exactly the same location in the sky year after year?

Most things that move must have stability factors built into the movement or drifting will be expected. This is true whether we drive a car, throw a ball or fire a bullet. In the particular case of driving a car we readily understand that we cannot just set the steering wheel and expect the car to go in any particular direction. Small bumps in the road, wind and passing vehicles all slightly upset the direction of movement. Therefore the steering wheel must be continually adjusted to stay on the road properly. This is also true concerning how heavenly bodies move. Throughout the solar system there are numerous forces involved in pulling on every object that exits. In addition, transient (i.e. temporary) forces develop when some unexpected object comes through. Comets are in this category and whenever one of them speeds through the inner solar system, their force of gravity will have an influence on all of the planets and asteroids in their vicinity. Usually these forces are very small and of no consequence but occasionally a large comet will arrive which could have an effect on other objects near its pathway. There was a case in 1979 of a large comet coming into the inner solar system and going directly into the Sun. While going into the Sun is not unusual for a comet (because many of them do go into the Sun), this particular one was very large. Howard-Koomen-Michels 1979X1 was larger than the Earth and hit the Sun in 1979. (22) If this comet had come within a million miles of Earth, the pull of its gravity would have been noticeable with respect to any moveable objects on the Earth but worse than that it might have modified the Earth's orbit slightly which could have resulted in total disaster. Therefore in order for the Earth to have stability, just as a car must have stability when it is driven, there must be an arrangement which can make minor corrections on an ongoing basis. The steering wheel of a car must be continually adjusted by the driver to keep the car on the road and the axis of inclination of the Earth must also be continually maintained or it too will change with devastating results.

Asteroids are relatively small objects when compared to either planets or moons and because of this their orbits can be more easily changed. The gravity of the planets – in particular, Jupiter – is credited with causing the occasional asteroid to drift out of the space between Jupiter and Mars and wonder into other areas. The gravitational influence of Jupiter gradually modifies the orbits of comets and occasionally shifts them into trajectories that carry them inward to the Earth. (29) As a result, asteroids are now found not only all the way out to Jupiter and beyond but throughout the inner solar system as well. This poses some concern for Earth because it is now well known that there are many that cross Earth's orbit. These are called Apollos and a small one actually passes quite close to the Earth almost every week. We would certainly prefer that all of them keep a safe distance from us. While the Earth is many times larger than an asteroid and consequently much harder to move, the occasional passing of such objects underscores the necessity of having stability factors in place. With this in mind it is clear that the system including the Earth and the Moon cannot be left unattended any more than the steering wheel of a car can be left unattended.

For a considerable time now the Moon has been recognized as the factor that stabilizes the Earth's Axial Tilt. (25) As the Moon orbits around the Earth it exerts a very strong gravitational pull on the Earth. This is the force that holds the Axial Tilt in its proper position and it does this by pulling on two of the Earth's bulges. It is the gravity of the Moon pulling on the Earth's bulges that continually generates the forces required to hold the Earth's Axial Tilt at the correct value.

The first bulge involved is the Earth's tidal bulge. The Moon's gravity causes the Earth to have tides. The tides are recognized as an important factor in the survival of coastal animal life but they don't get much credit as a stabilizing factor for the Earth's Axial Tilt. With reference to the diagram, the tidal bulge is shown exaggerated. While we are most familiar with the water tide there is also a land tide. While the land tide isn't very noticeable, it does exist and contributes to the overall tidal bulge. The tidal bulge is at its maximum 28 ½ degrees above the equator which is the angle that the Moon is above the Earth's equatorial plane (which includes the 23 ½ degrees of the Axial Tilt as well as the 5 degrees that the Moon is above the ecliptic (i.e. the Earth's orbital plane around the Sun)). Therefore as the Moon pulls on the tidal bulge it is actually tipping the Earth over further in the direction of increasing the Axial Tilt. If this situation wasn't countered by an offsetting force the Earth would gradually tip right over until the North Pole was pointing directly at the Sun (for a short period of time every year). The Axial Tilt would then be 90 degrees and the Earth would not be habitable.

The second bulge is the equatorial bulge. The Earth is basically a fluid body. Therefore when it rotates on its own axis it bulges at the equator. This bulge is quite significant and results in sea level being several kilometers further from the center of the Earth at the equator than it is at the North Pole. The equatorial bulge is quite flat however which means that it is not noticeable and in fact is spread out in such a way that the poles are only slightly flattened while the rest of the Earth is very close to spherical. While it is not noticeable it provides the second quantity of material or bulge on which the Moon's gravity can operate. In this case since the portion of the bulge nearest the Moon is below the Earth's orbital plane, the gravity of the Moon pulling on it operates to pull the Earth back upright. This offsets the Moon's pull on the tidal bulge. The system works because the equatorial bulge is fixed and the tidal bulge is adjustable. The equatorial bulge is always at the same location on the surface of the Earth and determined only by the Earth's spin. If the Earth should tip over a little too far the pull of the Moon on the equatorial bulge would therefore increase and provide the necessary restoring torque to bring the angle of inclination back to its proper value. If the system was upset the other way so that the Axial Tilt was reduced, the pull on the equatorial bulge would be reduced so the Earth would just tip over again until the proper angle was reached. Stability for the Axial Tilt is thereby provided and the Earth remains habitable with an optimal seasonal arrangement. Any other Axial Tilt would be less desirable and in fact any significant deviation from the present arrangement would make the Earth much less hospitable to life.

While the orbiting Moon is understood to provide orbital stability for the Earth directly, the Moon's involvement in generating and stabilizing the Axial Tilt is paramount in keeping that angle at exactly the right magnitude. The Earth is the only inner planet that has a large enough satellite to achieve axial tilt. Without the Moon's effect the Earth would tilt as much as 85 degrees off vertical. A tilt that large would be catastrophic because the seasons would not occur. Murray concluded that the influence of the Moon is very critical and he concluded that the forecast for a moonless Earth would have been very bleak for life. (58)

The several factors involved in providing axial stability include; the size of the Moon, the fluidity of the Earth, the rotational speed of the Earth, the orbital plane of the Moon and the distance of the Moon from the Earth. If any of these factors should change, the Axial Tilt would change or even drift right out of control. In this manner the very existence of the Moon is recognized as a necessary factor in order to have life on the Earth.

Mars has two very small moons which allow for chaotic tilt up to 60 degrees. Consequently its current 25 degrees will not be stable. (58) While the uselessness of Mars' moons as stability instruments is properly recognized, Mars could never have had an axial tilt in any event because an appropriate tidal bulge is required along with an equatorial bulge. Such features would be necessary in order for any planet with a moon for that moon to provide dynamic self-correcting axial stability. The absence of these factors on Mars, rule it out as a possible place for life to exist or to have ever existed. The same factors are necessary if any planet, anywhere, is to have life-enabling axial stability.

3.1.2   The Receding Moon

The Moon is receding from the Earth. This is caused by the way the gravity of the Moon pulls up the tidal bulge on the Earth. As discussed above, the pull of the Moon on the tidal bulge is one of the important factors in the establishment and stability of the angle of inclination. The equatorial bulge, as caused by the rotating Earth, is also important in that stability. As the Earth rotates it does so fairly rapidly. It is the speed of rotation that generates the equatorial bulge causing it to have the particular magnitude that it does. Since the equatorial bulge is caused by rotation, it is the same all the way around the Earth and is determined solely by the speed of rotation. By contrast, rotation speed does not influence the size of the tidal bulge but it does cause the maximum elevation of the bulge to occur slightly ahead of a line drawn through the center of the Earth and the center of the Moon. This is shown in the diagram entitled 'The Tidal Effect on the Moon'. As the Moon pulls up the tidal bulge, the Earth keeps rotating causing the bulge to reach its maximum height just after that particular location is directly under the Moon. Any time gravity pulls on something (if there is viscosity involved) it will always take a certain amount of time for the pulled object to respond. In this case during the response time, the Earth will have rotated around a little further so the maximum elevation of the tidal bulge is slightly ahead of the Moon. This slight delay has virtually no effect on the stability function but it does have an effect on the speed of the Moon. Since the tidal bulge is slightly ahead of the Moon it pulls the Moon forward in its orbit. The Moon is thereby being continually lifted a little higher in its orbit and a little further away from the Earth. The amount of this increase has been measured and while it is not very much per year, over long periods of time it adds up. The approximate annual increase in the distance is 3.8 cm. (30) Magnitudes of this size have no meaning whatsoever over the lifespan of a person but after a few million years they become very meaningful indeed. Clearly over the long term, the Earth is losing the Moon. It will not completely lose it. The Moon will continue to recede from the Earth and therefore take longer and longer to circle the Earth while the speed of rotation of the Earth will continue to decrease until the two are matched. At that time (assuming there is no friction loss) the Moon will be about 50% further away than it is now. The final result will be a day-night cycle which will be about 1100 hours long (i.e. 46, 24 hour days). (31) With this arrangement, a day will be about three weeks long and a night will be about three weeks long. During the day the Earth will overheat and during the night it will freeze solid.

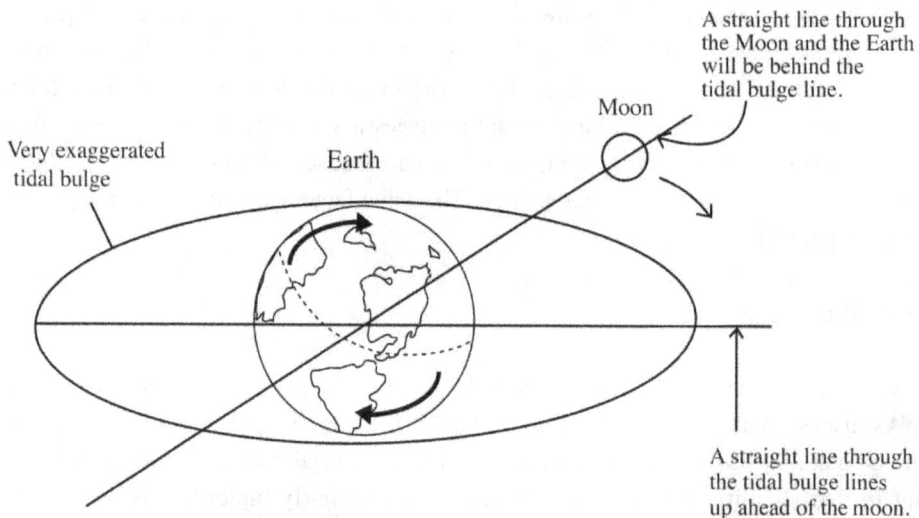

Very exaggerated tidal bulge

Earth

Moon

A straight line through the Moon and the Earth will be behind the tidal bulge line.

A straight line through the tidal bulge lines up ahead of the moon.

TIDAL EFFECT ON THE MOON

a) Moon pulls up the tidal bulge
b) Earth rotates before tidal bulge reaches maximum height
c) Tidal bulge is always ahead of the Moon
d) Tidal bulge pulls Moon forward and hence higher

This situation would be very similar to the present situation on the Moon which always has a night that is two weeks long and a day that is two weeks long. During mid-day on the Moon the temperature is about 107C (224F) and in the middle of the night the temperature is about -153C (-243F). (336,337)

While the surface of the Earth would freeze solid during the long night, during the daytime it would be expected to melt again except that it could be well into its three-week day before the surface would thaw out. Even though the Sun would be shining on the surface the increased albedo or reflectivity caused by the frozen surface would reflect a great deal of solar heat and the temperature might not rise appreciably before night set in once again. The reason for this is that several weeks of bitter cold would diminish the water vapour portion of the greenhouse gas inventory. This means that when solar heat was once again available, much less heat would be retained by the greenhouse effect so the temperature would not readily rise into the habitable range. This factor together with the reflective characteristic of the Earth's frozen surface could very well result in an Earth that simply wasn't habitable at all in spite of still being in the thermally-habitable zone of the Sun! Therefore whether we look forward or backward in time, the relatively short (i.e. a few tens of millions of years) habitable period for the Earth caused by the receding Moon generates serious questions with respect to the notion that the Earth is 4.5 billion years old.

As the Moon recedes, the rotation of the Earth slows down. Conservation of energy is thereby retained and basic physics is not violated. However a slowing Earth will have a smaller equatorial bulge. There will be less and less of an equatorial bulge for the Moon to pull on so the Earth will gradually tip over until the Sun shines almost straight down on the poles – first the North Pole and then 6 months later the South Pole. The opposite situation would have been in place if we look backwards in time. In this case the Earth would have had a faster spin rate with a larger equatorial bulge. The gravity of the Moon would have pulled on that greater equatorial bulge causing the Earth to be more upright resulting in a much smaller angle of inclination. More of the Sun's heat would have been consecrated on the Equator making the Earth much less hospitable for life and it would only have become reasonably hospitable just a few hundred million years ago as the rotation of the Earth slowed down producing a smaller equatorial bulge and a more appropriate angle of inclination.

Whenever an orbiting object is moved into a higher orbit it's forward speed slows down. Therefore as the rotation of the Earth continues to slow and the orbital speed of the Moon continues to decrease, the rotation of the Earth will be more and more synchronized with the orbital speed of the Moon until the Moon gradually slows to an apparent halt above some particular meridian of the Earth. At that time the Moon will appear stationary over one particular location and it will just stay there. The forward movement of the Moon in its orbit and the rotation of the Earth will be matched. On one side of the Earth the Moon will be visible and on the other side it will not be visible. At that time a reduced tidal bulge will still exist but it will not move around the Earth as it did before. It will still be useful to pull the Earth over and generate an angle of inclination but it will pull with a reduced force. In fact, since the Moon will be about 50% further from the Earth at that time its pull of gravity on the Earth will be reduced to about one-half of the present value. Unfortunately, at the same time the equatorial bulge will be reduced to almost nothing because the Earth will have slowed down from one rotation per day to about one rotation every six weeks. The pull of the Moon on the tidal bulge will therefore be much more dominant and the Earth will be pulled over so far that the North Pole will almost point at the Sun. The heat from the Sun will not be uniformly distributed. The loss of the balance between the pull on the equatorial bulge and the pull on the tidal bulge will mean that the Earth will not be habitable. While this development will not be in our favour and will require several hundred million years to happen, what are the implications of a Moon that was closer to the Earth in the past?

If we use the current rate of separation and calculate backwards, the Moon would have been touching the Earth about 2 billion years ago. This is only about one-half of the time that the Earth and the Moon are thought to have existed! The situation degenerates from there. If the friction created between the Earth and its oceans caused a loss of energy the time is considerably reduced. (possibly only one billion years.) (32) (Since this estimate was offered, friction loss has been measured to be much greater than expected so the time involved is much less. See discussion in section 3,1,3 The Decelerating Earth.)

With the Moon close to the Earth, the crust of the Earth would have been pulled up into a tidal bulge just as the oceans are now. Furthermore it would have been much easier to distort the crust at that time because it supposedly would have been much thinner. We only have a crust because the Earth has cooled off from its supposedly original molten state. As it continues to cool the crust should gradually become thicker. Several million years ago it would have been much thinner and much easier to deflect and fracture. Also at that remote time in the past the Earth would have been rotating at a much faster speed. Even now the tidal bulge races around the Earth at hundreds of miles per hour and there are always two tidal bulges every day. The situation long ago would therefore have been an Earth which was enduring two tidal bulges every day moving at thousands of miles per hour with a day that was much shorter. Therefore every few hours another tidal bulge would pass through. With the Moon much closer the magnitude of these bulges would have been miles high and the crust would have been extensively fractured into numerous pieces with ongoing volcanic activity everywhere. However, while the crust of the Earth is certainly fractured, it isn't fractured that much so we are left to wonder how the whole setup could possibly be billions of years old.

If the Moon is actually receding as believed, the width of the temporal 'window of life' that the Moon provides to make the Earth habitable, is not very wide and is probably less than one hundred million years in either direction from the present time. Such a relatively narrow window of possibly two hundred million years gives cause to suspect that there is something very wrong with the idea that the Earth is several billion years old.

3.1.3   The Decelerating Earth

Just when it appeared that things couldn't get worse they did. All of the above discussion is quite valid if we assume that all of the rotational energy of the Earth is applied to lifting the Moon into a higher orbit. Since the rotational energy of the Earth can be reasonably approximated and the energy needed to lift the Moon into a higher orbit can also be approximated, a valid calculation can be carried out to determine how much higher the Moon would be raised as the Earth's rotational energy is transferred to the orbital energy of the Moon. However, this straightforward approach ignores friction. Whenever something moves there is always friction involved. In the case of the Earth and the Moon the friction appears as the water in the ocean is sloshed around. As it moves past islands, around continents, up estuaries and into bays there is a loss of energy. The bulk of the Earth is also involved as the crust is slightly lifted up and then settles back down as the tidal bulge travels through. All of this means that a portion of the rotational energy of the Earth is not be available to lift the Moon so the Moon will not get nearly as high as it would if there wasn't any energy loss. However, the Earth will slow down simply because energy is lost and it will not matter how it is lost. As a result of these actions the rotational speed of the Earth should be slowing down at the present time and if it is slowing down enough to be measured, confirmation of the whole idea would be obtained. Coincidentally, the length of the day has been measured and has been determined to be increasing at about one second every five hundred days. 'Currently the Earth slows down at roughly 2 milliseconds per day. After 500 days the difference between the Earth rotation time and the atomic time would be 1 second. Instead of allowing this to happen, a leap second is inserted in the Atomic Clocks to bring the two times closer together.' (534) While this does not really sound like very much time, since there are only 86,400 seconds in a day, a constant rate of slowdown at the present rate, would only require 129,600 years for the Earth to slow down to one-half of its present speed. In reality, as the Earth slows down, one would expect some conservation of rotational momentum due to a shrinking equatorial bulge, so the actual time would probably be longer.

A problem will develop because as the Earth rotates slower and slower, the centrifugal force generating the equatorial bulge will diminish. However, the counter-balancing force caused by the tidal bulge will not diminish because the Moon will not be much further away than it is now. With a reduction in the equatorial bulge, there will be a reduction in the restoring force keeping the Earth's axial tilt from dramatically increasing. The Earth will just tip over. The habitability of the Earth will 'tip over' with it and sooner than we would like to contemplate, the Earth will not be habitable at all.

The centrifugal force involved with the generation of the equatorial bulge involves a square law where the speed factor is multiplied twice. This means that the rotational speed only has to drop to about three-quarters of the present speed before the equatorial bulge would be reduced to only a little more than one-half of its present size and would not be able to offset the tidal bulge at all! Therefore, if it took 65,000 years for the rotational speed to drop to one-half (as current measurements indicate), the Earth would be starting to tip over within a fraction of that much time and possibly within 20,000 years making the Earth increasingly less habitable well within the 129,000 year time frame! And just to make matters a little worse, an Earth that was rotating slower would have a more exaggerated tidal bulge helping the tip-over process a little more. Further, the stabilizing capability of the equatorial bulge goes through a maximum point as it's near side (i.e. nearest to the Moon) gets further from the Earth's orbital plane (i.e. as the Earth tips over a little more) but then reduces again as the far side of the bulge comes up and gets closer to the Moon. When this point is reached, tip-over will accelerate quickly. None of these realities bode well for the idea that the Earth has been here for 4.5 billion years because using the same information and calculating backwards, within a similar amount of time, (i.e. 20,000 years) there would also have been a loss of habitability. (i.e. the Earth would then have been sitting up straight.) To summarize, the hypothesized extreme age for the Earth does not correlate with the current observations that the rotation of the Earth is slowing down rapidly and days are getting longer. The old-age idea should therefore be set aside.

The understanding that the Earth's axial tilt will increase is based on measuring the rate of deceleration of the Earth's rotation. Prevailing wisdom, on the other hand, is based on historical measurements and declares just the opposite! Historical measurements and computer models indicate that the Earth will move from its present angle of 23.4 degrees down towards 23.1 degrees and then come back up again in a cyclical pattern. All of this will require a number of years with a full cycle requiring about 41,000 years. (540) A considerable amount of effort has gone into making these determinations and they involve the slight variations expected in the Earth's orbit around the Sun. The quality of the work is to be commended but no recognition seems to have been given to the rapidly-decelerating Earth! The prediction hereby being made is that the decelerating-Earth factor will presently overwhelm the cyclical factor and in the not-too-distant future the Earth will 'tip over' and become uninhabitable!

In this regard the Earth can be compared to a spinning top. Due to friction a spinning top will slow down and fall over. The Earth is slowing down due to friction – in this case the friction is the energy lost moving tidal water. It takes a lot of energy to move the water of the tides and this energy comes from the spinning Earth. However it will not spin forever because the energy being used to move the water of the tides will use up the rotational energy of the Earth and cause it to slow down. When it slows down to a critical rotation rate, the pull of the Moon on the tidal bulge will overwhelm the pull of the Moon on a reduced equatorial bulge and the Moon will consequently pull the Earth over making it uninhabitable. The time for this to happen will be within tens of thousands of years and not millions.

3.1.4   Transient Activity on the Moon

If the Moon was older than one billion years, one would expect that it would be a very quiet inert mass of rocky material by now. Any volatiles would have all out-gassed long ago and any substance that was too light for the gravity of the Moon to hold would have dissipated away into space. The Moon obviously is too small to retain an atmosphere so any gaseous substance that might appear on the surface would just drift away. With these seemingly basic factors in mind it was with considerable surprise to learn that the Moon is a very active place. That is, it was a surprise to those who had a preset opinion on the matter to the extent that reporting a lunar transient event was considered totally unacceptable even to the degree that doing so would result in being ignored or banned or having your reports and papers rejected. (15) After all, if one is to be a scientist and enjoy even minimal favour of his colleges one cannot go about upsetting the governing wisdom of the day. It would be interesting to know how many events have gone unreported just because it was considered for a long time to be unacceptable to do so. It is most unfortunate that science occasionally gets straggled by preset opinions to the extent that the actual picture is not available. The stigma about reporting on lunar transients has apparently been lifted and the change in outlook might be partially attributed to the report on lunar transients published by NASA just before the first manned lunar landing in the late 1960s. Short-term activities are referred to as transients and in preparation for the manned journeys to the Moon, NASA published; Chronological Catalogue of Reported Lunar Events. The events listed were reported from 1540 to 1967 and included 579 separate entries. This publication was preceded by 'The Craters of the Moon' by Patrick Moore and Peter Cattermole. An earlier writer named Charles Fort went against popular opinion and published in 1923 describing shadows, glows, points of light and moving clouds. Imagine, clouds on the Moon! Another writer, Frank Edwards, described several unusual sightings in a 1959 publication. Most of these reports were looked upon with disgust by many in the scientific community and this outlook seemed to prevail until the NASA publication became available.

Generally, the sightings were short-lived events and as such would have been unrepeatable. If an astronomer was not observing at exactly the correct time, the event would not have been seen. Of course if an independent observer saw something it very probably would never be reported in such a way to have it 'officially' recognized either.

Included in the transient events on the Moon are: light, coloured clouds, mysterious glows, and coloured shadows. Many of them are reported from the maria as well as certain craters including Aristarchus, Plato and Alphonsus with more than 300 associated with Aristarchus alone. (14)

All of these observations indicate activity, but if the Moon, being a relatively small heavenly body, is really very old, why should there be any activity going on at all. In particular, the clouds indicate some type of outgassing which means that there are still pockets of gas trapped within the Moon. The lights indicate some type of activity which is able to emit light and the one that comes most readily to mind would be hot magma showing itself through some temporary opening in the surface. However after several billion years why would there be any hot magma left at all? In general, where there is activity of this sort there very likely aren't any vast amounts of time involved. Such activity should have petered out long ago.

In 1178 an unusually massive transient event occurred on the Moon and was witnessed by a group of monks sitting out one evening 'observing' the Moon. In part they reported that ' ... suddenly the upper horn (i.e. It was a new Moon appearing as a thin crescent and the points were referred to as horns.) split in two. From the midpoint of this division a flaming torch sprang up, spewing out, over a considerable distance, fire, hot coals, and sparks. ... the Moon throbbed like a wounded snake ... This phenomenon was repeated a dozen times or more, the flame assuming various twisting shapes at random and then returning to normal.' (107) While the impact crater associated with this event was later named Giordano Bruno, the flame and sparks, while resembling a flaming jet of burning gas, are somewhat difficult to explain. One commentator offered that 'the flame and sparks were actually incandescent gases' and that the throbbing of the Moon 'like a wounded snake' could have been caused by the refraction of light through the gas cloud. This is a reasonable explanation. It could not have been flaming gas in the usual sense because there is no atmosphere containing oxygen on the Moon to support combustion. The greater point of interest however is that there was a pocket of some kind of gas just below the surface and that it must have been under a certain amount of pressure or it wouldn't have come out so readily. How long could a pocket of pressurized gas have been retained under the surface? Why wasn't this gas released many years earlier? After all, there are more than 200,000 craters on the Moon that are more than one mile in diameter indicating that the Moon has been repeatedly impact-shocked. (110) Admittedly, the crater formed during this event was quite large at 13 miles in diameter (110) indicating that the impactor was quite large and would have struck the surface with great force. A fissure must have been opened to a pocket of gas that had been trapped below the surface but how could it have remained trapped for billions of years? As soon as the surface was struck, the gas came spewing out in great quantities so it must have been under significant pressure. Why didn't it come out long ago? After all, the Moon has been repeatedly shaken and fractured by one impact after another year after year? Why didn't it escape when all of the mass concentrations (mascons, see chapter 7) came crashing in? At the very minimum the entire situation is suggestive that the Moon cannot be very old!

3.1.5 Temperature of the Moon

The Moon is still hot! This, apparently, was quite a surprising discovery – that is surprising if you expected that the Moon was old, dead and cold andit should have cooled more rapidly than the Earth or the other planets. Lunar mapping by the Clementine satellite showed that part of the Moon is still molten. Indeed the first lunar astronauts measured a higher heat flow from the Moon than expected, indicating that the Moon's interior is much hotter than had been predicted. The logical explanation is that the Moon has not had time to cool and is therefore not nearly 4.6 billion years old. (130) Other commentators also recognize that the heat flow from the Moon is much greater than expected. 'Some theorists expected a slight heat flow due to internal radioactivity, but the measured heat flow is much greater than expected even from this process.' (338) Very simply put, a hot Moon cannot be an old Moon.

3.1.6   Lunar Dust

In preparation for going to the Moon a lunar lander was designed. Of course when something is designed every possible anticipated circumstance must be taken into account. In this case since the Moon was thought to be very old, there would have been time for a layer of dust to accumulate because of the constant bombardment of small bits of rock coming in from space. This bombardment would have broken up the surface and a layer of dust would likely cover the entire Moon. A thin layer of dust would seemingly not cause a problem but some commentators (including Carl Sagan) confidently predicted that the thickness of the dust layer would be at least 50 feet. It really would not be possible to deal with a layer of dust that thick but the possibility of a layer of dust was recognized and the landing pads of the lunar Eagle Lander module included large dish-like pans on the bottom. This would prevent the craft from - hopefully - sinking down too far. Even as the module descended to the surface there was considerable trepidation that they might sink in too far. Surprisingly and to the great relief of the astronauts, the Eagle Lander touched down firmly with a slight jolt. This meant that there wasn't any dust. When they looked out they could see that the four landing pads, which were 2.5 feet in diameter, (339) had not sunk into the surface at all. They were sitting right on top. When the astronauts came out and walked around, they did not sink either but only left footprints similar to what would be left if a person walked across a sandy beach near the waterline. There was really no significant layer of dust on the surface of the Moon at all.

Even if all of the material right down through the regolith (i.e. the surface layer of unconsolidated material on the Moon) is considered there really doesn't seem to be enough to justify a 'billions of years' explanation. Subsequent landings included rovers which enabled the explorers to travel much farther from the spacecraft and gather more information and the rover wheels were designed to only deal with a firm surface. If a great deal of dust was expected on an ancient Moon and no dust was found would this mean that the Moon isn't really ancient at all or is there something more subtle involved?

There are two factors to consider. One is that the Moon is (thought to be) continually bombarded by rocky objects – both large and small. The other is that there is hardly any layer of dust. These two factors can only be reconciled if the Moon is not very old – not even a million years old and more likely only a few thousand years old! Uncertainty does remain however and relates to the paucity of observations regarding the ongoing arrival rate of space material as well as the historical arrival rate of space material. In particular the evidence for an asteroid shower on the Moon (discussed in chapter 7) is suggestive that the entire current accumulation of lunar dust could have substantially been deposited within a very brief period of time. This means that the layer of dust on the Moon does not provide any support at all for a lengthy period of billions of years.

3.1.7 The Uncratered Maria

The near side of the Moon has very large areas referred to as maria. These areas are volcanic in origin meaning that sub-surface material has poured up onto the surface from down below. The maria are so extensive in area that they cover thousands of square miles and occupy more than 50% of the near side. The problem with respect to time is that they have very few impact marks.

This could indicate that the maria formed when a few very large asteroids hit the Moon and broke through the crust into the molten interior. Several of them seem to have left indications of their coming as concentrations of mass below the surface. These are called mascons and were discovered during one of the missions to the Moon. One of the maria does not have an associated mascon. Either the other impactors opened the crust of the Moon enough to cause a mascon-free maria to form or it is possible that one of them was large enough and the material below it was molten enough to allow it to sink downward and settle close to the geological center. In this case it would not currently be recognizable as a mass concentration and it would not be contributing to the gravitational lop-sidedness of the Moon.

The monster Hellas crater on Mars does not have an associated mascon either while the other large impact sites do have one and the same reasoning could apply there. (344) The Hellas impactor would have been a very large object. It too might have just settled down right to the center and only left the crater as evidence of its coming.

Our current interest relates to time and with the almost total lack of impact marks in the maria it is apparent that the entire development must have happened only recently. Otherwise the surfaces of the maria would be riddled with impact marks like the rest of the lunar surface. (We recall that the lunar surface has more than 200.000 impact marks more than one kilometer in diameter. (110)) Consequently, it is difficult to imagine that the maria were formed a very long time ago and it is even more difficult to imagine that they were formed more than a few thousand years ago.

3.2 The Sun

3.2.1 The Faint Young Sun Paradox

The problem called the 'Faint young Sun Paradox' recognizes that the Sun was once cooler (i.e. fainter) than it is now. This idea is derived from nuclear theory which teaches that 4.5 billion years ago the Sun would have been only about 75% as bright as it is now. According to certain commentators (i.e. Newman and Rood0 over the supposed 4.5 billion year history of the Sun its luminosity should have increased by about 25%. (33) If this was the case, would life on the Earth have been possible? This is not to suggest that there was life at such a remote time but rather to wonder if life would be possible at the present time if that was the case so very long ago.

Temperature is the primary concern involving the possibility of life. Currently the average surface temperature of the Earth is about +15C. (51) This is about half-way between the freezing point of water and the body temperature of most mammals (or the temperature at which most seeds can germinate). While this temperature appears to be quite satisfactory it is understood that deviation either way would be disasterous. A deviation of a few degrees would not seem to be significant but there is currently widespread concern that even a deviation of two degrees would have serious consequences.

In order for the Earth to be hospitable to life it must not only have a particular average temperature but it must also have temperature stability. The first priority to achieve this is a stable orbit. The orbit must be stable and repeatable year after year. In this regard the orbit of the Earth has often been referred to as a 'Goldilock's orbit – it is just right'. (186) It is also recognized that deviation is not permissible to the point that even a 5% deviation either way is recognized as disaster. (24) Such a small amount does not really seem like very much particularly when compared to other planets. Mars, for example has an orbit which is quite elliptical allowing Mars to drift out until it is about 10% further away from the Sun than its average distance. Similarly it drifts in about the same amount. A deviation of plus or minus 10% is too much because the amount of heat received from the Sun varies with the square of the distance from the Sun and the variation of temperature on the surface of Mars is much more than plus or minus 10%. This is far too much variation for life to have ever happened – even if that average temperature was above freezing.

The Earth has a slightly elliptical orbit which would suggest that its temperature would also vary. However in the case of the Earth the slight variation in the distance to the Sun works synergistically with the angle of inclination and the types of surface material of the Earth (i.e. water and land) to provide optimum heat management. In this manner, the distance variation works in our favor but all of the other factors (i.e. surface material, rotation speed, lunar distance, lunar size, and the Moon's orbital plane) must also be in place for this to happen.

The temperature of the Earth is determined by two main factors. One is obviously the heat that is received from the Sun and the other is the complement of greenhouse gases. Clearly the Sun is the actual source of our heat but the characteristic of the greenhouse gases enables some heat to be retained that would otherwise be lost. This arrangement keeps the surface temperature of the Earth right where it is and right where we need it to be. There are several gases that have a greenhouse effect but there are two that have by far the greatest influence. Carbon dioxide is very influential but water vapor is even more influential. While there are several factors that determine how much carbon dioxide the atmosphere includes, temperature alone determines the amount of water vapor. Warm air can hold more water vapor and cold air holds less.

Now if the Sun was only 75% as hot, it would be equivalent to having the Earth about 12% further from it. However the maximum deviation that can be tolerated and keep the Earth in the thermally-habitable zone is given as being between 1% and 5%. A 12% outward deviation would therefore result in the surface of the Earth being well below freezing. This is not welcome news because if this was true, there would have been ice on every body of water on the entire Earth. If all of the water was in a frozen state it would mean that there would have been very little water vapour in the air. Water vapor is, of course, directly necessary for life to exist but it is also necessary as a greenhouse gas to keep the surface temperature in a livable range. In fact, water vapor is the most important greenhouse gas followed by carbon dioxide. (Clouds also have a greenhouse effect which is in between water vapor and CO2.) The further complication involved in being below freezing is that there would be much more ice on the surface of the Earth. Ice reflects a lot more heat and light than an unfrozen surface would and this means that much of the incoming heat from the Sun would be reflected away before it could contribute to any warming effect. In other words the habitability of the Earth would be dramatically reduced if the average surface temperature was below freezing.

Let us suppose for a minute that the following relationship is valid.

Direct heat from the Sun	8,000 units
Greenhouse heat (i.e. trapped solar heat)	2,000 units
Total heat needed to keep surface temperature at +15C;	10,000 units

These quantities are being offered for illustration purposes. With these quantities in mind and the understanding that the Sun was originally only 75% as bright as it is now we would have a relationship as follows.

Direct heat from the Sun	6000 units
Greenhouse heat (assuming that water vapour and clouds are missing)	500 units
Total heat available	6500 units

Recalling that a 5% deviation would spell disaster we see that we are short 3500 units of heat or about one third of what we require. The Earth under such conditions would have had a surface temperature well below freezing and all of the oceans as well as the land would have been frozen solid. Actually the situation might not have been quite this dismal because at that remote time the Earth would have been cooling down and just starting to form a crust. If we allow that it would have taken a billion years to form a crust that was a few thousand feet thick the output from the Sun would have supposedly increased somewhat from what it was at the beginning. If the increase was reasonably linear over the entire time, then the Sun's output might have been about 80% of what it is now. This would be about 6400 units. The situation would therefore appear as follows.

Direct heat from the Sun 6400 units	
Greenhouse heat 500 units	
Total heat available	6900 units

There would still have been a significant shortfall. The surface temperature of the Earth would have been well below freezing even if heat was still coming from the interior. It means that the surface would have been frozen but a frozen surface brings another aggravating factor.

Albedo is a term used to identify the percentage of incident solar energy that is reflected from a surface. For example the ocean has an albedo of 0.2. It absorbs 80% of the incoming energy and reflects 20%. Ice is just the opposite. Ice will reflect about 80% of incoming energy. A world covered with ice would reflect 80% of the solar energy that was incident upon it. Returning to the above relationships this means that there would only have been about 20% of 6400 units received by the Earth as useful energy. The rest would have been reflected away. Therefore we would have had the following.

Direct heat from the Sun	1280 units
Greenhouse gas	500 units
Total heat available	1780 units

Clearly a situation like this is not workable and the Earth would have been frozen solid. Even if we allow that the crust wasn't thick enough until 2 billion years had passed we would still have a situation that would not be the least bit hospitable to life and the Earth would have been uninhabitable.

Further, suppose that the Sun's heat increased steadily until it reached its present level but if the Earth had been in a frozen state, the total heat from the Sun plus the negligible greenhouse gas effect would still not have been quite enough to bring our surface temperature to above freezing. In this case the following situation might have been in place.

Direct heat from the Sun	8000 units
Greenhouse effect	500 units
Total heat available	8500 units

Since we need 10,000 units to be viable and the surface had historically been frozen, the surface would simply remain frozen. In such a case the albedo effect would have still been active and the direct heat from the Sun would effectively have only been about 20% of 8000 units so the situation would have been closer to the following.

Direct heat from the Sun	20% of 8000	1600 units
Greenhouse gas effect		500 units
Total heat available		2100 units

This is only one-quarter of what we need and illustrates that there is no doubt that if the Earth had ever become frozen solid, it would still be frozen solid. Further it would remain frozen solid for another few billion years. This is a totally dismal situation to contemplate and raises a serious doubt about whether or not the Earth is actually 4.5 billion years old.

3.2.2   A Cooling Sun

While theories of nuclear reaction indicated that the Sun would have warmed up over the long time frame of 4.5 billion years, what is one to think if the Sun is actually cooling off instead? While the heating scenario is theoretical is there any measurement to confirm this or is there any to suggest otherwise? 'A meeting of the US Geophysical Union in 1986 brought together a number of independent scientific studies which all agreed that the Sun has steadily declined since 1979. All types of measurements including satellites, rockets and balloons all showed that the Sun is getting dimmer. (35) This would appear to contradict the idea that the output of the Sun is increasing but it represents a very short time frame from which to draw any firm conclusions. Over the short term the Sun's output is expected to increase and decrease due to solar flares. (The solar flare being slightly darker than the rest of the surface of the Sun actually reduces the output of the Sun) This further complicates obtaining measurements that show the long term activity and makes it virtually impossible to get a valid long-term picture. It never-the-less casts a shadow of doubt on the 4.5 billion year idea.

However, there is an even greater consequence involved. While the Sun is observed to be cooling down, the Earth is observed to be heating up! (306) What would be the consequences for the habitability of the Earth if the Sun stopped cooling down and heated up again instead? What does this say about the thermal stability of the Sun over a time period of 4.5 billion years?

3.2.3   The Formation of the Sun

The historically-dominant idea of how the Sun was formed involves a large cloud of dust. As this cloud of dust collapsed, (i.e. due to gravity - but unfortunately there would not have been enough gravity to cause collapse.) (305) it started to rotate and this explains the spinning activity of both the Sun and the planets. Of course the observation that the planets have about 1% of the material of the solar system and 99% of the angular momentum compared to the Sun which has 99% of the material and only 1% of the angular momentum had to be overlooked or explained away. So it was suggested that the original angular momentum of the Sun must have been swept away soon after the Sun took its present shape. This type of explanation is somewhat ingenious but not really very scientific.

The further difficulty is getting a cloud of gas to collapse in the first place. The great gas clouds in space are expanding not contracting. Explanations for the initial formation of stars is missing. ... The conclusion is that stars will not spontaneously form in space because the dominant force is outward and this will forbid collapse. Instead gas clouds expand and dissipate outward. (244) Even if the cloud of gas is relatively cool, it is still expected to expand. It is clear that the outward push due to thermal motion of the molecules, even at low temperature, is greater than the inward pull due to gravity. (245) In order for a cloud of gas to contract in space it must be brought down to a certain critical size called the Jeans length, after Sir James Jeans who initially developed the mathematics. (246)

However the whole idea of gravitational collapse of a giant cloud of gas ran into insurmountable difficulties when it was observed that many stars are actually two closely-connected luminous masses instead of being a single body of material like our Sun. The problem related to this observation is that the ages of the two objects (theoretically) should be the same but (theoretically) they aren't. 'There is a class of binaries known as semi-detached systems in which the stars almost touch each other. These systems have a large secondary of lower mass and a small primary of higher mass. This means that the larger secondary would appear is at a later stage of evolution than the smaller primary. Computer models predict that the more massive star should evolve more quickly than the less massive. So why is the star of lower mass, larger than the one with most of the mass? (36)

The Laplace Nebular Hypothesis of Solar System Formation ran into even more trouble when other solar systems were discovered. With many of these far-away systems the arrangement of the planets is totally different from what Laplace's theory predicted. The Nebular Hypothesis predicted that rocky planets like Mars, Earth, Venus and Mercury would form nearest their host star. Gaseous planets like Jupiter and Saturn would form further away. This seemed to be valid because that is the way that the solar system is formed. Then other solar systems were discovered and their formation is completely different. Large gas giants were right in close to their host sun. In some cases they were so close that they circle their host sun in a matter of days (or even hours) and this discovery disabled the Nebular Hypothesis. (37) No planet could ever form so close to its host star. (e.g. Kepler 70b orbits Kepler 70 at 1/130 of the distance of Earth from the Sun. (150)) In other words we do not have any idea how the solar system formed so how would we have any idea about how old it is?

3.2.4   Nuclear Reactions

If the Sun is being powered by nuclear reactions, it is understood that there would be enough fuel in the Sun to keep it hot for billions of years. In order to confirm that this is happening we require a combination of theory and supportive observation. When nuclear reactions take place they supposedly release very small particles called neutrinos. Most atomic particles are detectable because they carry an electric charge or because they have detectable mass. In this regard the neutrino poses a challenge because it does not have an electric charge and its mass is extremely small. However detection equipment has been set up and a certain number of detections have been made. Hopefully they were actually neutrinos and not some other form of radiation like cosmic radiation. It is expected that cosmic radiation would be blocked by several hundred feet of rock but this too is theoretical and difficult to confirm. We cannot just declare that a few thousand feet of rock is sufficient to completely block all forms of radiation that we do not want and then settle back and assume that the problem has been solved. Further, there is the question of direction. Which direction did the detected particles come from? Of course, we only want the ones from the Sun which really presents a very small window to leave open for detection. The total angle that the Sun presents in the sky is so very small that to set up an experiment to block out particles from all other directions is a very serious challenge of its own. This means that there are two types of undesirable radiation to be dealt with and the shielding apparatus would reasonably appear like a solid sphere of material capable of blocking all kinds of radiation. In one side of this shield there would be a very small hole pointing at the Sun. Only radiation from the Sun would be able to get through this hole. There is still the question of cosmic radiation so we should insert a cosmic radiation block in this opening to make sure that no cosmic radiation was able to get through. We understand that neutrinos will pass right through great amounts of material because to a small atomic particle like a neutrino solid material is more than 99.99999% open space. In recognition of this reality one commentator offered; neutrinos are so incredibly small that the entire Earth is transparent to them. So most of the Sun's neutrinos go straight through without interacting with anything. (72) In order to solve the direction problem, either a neutrino block must be set up or some other reliable way to determine direction must be set up. The direction detector must be able to exclusively recognize the area of the sky that is occupied by the Sun because the other 300,000,000,000 stars in the galaxy might also be emitting vast numbers of neutrinos with the expectation that all of space would be literally full of them. They would be sweeping past and through the Earth in unfathomable numbers. How would a neutrino detector distinguish those coming from the Sun?

But let us assume that neutrinos from the direction of the Sun are detected. Then, how could it be determined that they were coming from the core of the Sun and not from solar flares on the surface? (39) Only neutrinos from the core would support the idea that the Sun is powered by nuclear reactions. Then once it has been confirmed that neutrinos were coming from the core of the Sun, in order to acquire information concerning the age of the Sun, we need to know how long the neutrino-producing reactions have been proceeding. Just because such reactions are happening now in no way identifies how long they might have been happening. On the other hand, if neutrinos could be detected as having come from some particular remote star, this would suggest that the neutrino-producing reactions were happening at the time when those neutrinos left to come to us. For example, if they came from a star that was ten billion light-years away the reactions in that star would have happened ten billion years ago. (Actually the neutrino-producing reactions would have been happening longer ago than ten billion years because neutrinos do not travel quite as fast as the speed of light.) The star could, of course, be long gone by now because the light we would be receiving would be ten billion years old but at least it would be suggestive that the reactions might have been going on for a long time. It also means that nobody knows what is going on out there now, including whether or not remote objects in the universe still exist.

One more question deserves an answer. How did the nuclear reaction process start in the first place? An astrophysicist with Harvard University (i.e. Cameron) calculated the maximum temperature obtainable by the Standard Evolutionary Collapsing Gas Cloud Theory of Star Formation as one million degrees Kelvin. This would be much too cool to initiate hydrogen fusion. (329) Such embarrassments must be swept aside as soon as possible as illustrated by the following. The large uncertainty about the way in which nuclear reactions turn on is not presently understood. (329) Which theories should we accept and which ones should we ignore? The uncertainties seem to outnumber the certainties with this entire situation which makes it difficult to determine with any degree of validity that the Sun is even being powered by nuclear reactions or for how long this might have been happening. This leaves the 4.5 billion-year idea without much support.

3.2.5 Gravitational Collapse

If the Sun isn't powered by nuclear reactions is there any other way that it could be powered? Gravitational collapse is one such possibility but this would only keep the Sun operating for about twenty million years instead of for billions of years. (323) It was reported a few years ago that certain measurements confirmed that the Sun was shrinking. It appeared to be collapsing inwards and the reason for this would be its own gravity. The problem here is to obtain valid measurements that show the long term trend. While we understand that the Sun is a very large object it only occupies a very small angle when viewed from Earth. Therefore to measure the extremely minor change in diameter that would be sufficient to verify that it is collapsing at a rate that would explain its thermal output seems like an insurmountable challenge. All of the Sun's current heat could be produced by a contraction rate of only 0.02 ft. per hour. Tthis is much too small to be seen because it only amounts to about 3 miles/century. (38) It is certainly possible that the Sun could be powered by gravitational collapse but obtaining valid measurements to confirm this is a most difficult task. Never-the-less contraction of the Sun has reportedly been observed but the rate was much higher than would be necessary to explain the heat output (which undoubtedly made the measurement process somewhat easier). Moreover the Sun was observed to be pulsating which implies that the interior temperature would not be high enough to enable nuclear reactions to continue! (324) These types of observations raise a serious doubt about the age of the Sun and, by association, the age of the Earth.

3.2.6 Solar Mass Stability

Another factor directly involved with the Earth's orbit and hence the age of the Earth is the mass of the Sun. In order for the Earth to have a stable orbit the mass of the Sun must be constant. However, how stable is the mass of the Sun because it might be constantly changing. Material is continually being added to the Sun's mass and material is constantly leaving. The material being added includes comets and asteroids. These objects keep pouring into the Sun all the time. Sometimes the mass being added is substantial as in 1979 when the giant comet Howard-Koomen-Michels 1979X1 crashed into the Sun. As mentioned earlier, the head of this comet was as big as the Earth. Comets continually crash into the Sun and apparently they arrive on a regular basis. Asteroid material is also involved. None of this should be any surprise because the gravity of the Sun is enormous and falling into the Sun would seem to be the default option. Even the gravity of the Earth attracts a great amount of material on a regular basis and thousands of tons are understood to impact the Earth every month. (57) Most of this seems to be rocky material but there is some evidence that comet material is also involved. (43) How much more material does the Sun attract? In the case of both the Sun and the Earth material also leaves. The Earth is continually losing hydrogen from the upper atmosphere. (59) The amount has been estimated and also involves tons of material every year. The Sun also loses mass as some of its material leaves. In this case some material is ejected from the solar flares. Other material leaves by the solar wind. Is the amount of material that leaves, and the amount of material that arrives, in balance? Since the Sun has such an enormous mass (2×10^{30} kg) it would take time to significantly change that mass. However if the period of interest is 4.5 billion years the question takes on more significance because if the Sun had even 1% less mass in the distant past, the orbit of the Earth would have been too far from the Sun and the Earth would have been frozen solid. (It simply is not allowable to have the Earth outside of the Goldilock's zone for any reason whatsoever.) If, as time went by, the mass of the Sun increased enough to bring the Earth into the Goldilock's zone it would still be in that frozen-solid state and would have remained there right to the present time. (i.e. it would have been a 'snowball Earth') If the Sun continues to accumulate mass, the Earth will continue to spiral inward, making the habitability time for the Earth limited because it would soon be beyond the 5% limit mentioned earlier.

Whenever very long time spans are used to try to explain something there are invariably factors involved that become quite significant whereas over short time spans they would hardly have been recognized. Over a time span of 4.5 billion years very small repeated changes can become overwhelming and this simply points out that recognizing long time spans like 4.5 billion years comes with a large contingent of complications.

3.3 The Earth

3.3.1 Earth's Orbital Stability

Repeatedly throughout the literature, the orbit of the Earth is referred to as a 'Goldilock's Orbit – it is just right'. (186) This is most fortuitous because we certainly do need an orbit that is dependable, repeatable and stable over the long term. And there does not seem to be any room for variance. Certain commentators refer to this stringent requirement as a 'bottleneck'. 'To be habitable, a planet must not only maintain a circular orbit within the Goldilock's zone but must take the right orbit within this zone. According to former NASA astronomer Michael Hart's computer simulations: "If the Earth's orbit were only 5% smaller than it actually is, during the early stages of Earth's history there would have been a 'runaway greenhouse effect', and temperatures would have gone up until the oceans boiled away entirely." On the other hand, he found that "If the Earth-Sun distance were as little as 1% larger, there would have been runaway glaciations on Earth about 2 billion years ago. The Earth's oceans would have frozen over entirely and remained so ever since, with a mean global temperature of - 50F.' (42)

While orbital stability is required it isn't just stability, it is long-term stability. This is very difficult criteria to meet because there are so many factors involved with stability. As pointed out above, the Moon is very much involved in the stability of the Earth's orbit but the Moon is only providing relatively short-term stability because it is receding from the Earth at a rate that has long-term significance. The window of stability in this case might not seem really that short when compared to every day experience but 100 million years is a far cry from the 4.5 billion years that is so widely assumed for the Earth's existence. If we look both ways in time from the present, the total 'window' might only be 200 million years wide. This would only take us 100 million years into the past which is hardly enough time for the last major asteroid impact (i.e. the Chicxulub impact has been pegged at 65 million years ago) to have occurred. All of this ignores friction which is lengthening the day by about one minute over the life-span of a person giving the Earth in reality probably less than 50,000 years until it 'tips over' and becomes uninhabitable. (refer to 3,1,2 above) These realities raise a serious doubt concerning just how long ago the asteroids actually hit the Earth.

3.3.2   Erosion

While the exposed crust of the Earth appears quite solid it still weathers and erodes. Both rain and wind – especially wind that has embedded in it erosive agents like sand - will eat away at exposed rock and gradually wear it down. How long could it last? The exposed above-sea-level material of the Earth also includes soft material like topsoil and clay. These materials erode much more easily than solid rock. How long could they last?

We understand that the average depth of the ocean is about 2 1/2 miles. This is much greater than the average elevation of all landforms above sea level. In other words the world is mostly under water at the present time. As the ocean continues to heat up and expand and as the glaciers of the world continue to melt, ocean level will rise. This is alarming but in reality it will only involve an increase in ocean level of a little more than 200 feet. But it is alarming because it is well understood that a great deal of habitable land is less than 200 hundred feet above sea level. The area of land that is much greater than this above sea level comprises a relatively small percentage of the total. How does all of this stack up against the reality that erosion is occurring at a rapid rate - even rapid on a scale of one thousand years? What would the accumulative effect be after 10,000 years or 100,000 years? We are instructed to believe that several hundred million years are involved in the history of the Earth but what would be the effect of erosion after a few hundred million years? Would any portion of the continental land masses currently above sea level, still be exposed? We cannot have a world that is continually eroding at a rate that will remove the entire continental volume in only about 14 million years (321) and a world that is several hundred million years old at the same time. The two ideas are incompatible. Based on the reality that erosion is observable, the idea that the Earth is 4.5 billion years old is questionable and should be set aside! In fact we do not even need to know the magnitude of the erosion rate. Simply the fact that erosion is taking place at a measureable and significant rate and that 4.5 billion years is in the offing is enough to recognize that either one or both of these ideas are invalid. Unfortunately for the long-time view, erosion is observable while 4.5 billion years is an abstract idea.

This basic reality has been countered by declaring that the mountains formed quite recently and therefore would not have significantly eroded since they were formed. While this seems like a valid and appropriate response, it does not say anything about the erosion that would have occurred prior to mountain formation. If the mountains formed recently it means that all of the rest of the exposed land had been eroding for several billion years so that by the time the mountains formed there would not have been any exposed land at all!

The mountain-forming scenario should be reviewed at this point. All of the mountain ranges of the world are formed from sedimentary rock. Sedimentary rock is rock that has been formed from material that was deposited by water. It is also true that wind will form sedimentary rock but rock that has been formed this way usually consists of very thin layers. By contrast the layers of sedimentary rock in the mountains are hundreds of feet thick. There is no conceivable way that wind formed any of these structures.

The layers of sedimentary rock in the mountain ranges are invariably thick. This means that massive water flow was involved. It would have required massive water flow to form a layer of rock even one hundred feet thick but the layers in the mountains are invariably much thicker than this. They are, in fact, commonly several thousand feet thick. The water movement that caused such formations to form would have been immense and far beyond everyday experience and even beyond the imagination of most of us. How can we visualize water movement which could deposit a layer of material that thick? It had to entrain and transport all of that material for an unknown distance. The magnitude of the water flow means that it would have involved more than one or two continents. It would have involved all of the continents. A water flow of such magnitude would not have stopped at any continental boundary but would have been a globe-encircling water flow which could well have encircled the Earth more than once.

This type of activity is exactly what would be expected when an asteroid impacts the Earth. Massive waves would have been generated and most of them would have encircled the world – possibly repeatedly - before dissipating their energy. Even much smaller waves could encircle the Earth. This is clear from the results of the Krakatoa explosion in 1883. Krakatoa created several monster waves with at least one being 130 feet high. All of them would have traveled at tsunami speed (i.e. 700 mph) across the sea. The tsunamis swept over the coast of Java and Sumatra, destroying most of the settlements and killing thousands of people. (353) One of the waves resulting from that explosion in Indonesia made it all the way to the English Channel. While it started off as a large wave, it would have been dwarfed by the waves resulting from an asteroid impact. These waves could well have started off several miles high because the size of the originating asteroids would have been several miles in diameter and moving into the ocean at thousands of miles per hour. The great tsunamis produced by these monsters would have over-run any land-form simply because of their speed. At tsunami speed, which is commonly upwards of four hundred miles per hour, the kinetic energy of the forward movement could readily have caused them to climb right over a barrier even if it was more than a mile high. The incredible magnitude of the water movement would have ensured that continent-crossing and mountain-climbing would have been guaranteed. Entraining the vast quantities of material which have resulted in the sedimentary layers in the mountains would have been well within the capabilities of such water movements.

The other factor which is evident from this scenario is that it means that all of the mountain ranges of the world would have formed within a single one-year time period. This reality becomes evident from the shape of the sedimentary layers. They are bent into basic sine waves where there is an upward bending arc and a downward bending arc. Layers of rock thousands of feet thick are bent into these anticlines and synclines. Of course every imaginable assembly of rock is also involved in most mountain formations but the up and down sine-wave formations are clearly evident. The only way that such formations could have formed is if they were formed before the deposited material became too stiff. It would not have taken very long for stiffness to have developed. As soon as the layers were placed they would have started to harden and within a short space of time they would have been too hard to bend without breaking. All of the mountain ranges of the entire Earth must therefore have been formed in a very short period of time. There isn't any way around this because if rock several thousand feet thick was forced to bend, even long before it was absolutely solid, it would have fractured. There is certainly plenty of fractured rock in the mountains but the sine wave formations are clearly evident and demand a rapid-formation explanation. Any other explanation must simply ignore the rapid drying out and hardening reality.

Now if we return to the original situation where the exposed material of the Earth is being eroded we have a dilemma much worse than we had before. While the mountains apparently have been formed relatively recently this means that prior to mountain formation, the surface of the Earth would have been quite flat and the highest elevation above sea level would only have been a few hundred feet at the most. If the Earth had been here for several billion years why weren't these low-lying exposed areas completely eroded? How is it that there was any land left above water at all? If the mountains only formed recently, the surface of the Earth prior to their formation must have been mountainless with only relatively low land existing. This low-lying land would supposedly have been in existence for several billion years and would therefore have been eroding for several billion years but such a scenario is not realistic. Why did it exist at all? Eroding land – even if it was eroding at some minimal rate – could not have existed for several billion years. This casts the entire idea of an Earth that is several billion years old, into doubt. Either the Earth has not existed for nearly that long or there really wasn't any land above water until very recently. It was just one world-encircling ocean from pole to pole. The only creatures to be in existence would have been the creatures of the sea. There would not have been any point in evolving into land animals because there would not have been any land to evolve onto. On the other hand if there had been exposed dry land much earlier, any land-based animals that had existed before the last bit of land eroded and disappeared would have died out with no hope of being replaced. Something is dramatically wrong with such a scenario. The length of time that the Earth is declared to have existed cannot be right because erosion really does take place. The most obvious conclusion is that the Earth is not 4.5 billion years old at all!

Just when things couldn't possibly get any worse they did. Hundreds of millions of years ago the Moon would have been much closer to the Earth as explained earlier. This is determined from current measurements which tell us that the Moon is receding from the Earth. The Moon is receding from the Earth due to the tidal effect of the Moon on the Earth. (127) The Moon pulls up the tides but because the Earth is rotating at a fairly rapid rate, the tidal bulge does not reach maximum until it is slightly ahead of a line drawn through the centers of both the Moon and the Earth. This puts tidal maximum slightly ahead of the Moon even though they are both moving in the same direction. But now the tidal bulge pulls on the Moon and actually pulls it forward in its orbit. It accelerates the Moon. This causes to Moon to continually move higher in its orbit around the Earth and as time goes by the Moon will get farther and farther away from the Earth. It also means that in the distant past the Moon was closer to the Earth and pulled up tides that were much greater than they are at the present time. Time is significant in this situation and calculations reveal that the Moon would have actually been touching the Earth about two billion years ago. The rate at which the Moon is receding is measurable and indicates that the Moon would have been in contact with the Earth only 2 billion years ago. Also the friction between the Earth and its oceans results in more energy being dissipated. This further reduces the time back to Earth-Moon contact to only one billion years. (127) This is not good news for either the erosion scenario or the ancient-Earth scenario. Monster tides and low lying land are a very bad combination. The tides would have repeatedly swept right over the land and in fact would have been the primary means of erosion. Rain and wind would have had minimal effect by comparison. Also, since the Earth would have been spinning much faster at that time, the tides would have come much more frequently. With our present schedule, the tide peaks about every twelve hours. Back in that ancient time it would have been every one or two hours. A further aggravation recognizes that the tides would have been moving faster. At the present time tidal speed is more than 500 mph. Back then, the speed would have been thousands of miles per hour. So the tidal situation involved tides every few hours traveling several thousand mph and peaking out many times higher than present tides peak. How would any exposed land – either soft or hard – have withstood such abuse? From this factor alone it is clear that the idea that the Earth is 4.5 billion years old is totally false and should be set aside.

While a nearby Moon and a rapidly rotating Earth mean that enormous tides go by frequently, it also means that the temperature of the continental crust would have been elevated due to the internal friction of the Earth's material. (This is referred to as the mechanical equivalent of heat where mechanical movement results in material warming up.) This would have raised the temperature of the Earth's crust by 1000 degrees Celsius. (352) This means that a nearby Moon would have caused the surface layers of the Earth to have been liquid. Why then wouldn't the continental crust – being slightly lighter than the oceanic crust – have simply spread out evenly over the entire Earth. Being lighter it would have floated on the heavier material underneath and being liquid it would have spread out. In this case there never would have been any concentration of continental material to enable the continents to stick up in the first place!

Erosion makes one more contribution to the time element because material that erodes from the land enters the ocean and never returns. Sediment from rivers continually brings material to the ocean. Wind-blown dust is another factor. Of probably greater consequence material entering the atmosphere from space is continually contributing to the accumulation of material in the ocean. The rates for all of these material contributions have been estimated and indicate that the ocean basins would be completely full of material in only a few million years. However before the ocean basins were even 10% full, enough ocean water would have been displaced to completely cover the land. It would have been a water-covered Earth. The great average depth of the ocean at more than two miles compared to the relatively-low average land elevation of less than five hundred feet means that only a small fraction of the ocean would have to be filled before most of the land would have been covered. With a scenario like that, massive long-traveling waves would have attacked the remaining exposed land and would have hastened the erosion process. Therefore due to sediment accumulation alone, exposed land could never have remained exposed for billions of years!

3.3.3   The Crust of the Earth

The Earth is a large sphere of hot molten material covered by a relatively thin crust. The crust has supposedly formed as the Earth cooled from some previous state where it was hotter than it is now. The length of time required to cool the originally-molten Earth would be unknown partly because the starting temperature for the cooling process would also be unknown. The initial state would supposedly have been at least red hot because the interior of the Earth is still red hot at the present time. Of course even 'red hot' does not indicate any particular temperature but the radiation rate would likely have depended on the temperature colour. Hotter would just radiate faster. Various commentators have weighed in on the matter and concluded that it would have likely taken at least one billion years for the Earth to cool enough for a crust to just start to form. (128) Even after the crust actually started to form it would still have been red hot and water - if there was any (Water is mostly oxygen and it has been hypothesized that the early Earth was void of oxygen and did not have any until certain microbes 'evolved.' When cyanobacteria developed (about 3.5 billion years ago) oxygen was produced and began to fill the atmosphere. (162)) Any other volatile material would also have remained in suspension in the atmosphere. The ongoing redness would have enabled the cooling process to proceed almost as rapidly as it did in the molten state.

The rate of formation of the crust after it initially started to appear would expectedly have been non-linear because as the crust thickened the rate of heat loss would have decreased. Cooling would obviously have continued but with a thickening crust it would have slowed down somewhat. The necessity would have been that the thickness of the crust increase to the point where the surface temperature was well within the animal comfort zone. Animals cannot walk on a hot surface and plants cannot grow on a hot surface. Also water will not remain on a hot surface but will just evaporate. Even at the present time the surface of the Earth will dry out very rapidly if it is exposed to direct sunlight during the summer. In fact, if the surface of the Earth was 'warm' on its own, moisture retention would be virtually impossible with the result that plant life would not thrive at all.

One approach to estimating the time required for the surface of the Earth to get cool enough to be usable by both plant and animal life can be recovered by noting the temperature profile of the Earth's crust at the present time. While the thickness of the Earth's crust is not really known certain commentators have offered that the continents are about twenty miles thick while the ocean floor is much thinner. (129) Thickness, in this context, is not a very precise term but will allow the discussion to proceed. Various instruments have been developed to try to measure the thickness of the crust but the only sure way would be to drill a hole as deep as possible and measure the temperature directly. This has been attempted but it must be recognized that going all the way through would not really be feasible because a drill would not be able to continue operating where rock was red hot and its viscosity was low. As the viscosity of the rock decreases with greater depths and increasing temperature, a drilled hole would not remain open. It would flex and distort and soon disappear. Something similar to this happens when holes are drilled down through the great glacier on Antarctica. Ice is not reliable as a solid and soon a drilled hole will become distorted and might disappear altogether.

The 'bit became too hot' was the reason that the Russian drilling team terminated their deep borehole drilling project on the Kola Peninsula near Finland. They were forced to stop at about the 12 kilometer depth (7.2 miles) when the temperature was about 180C (356F). (335) Commendably they had persisted for about twenty years but had to abandon the work due to the temperature limitations of the equipment. (6)

It will also be understood that the crust would not have exactly the same thickness everywhere. It would be expected to vary and the variation could be several percentage points. Never-the-less it would seem reasonable to recognize that the thickness of the crust of the Earth could be about 20 miles but for our present discussion we will be a little more conservative and use a ten-mile thickness.

Mines are occasionally sunk to a considerable depth into the Earth and one of the difficulties of working in a deep mine is the temperature. One of the deepest mines ever operated was the nickel mine near Sudbury Canada. This mine eventually went down for more than one mile but even at the one mile depth the temperature was too hot for the workers without having cool air piped in. A similar temperature on the surface of the Earth would inhibit both plant and animal life from being able to survive. The ground would just dry out and water would never stay in any reservoir. If the surface temperature of the Earth was similar to the temperature of a deep mine, the Earth would not be habitable.

Suppose that it only required 500 million years for a crust to initially form and that it has been thickening at a constant rate ever since. Further, if the crust is only ten miles thick but the temperature in a mine one mile deep is too hot, it would be suggestive that nine miles of crust would not be enough to allow the surface of the Earth to be at a temperature that enables the Earth to be habitable. If the age of the Earth is 4.5 billion years and it took about 500 million years for the crust to start forming, the crust has only been forming for about 4 billion years. During that time it has formed (for our discussion purposes) to a depth of 10 miles but when we are only 10% of the way through it is too hot. This is implies that for 90% of the 4 billion years, the crust was still too hot for either plant or animal life. 90% is only about 400 million years ago. Did all of the plant and animal life on the Earth develop in just 400 million years? We could even offer another step of conservatism and allow that it was actually one billion years ago that the surface of the Earth became cool enough for plant and animal life to thrive. While these are only numbers for discussion purposes, they illustrate that the thickness of the crust and the idea that it formed from a molten state by cooling, are important factors with respect to the length of time that the Earth has been habitable.

Currently the average surface temperature of the Earth is about +15C (51) and it is observed to be rising. The increase expected during the next few years would only be a few degrees but if this happens it is anticipated that it will be a catastrophe for the entire world.

Since the interior of the Earth is very hot, the Earth is losing heat. How does this heat loss affect the surface temperature at the present time? What if the crust was only one-half as thick as it is now? Would our surface temperature be a few degrees higher and would that be a disaster similar to one expected due to the increase in surface temperature expected on the Earth in the near future due to increasing amounts of greenhouse gas? While such a question is only of academic interest to the current inhabitants of the Earth, it is an important factor with respect to the length of time that the Earth could possibly have been habitable.

The crust under the ocean is a slightly different type of material than the crust under the land areas. The ocean crust is heavier. This enables the land crust to float up higher and be elevated slightly above the ocean. Actually it is just barely higher but it is obviously necessary that this be the case. If continental crust was as heavy as ocean crust, continental crust would sink and ocean crust would rise. As it is, the ocean is very deep at an average depth of approximately 2.5 miles and except for mountains the continents barely rise out of the ocean with an average elevation of about one-fifth that amount. (250) In comparison to the great diameter of the Earth a shift of as little as 0.25% in either of these two factors would result in all land areas being submerged under water and this would make the Earth completely uninhabitable for many types of animal life. Therefore the ocean crust and the continental crust must not only have a slightly different density, the difference must be maintained over the long term. However without continuous recycling of materials through non-stop plate tectonic activity, granitic magmas would not give us our high-standing land masses for very long and we would soon have a watery world. No other body in our solar system apparently has such regular large-scale tectonic cycles to regulate crustal growth. (250)

One wonders how the situation developed in the first place with small amounts of comparatively light material floating on a giant ball of heavier material. The amount of continental material that exists in comparison to the rest of the Earth is only a very small fraction of 1%. This appears as a curious arrangement on its own but long-term stability is also required and if the Earth is to be 4.5 billion years old, stability must be maintained for that entire time.

While the crust of the Earth is a curiosity with respect to its density, it is also curious with respect to its contents. In order to float on top of the remainder of the Earth's material it must be lighter. However it contains numerous heavy elements including; gold, silver, lead and iron. If the crust of the Earth was once molten, these heavy elements would have sunk to the very center of the Earth. Gold, however, has been found right on the surface. How is this to be explained?

If a planetary body were to experience a liquid phase at some time in its history, it is understood that denser material would sink. Many planetary bodies in our solar system were once liquid and this allowed denser materials to sink to their centers. (340) This understanding has now been applied to asteroids as well. It was reported in a recent issue of Nature (i.e. the magazine Nature) that for at least for two of our solar system's largest asteroids, melting was dramatic. (340) The find suggests that more than 50 percent of each object was once liquid. (340)

Gold and lead are among the heaviest elements known. In order to capture and keep such heavy elements from sinking, the support structure in which they are found must have been in place in a solid state ever since these elements arrived. This means that the Earth could never have experienced a time when it was entirely molten and that the crust of the Earth has always been in place just as we currently know it. Therefore we cannot talk about a time when the Earth cooled down enough for a crust to form because the crust must have always been here to support all of the heavy elements. This casts a serious doubt on the idea that the Earth is 4.5 billion years old because the crust must have formed instantaneously in order for the heaver elements that are embedded in it to have been kept from sinking to the very center of the Earth! (We must note that volcanic activity cannot account for the presence of heavy metals near the surface because volcanic activity never involves the material near the center of the Earth.)

As with the tidal friction and heat problem, if the Earth had ever been liquid for any reason there is no cause to expect that continental material would have clumped together to enable continents to be formed. Rather it would have simply spread out evenly over the entire globe. In this case any water that showed up would have also spread out and we would now have a water-covered Earth!

There is another aspect of the crust of the Earth that is relevant to the long-term time question and this relates to the stones found on the surface. In many areas where the bedrock of the Earth is exposed, forests are growing. The trees of these forests drop their leaves every year as well as some branches. Occasionally a tree will fall over as well. All of this material contributes to the build-up of a layer of topsoil. However the stones on these forest floors are still exposed! How could this be possible if leaves have been falling for 100,000 years? Perhaps these forests did not form until recently but how could that be? Trees will grow where there is seemingly no soil at all and are commonly found growing on near-vertical cliff faces with their roots wedged into a few small cracks. The entire scene speaks of recency and casts one more doubt on the idea that the Earth is very ancient. If it was, all of the 3,000,000 square miles of the Laurentian Plateau (The Canadian Shield) would be covered in a layer of topsoil many meters thick by now and there would not be any exposed rocks on the forest floors at all.

3.3.4   The Asteroids and a Life-Filled Earth

3.3.4.1   The Major Impacts

If the Earth is several billion years old it seems reasonable to expect that the asteroids that have hit the Earth would have come at different times and these times would have involved large intervals of millions of years. This expectation is confirmed by the various listings of major impacts where each major impact has a date assigned to it. These dates have been determined by an analysis of rock samples taken from the site and thought somehow to represent the time of impact. Apparently the glass nodules that accompany many impact locations are also used as dating material. The following list of impacts includes the dates that have been assigned. Crater diameters are measurable and the list shows the diameters that have been measured for each of these craters. This particular list is similar to other lists commonly given and starts with the largest 'confirmed' impact. (278) There are several other suspected sites which have not yet made it to any 'confirmed' list and of course there are sites which have not even been identified yet.

Major Impacts					
Location	Diameter(km)	Age(millions)	Location	Diameter(km)	Age(millions)
Vredefort	300	2033	Sudbury	250	1849
Chicxulub	180	65	Popigai	90	35
Acraman	90	580	Manicouagan	85	215
Morokweng	70	145	Kara	65	70
Lake Tai	65	375	Beaverhead	60	600
Tookoonooka	55	120	Charlvoix	55	342
Silian Ring	52	377	Karakud	52	25
Montagnais	45	50	Araguaainha	40	244
Chesapeake Bay	80	35	Mjolnir	40	142

Major Impacts					
Location	Diameter(km)	Age(millions)	Location	Diameter(km)	Age(millions)
Puchezh-Katunki	40	167	Saint Martin	40	220
Woodleigh	40	364	Carswell	39	115
Clearwater West	36	290	Manson	35	74
Slate Islands	30	450	Yarrabubba	30	1800

At each of these impact sites crustal material would be available for dating as well as possibly asteroid material. Both of these materials would have already have had an age prior to impact so how would dating them tell us when the impact occurred? How is one to know with any certainty that the 'age' of any particular sample of rock or other material recovered from an impact site is indicative of the time of the impact? All of the material in the neighbourhood would have had an age just before the asteroid touched down so it would not be the least bit clear how dating rocks from the area would tell us when the impact happened.

The theories of rock dating – in particular potassium-argon – declare that if a sample of rock is 'reset to zero', we can tell how long it has been since the 'reset' (or melting) occurred. This has proven to be a very unreliable procedure and rocky material that was molten only last week has been found to be several million years old this week. (279) The procedure also assumes that melting would have occurred at impact and that the particular sample in hand had melted and been reset to zero at impact time. The presumptions are thereby compounded and one is left wondering if there is any credibility in the age declarations at all.

Before leaving this topic we must recall that the current rate of the recession of the Moon from the Earth only allows a time frame of, at the most, a few hundreds of millions of years back to when the Moon would have been in extremely close proximity to the Earth and would have been distorting the crust of the Earth so badly that it would have practically been fluid. The tidal bulge of the Earth that would have extended upwards towards the Moon at that time would have been hundreds of miles high. What does this say about impacts that are dated to two billion years ago?

3.3.4.2   The Hazards of a Single Impact

From the crater diameter the size of the asteroid that made it is assumed and, while not included in the list will usually be thought of as being about 10% of the crater diameter. This is not known but only surmised. As pointed out in the previous chapter, it is more likely that for large impacts, the size of the impactor was closer to the size of the crater because the Earth is a fluid body covered by a very thin (comparatively speaking) crust. If an asteroid were any appreciable portion of the thickness of the crust, it's very high speed and mass would enable it to punch right through the crust and plunge into the interior before slowing down. In order to form a crater in the usual sense, the impacting object must slow down and stop. If it didn't slow down and stop partway through the crust there is no reason to assume that its size would relate to the crater size the same way that a much smaller one would that couldn't get through the crust at all. This means that the size of a large impactor could be an appreciable fraction of the size of its crater instead of only a small fraction. On the other hand if an asteroid was only a small fraction of the thickness of the crust it would be like hitting a solid barrier and most of the energy of the asteroid would be dissipated right at the impact site. However if the asteroid could punch through the crust, its energy would be dissipated far from the impact site and a size calculation could not proceed based on the crater size. Strictly speaking, only very small asteroids form craters. Large asteroids leave a basically-flat surface with only a rim appearing. The flatness (actually the curvature of the impacted body) indicates that the impactor did punch through and enabled molten interior material to rise up into the opening and basically make it almost disappear – like the huge impact sites on Mars, Mercury and the Moon. Otherwise there would be a classic bowl-shaped formation - probably partly filled in – like the Brent impact site in Ontario. Punch-through means that most of the energy of the asteroid would be dissipated far from the impact site making any calculation based on local evidence invalid. However, even using conventional reasoning with respect to the relationship between crater size and asteroid size all of the impacts listed would have caused world-wide flooding, darkness, volcanic activity, and chilling, all of which would have persisted for months and all of which would have reduced the habitability of the Earth for as long as they lasted.

It has been estimated that the total devastation from a single asteroid, approximately ten miles in diameter, would be sufficient to decimate life in an entire hemisphere. (315) This is suggestive that if an even bigger one struck the Earth, the devastation would be sufficient to decimate life on the entire Earth. Instant local death would result from the immediate effects of the impact. Soon-to-follow death would result from the high-pressure steel-plate effect of the atmospheric pressure wave, the earthquakes, the numerous globe-encircling water waves, the shower of rocks, and the mountain-forming crustal deflections. A massive cloud of dust and debris from the impact would rise high up into the atmosphere and spread out over the entire Earth. The impact cloud formed thereby would be augmented by volcanic clouds. Soon, the entire Earth would be enveloped in thick light-blocking cloud. Death during the next few days would result from choking on an atmosphere which was full of dust and particles of rock. There would not be any clean air to breath over the whole Earth. A post-impact winter would set in right away. The cloud would bring darkness and because the energy from the Sun would be blocked from reaching the ground, the surface of the Earth would become cold. The cold would kill plants and freeze drinking water as well as any unprotected animal life. Starvation would follow. Knowledgeable commentators state that between 75% and 96% of all plant and animal life on the Earth would die following the impact of a single large asteroid. (316) (19) For example, the Chicxulub asteroid impact event is claimed to have resulted in a 'mass extinction, which included the dinosaurs. (317) Another commentator offered that this is what happened when the dinosaurs were killed along with 70 percent of all of the other animals and plants. (18) On the other hand, from the expected results of the Chicxulub impact, it is a wonder that any life is left on the Earth at all. Between the direct devastation of the impact and the following post-impact winter, how did anything survive? A survival scenario following a major impact has never been offered!

3.3.4.3   The Problem with Repeated Impacts

Even small asteroids are feared much more than their size would suggest. The impacts of objects larger than one kilometer are large enough to cause climate change that would kill most of the animals but would not cause mass extinctions. (314) Asteroids which are over one kilometer in size and potential causes of mass extinctions and they are expected to strike about once every half million to ten million years. (315) If the generalization that there is a ten to one ratio between asteroid diameter and crater size, (which is probably reasonable for small asteroids at least) the evidence included in the Earth Impact Database (EID) indicates that the Earth has experienced at least 70 impacts involving objects greater than one km. diameter. (2)

With respect to the above declarations, an estimate can be carried out to determine how many animals would survive the repeated impacts that the Earth has endured. For the sake of discussion, if it could be allowed that there would be, more conservatively, only an 80% extinction accompanying all asteroid impacts which produced craters over 250 km. in diameter, there would have been four (reading from the EID) 80% extinctions. (313, 314) Further, if a 50% extinction accompanied all asteroid impacts which produced craters between 150 and 250 km. in diameter there would have been another mass extinction. (This is conservative in consideration of the 70% quoted above.) If all eleven of the craters between 50 and 150 km. diameter indicate 30% extinctions and the 17, between 25 and 50 km. diameter indicate 15% extinctions, an estimate can be carried out. The Manson, Iowa, crater is in this category at 35 km. diameter, and is considered to have been formed by a stony meteorite about one and one-half miles in diameter. (312) An electromagnetic pulse would have moved away from the point of impact at the speed of light, and would instantly have ignited anything that would burn within 150 miles of the impact (i.e. most of Iowa). The shock wave would have toppled trees for hundreds of miles and killed animals within the same distance. (311) This description, while consistent with others, does not mention any fatalities resulting from ongoing effects such as the dust cloud.

While these are the effects of the impactors from the 'known' list, it is highly probable that there are crater sites which have not yet been discovered. Further, the Earth Impact Database indicates that there are at least 40 more with diameters between 5 and 10 km. The following calculation will not include any entry to recognize these last two factors, which would only make things worse.

The oceans of the world account for approximately 70% of its surface area. Therefore it would be reasonable to expect that impacts would also have occurred in the oceans. In particular a crater was recently discovered under the Indian Ocean. The Burkle Crater is only 18 miles in diameter and is well buried below the bottom of the sea but it makes one wonder how many more there are on the ocean floor. (309) Further, the ocean floor includes numerous areas called igneous provinces, many of which are much larger than the largest craters so far discovered. These are places where molten lava from the interior has come up and spread around. (310) Do any of these provinces cover an impact site? (The igneous provinces have some similarity to the Maria on the Moon, where, because they have associated mascons seem to be covering impact sites.)

While any of these additional impacts would also have resulted in much destruction, the following calculation will ignore them. The portion of surviving species from an 80% extinction is 20%. Similarly from a 50% extinction the surviving species are also 50%. From 30%, 70% remains and from 15%, 85% remains. Therefore the calculation appears as follows.

20% (Tanitz Basin, Kazakhstan, 350 km.)
20% (Sudbury, Canada, 250 km.)
20% (Vredefort, South Africa, 300 km.)
20% (Czechoslovakia, 320 km.)

These four events would therefore leave 0.2 x 0.2 x 0.2 x 0.2 = 0.0016 or 0.16%. This means that only a fraction of 1% of all animal life on the Earth would be left as a result of the four largest impacts. Between 50 and 150 km. diameter there is only one event. Therefore from the above assumption of 50% survival for this category, only 50% of all animal life would be left after the asteroid in this category hit the Earth.

50% (Chicxulub, Mexico, 170 km.)

It is noted that this particular event is credited with completely wiping out all dinosaur life. A city-size meteorite slamming would have generated a global cataclysm. Giant waves would have inundated shorelines, and fine dust would have blotted out the Sun casting the world into darkness. This would have been a mass extinction, which included the dinosaurs. (308) The 50% being used for the calculation is therefore very conservative, as mentioned.

There are twelve craters between 50 and 150 km. diameter, to which a 30% extinction was assigned. Therefore, for this part of the calculation we have the following.

70% (Acraman, Australia, 90 km.)
70% (Beaverhead, USA, 60 km.)
70% (Charlevoix, Quebec, 54 km.)
70% (Chesapeake Bay, 90 km.)
70% (Kara, Russia, 65 km.)
70% (Kara-Kul, Tajikstan, 52 km.)
70% (Morokweng, South Africa, 70 km.)
70% (Puchezh-Kalunki, Russia, 80 km.)
70% (Tookoonooka, Australia, 55 km.)
70% (Manitouigan, Quebec, 100 km.)
70% (Popigai, Russia, 100 km.)
70% (Ries, S. Bavaria, 80 km.)

These twelve events only leave 0.7 x 0.7 x 0.7 x 0.7 x 0.7 x 0.7 x 0.7 x 0.7 x 0.7 x 0.7 x 0.7 x 0.7 = 0.0138 which is only 1.38% of all animal life remaining. In a similar manner, the 18 impact craters between 25 and 50 km. would only leave a small portion surviving.

There are also eighteen craters between 25 and 50 km in diameter.

85% (Araguainha, Brazil, 40 km.)
85% (Carswell, Canada, 39 km.)
85% (Clearwater East, Canada, 26 km.)
85% (Clearwater West, Canada, 36 km.)
85% (Kamensk, Russia, 25 km.)
85% (Keurusselka, Finland, 30 km.)
85% (Manson, USA, 35 km.)
85% (Mistastin, Canada, 28 km.)
85% (Mjoinir, Norway, 40 km.)
85% (Montagnais, Canada, 45 km.)
85% (Saint Martin, Canada, 40 km.)
85% (Shoemaker, Australia, 30 km.)
85% (Slate Islands, Canada, 30 km.)
85% (Steen River, Canada, 25 km.)
85% (Strangways, Australia, 25 km.)
85% (Woodleigh, Australia, 40 km.)
85% (Yarrabubba, Australia, 30 km.)
85% (Sahara, Eygpt, 31 km.)

Therefore, these 18 events would only leave 5.28% of all animal life remaining.

When all of the above events are considered together, only 5.28% of 1.38% of 50% of 0.16% would remain. This is approximately 0.00005%. While this does seem pretty pessimistic there is widespread recognition that by far the vast majority of creatures that once lived are no longer living. No one knows how many species of organisms have existed since life began. Thirty billion has been suggested but the number could be much higher even as high as 4,000 billion. Therefore 99.99 percent of all species that have ever lived are no longer with us. (538)

In everyday language this is less than one ten-thousandth of the original number of creatures. If this calculation is one thousand times too pessimistic for whatever reason, it would still leave only 0.05%. Since there were at least four impacts even greater than Chicxulub, the above calculation does seem to line up with the following comment. Every 300,000 years, a one-to-two-kilometer-wide asteroid crashes, initiating a short-duration global winter. Every 100 million years an even bigger one hits and produces world-wide calamities. There are 14 known mass extinctions in Earth's geological record. (19) One is inclined to ask how many times this sort of thing can happen and still have anything left. Of course, a different set of assumptions could be made and a different result would be obtained. It could be argued that some of these impacts occurred prior to any animals being on the Earth. One commentator offsets this idea by declaring that evidence gathered from around the world related to earlier (i.e. earlier than Chicxulub) mass-extinction episodes, 230, 365, and 445 million years ago. (316) On the other hand, even more impacts than those listed above were probably involved because it is presumptuous to think that all impact sites are currently known. In particular, the Earth is mostly covered by water. While a few craters have been found underwater, why wouldn't there be even more underwater than there are on land? In addition, more craters will certainly be discovered on land. In particular, the Chiczulub Crater and the Chesapeake Bay Crater are very recent discoveries. Unknown sites, like these two, might be so completely buried that they will not be found until some totally unrelated activity leads to their discovery, even though they may be as large as Chicxulub. No matter how many craters were actually involved in mass extinctions, the fact will still remain that following every major impact there would be very few, if any, survivors. Then, when the next impact happened, only a small percentage of that number would be left and so on. The so-far-unmentioned problem is that the surviving cohort would not include any large animals but would be dominated by vermin and other small creatures which, simply because of their size, were better able to avoid destruction.

Since the Earth has experienced several impacts, which are in the very large category, life must have been extinguished or virtually extinguished, several times. This is suggestive that life did not really result from evolution because there would not have been enough time for a diverse range of species to have evolved prior to the first major impact. Then, starting with the surviving cohort from the first impact, another diversity of species had to be in place prior to the second impact and so on. Mass extinctions every 50 or 100 million years would have made it very difficult for evolution to proceed properly, partly because, after every one of them, a different mixture of survivors would have been the starting point for the next period. On the other hand, if life on Earth was originally created, the setback of numerous extinctions would have required virtually total creation all over again every time. Whatever explanation is accepted must account for the present diversity of life on the Earth. In fact it must account for the known diversity which was even greater in the immediate past because of the numerous species that are known to have become extinct only recently.

There are several very large impact sites on the Earth. With each of these impacts, the "Window of Life" would have closed, if not from the impact events directly, at least from the subsequent post-impact winter. Could it have closed repeatedly? The probability that a diversity of life could have spontaneously developed prior to the first impact, is mathematically virtually zero, so the possibility that it could redevelop after subsequent impacts is just another zero. Any declaration that life repeatedly redeveloped into a multitude of different species from a few vermin-like and other stragglers from a major impact, is preposterous. There is no conceivable way that animal life could survive the many mass extinctions that wiped out millions of species. (320) Life is just too complex. Could a great diversity of life forms have been repeatedly recreated? Not very likely!. Therefore it is concluded that; THERE HAS ONLY BEEN ONE MASS EXTINCTION OF LIFE ON EARTH.

This is very bad news for the old-age scenario but it gets worse from here. The world-wide flooding would have involved continent-crossing tsunamis repeatedly encircling the Earth. Soil, forests and anything that was movable would be swept away. Land-based animals would not survive and it would be months before there would be any possibility of their survival. Sooner or later the water activity would peter out and things would settle down. However by then the Earth would be settling into an irreversible deep-freeze which would be caused by the reality that the greenhouse gas temperature-regulation mechanism had been irreversibly upset.

3.3.4.4 Upsetting the Greenhouse Effect

The atmosphere has – like the covering on a greenhouse - a greenhouse effect. A portion of the energy from the Sun that is radiating up from the surface of the Earth is re-radiated back to the surface of the Earth. It effectively becomes trapped in the lower atmosphere and helps to keep the surface temperature of the Earth in the habitable range. If this energy was allowed to radiate into space, the surface temperature would drop and the Earth would not be habitable. Of course all of our energy comes from the Sun but when some of it contacts the surface of the Earth its frequency is changed and it becomes heat energy instead of light energy. Then, as it radiates up from the surface it cannot easily get back through the atmosphere because the greenhouse gases re-radiate it right back to Earth. Consequently heat is trapped in between the lower atmosphere and the surface. This is very much in our favour and without this heat control factor the surface of the Earth would seriously chill at night. This actually happens at the present time over the Sahara Desert because the greenhouse effect (i.e. in this case water vapour and clouds) in the atmosphere over the Sahara is depleted.

While the atmosphere of the Earth consists mostly of oxygen and nitrogen it also includes relatively small quantities of several other gases some of which are referred to as greenhouse gases. These particular gases have the peculiar characteristic of being able to reflect heat energy back to the surface of the Earth. Several components of the atmosphere actually contribute to the greenhouse effect. The most influential one is water vapor. The second is clouds and the third one is carbon dioxide. (CO_2) (303) There are several others which collectively only make up a small portion of the total effect.

CO_2 is currently getting considerable attention because it has been measured to be increasing continually since the industrial revolution and has now reached a level that is causing alarm. The historically high level is being blamed for an increase in the average surface temperature of the Earth. While the increase is really quite small and only in the range of a few degrees, it is thought that such an increase will cause great havoc around the entire Earth. Ocean water will warm up and hence sea level will rise because of expansion. More of the great glaciers on Antarctica and Greenland will melt causing an even greater rise in sea level. The warmer water might release some of its CO_2 inventory which will only aggravate the situation even further. But possibly the most aggravating result would be an increase in the amount of water vapour in the atmosphere because this amount is solely dependent on temperature. Therefore an increase in temperature will result in an increase in the amount of water vapor. A situation like this can be self-perpetuating and only make the changes that occur even greater. (34)

The average surface temperature of the Earth at the present time is about +15C (+59F) (51). There are two sources of heat that cause the temperature to be exactly where it is. The first of these is the direct heat from the Sun. The second is the characteristic of the greenhouse gas inventory to retain a certain portion of this heat and to prevent it from being radiated back into space. Of course all of our heat comes from the Sun (except a very small amount which comes from the interior of the Earth). The necessity is to retain some of this heat and thereby keep the average surface temperature at a reasonable level. Over most of the Earth this works quite well but there are a few exceptions. One of these exceptions occurs over the Sahara Desert. The Sahara Desert is reasonably close to the equator and becomes quite hot during the day especially during the summer when the Sun is directly overhead. The surface material of the Sahara is mostly sand, stone and bare rock. Such materials are easily heated. Therefore during the day they heat up. Since there is a lack of water reservoirs on the Sahara there is very little water to produce cloud to shade the surface. All of the heat of the Sun therefore reaches the surface and heats it up. The opposite happens at night. At night it cools off and the temperature often goes right down to freezing and even below. The main reason for this is the diminished greenhouse gas factor in the atmosphere over the Sahara so very little of the heat that radiates up from the surface is reflected back to the surface to help keep it a little warmer. The missing factors in this case are water vapour and clouds – the most influential greenhouse factors. Actually a similar situation can develop in any number of other locations where it can be observed that a cloudy night will not usually be as chilly as a clear night.

In the case of water vapor, as with CO2, more moisture means higher temperature. Within a fairly small range, this is acceptable but the problem with water vapor is that it produces a self-perpetuating viscous cycle. This means that the worse it gets, the worse it gets. This type of activity is also called a positive-feedback loop where the results make the cause even more exaggerated. This could become a very serious problem for the Earth.

The reverse is also true. If the Earth should become chilled for any reason whatsoever the air would lose some of its ability to retain moisture. Therefore the greenhouse effect of the water vapor in the air would be diminished. This would cause more cooling which would result in less moisture in the air and even more cooling. The end result would only occur when the air could not really lose any more moisture. At that point the temperature of the Earth would settle out. Unfortunately this temperature would be well below the comfort zone for the survival of animal life and the Earth would have become uninhabitable.

Water vapour, as a greenhouse gas, is a Window of Life. A certain quantity is required to help keep the Earth in the comfort zone. If there was too much, the Earth would overheat. If there wasn't enough, the Earth would chill. The self-perpetuating results of both over-heating and over-chilling mean that this window is not very wide and could easily close making life on Earth impossible! Based on this recognition of the greenhouse gas effect, it will be appreciated that any introduction of significant prolonged cold on a world-wide basis must absolutely be offset by prolonged heat – also on a world-wide basis.

A single major asteroid impact would produce a globe-enveloping cloud. It would also stir up numerous volcanoes, the clouds from which would augment the impact cloud causing further cooling. This would be a recipe for disaster because the cooling effect of the clouds would reduce the overall amount of water vapor in the air. With the loss of this important greenhouse gas, the Earth would chill to well below freezing within a short time. (51) While any underwater volcanic activity would have warmed the ocean partially offsetting this loss, it is not likely that a single impact would stir up enough activity to offset the deep chill caused by the world-wide cloud cover. The cold resulting from the dust cloud would therefore dominate and following a single major impact the world would be subjected to a drawn-out world-wide post-impact winter where everything would become frozen solid. Unfortunately with the loss of the greenhouse effect of water vapor and the increased albedo effect of a frozen Earth there would be no escape from this condition and it would persist indefinitely. The greenhouse effect can only be upset at our peril. This causes one to wonder how there could have been numerous widely-spaced asteroid hits over millions and billions of years if the Earth became uninhabitable indefinitely after every hit! It also casts a doubt on the whole idea of millions and billions of years.

3.3.4.5 Asteroid Location

The vast bulk of the asteroids are located between the orbits of Mars and Jupiter. It is understood that the gravity of Jupiter has an upsetting effect on the asteroids causing some of them to be relocated to orbits outside of this area. Consequently asteroids are currently found throughout the inner solar system from inside the orbit of Mercury to outside the orbit of Jupiter. Over time, the gravitational influence of Jupiter modifies their orbits shifting some of them into trajectories that carry them inward towards the Earth. It has also been declared that the asteroids formed long ago about the same time that the planets formed. The asteroids themselves formed about 4.6 billion years ago, at the same time the planets were emerging from the solar nebula. Of course any asteroid that comes within range of any of the inner planets would also have its orbit modified. This can readily be seen by viewing some of the animations that have been posted by NASA concerning asteroids that have come near the Earth. However, if the asteroids are continually being relocated why, if billions of years are involved, do the bulk of them still orbit between Mars and Jupiter? The reality of the asteroids' location casts a serious doubt on the declaration that the solar system, and everything in it, is 4.5 billion years old.

3.3.4.6 Asteroid Features

Both the impact marks and the generally-soft rounded features of the asteroids indicates that they were once molten. One of them has retained an even more definitive feature of a splash that froze in time. Several globules of material are supported on stems above the general surface of the asteroid. Both the stems and the globules on the ends are quite large and would measure several hundred yards in diameter. If this asteroid was ever struck by another one it is doubtful if the stems would be able to support the globules and keep them from crashing down to the surface. If this asteroid has existed for long eons of time, one would expect that it would have been struck several times by now. If it was struck however the splash feature would have been destroyed. (537) What does this type of evidence say about the age of the solar system?

3.3.5 The Magnetic Field

The Earth is a magnet. Therefore it has a magnetic field. The term 'field' implies that there is some influence remote from the object, which is causing the field. The Earth also has gravity. The force of gravity is similar to a magnet because there is an effect far away from the Earth itself. While nobody has been able to explain this remote effect, the magnitude of it can be calculated. It is understood that the force of gravity between the Earth and the Moon keeps the Moon in orbit around the Earth. Similarly, there is a force of gravity between the Earth and the Sun which is recognized as keeping the Earth in orbit around the Sun. We have Isaac Newton and

Albert Einstein to thank for the mathematics involved but we do not have anybody to thank yet for explaining just how the force of gravity works. The theories which have been advanced so far are very tenuous but this is not important compared to the fact that the magnitude of the forces can be calculated.

As the force of gravity acts remote from the Earth and can be accurately calculated, the magnetic field of the Earth also operates remote from the Earth and its magnitude can also be calculated. While gravity is credited with holding everything in place on the surface of the Earth, the magnetic field of the Earth gets very little attention. It is handy for operating compasses. If a person is away from well-known areas, knowing which way is north will be helpful in determining which way to go. A compass and the magnetic field might therefore be helpful for our survival if we are far away from home but there is another way in which the magnetic field is helpful all of the time.

The magnetic field of the Earth is quite strong. It therefore extends well out into space and will have an effect on all incoming particles, which might be influenced by a field of this nature. (i.e. charged particles) Unfortunately the Earth is the target for a lot of particles from space – in particular subatomic particles. These particles are atoms but they are not complete atoms and this is the reason that the magnetic field is so important.

Since these particles are very small (i.e. atom sized) and move very fast they do, in many cases, come right through the atmosphere and go right into the Earth. They could go right through a person on the way. They might interact with the human body at the atomic level. They might crash right through the surface layers of the body and damage some part of the complicated structure which causes the cells in the body to divide. If this happens, the cells might not divide properly. Tumours could result, which would be a very unwelcome development.

The magnetic field of the Earth has an influence on many of these unwanted particles from space and actually traps them high above the Earth. In this way most of them do not actually get to the surface of the Earth at all. The magnetic field has prevented them from getting here and in this way it is acting like a giant shield. It is definitely safer to live on the Earth when this shield is operating. It is also to our benefit that it is transparent so we can still see the stars and sunlight can get to the surface of the Earth.

These little particles, which would have come towards Earth, become trapped high above the atmosphere in a region, which is called the Van Allen Radiation Belt. It is the magnetic field of the Earth, which causes this radiation belt to exist. If the magnetic field disappeared, the Van Allen Radiation Belt would also disappear. In such a case, the Earth would be fully exposed to all incoming radiation from space and the harmful effects that this would bring.

3.3.5.1   A Dying Field

Unfortunately, the magnetic field of the Earth is dying off. Measurements indicate that the overall strength of the field is falling. This is not good news, in particular since, at the current rate of falloff, it will be reduced to half of its present value in about 1400 years. (251) By 4000 A.D. the Earth's magnetic field will have disappeared. (252) The magnetic field of the Earth is therefore both a circumstances window and a temporal window and as a temporal window, it is rapidly shifting towards closure. Since the magnetic field is a necessary component of habitability, how does its transient nature relate to an Earth that is 4.5 billion years old? How has the magnetic field been maintained for so long? Why is it just dying off at the present time? Clearly a few thousand years is a very small portion of 4.5 billion years and one is left wondering what peculiar circumstance could have been in place to maintain a magnetic field for such a long time and then allow it to die off so quickly at the present time? Even more importantly from the standpoint of very long times, the current rate of decay indicates that the strength of the magnetic field would have been much too high in the very recent past. The currents that would produce a magnetic field would be operating in the core of the Earth. However the present rate of reduction in the strength of the magnetic field indicates that a few thousand years ago the currents that produce the magnetic field would have been so strong that the whole Earth would have been destroyed by the heat energy produced by these currents. (318) Therefore the 'window of life' provided by the magnetic field and the currents that produce it can only be a few thousand years wide. How then could life have existed on the Earth for more than a million years and by association how could the Earth be 4.5 billion years old?

3.3.5.2 The Theory of Plate Tectonics

To deal with the problem of a dying magnetic field, it has commonly been declared that it has frequently reversed and that the current reduction in magnetic field strength is because another reversal is underway. Evidence supporting this conclusion involves magnetic stripes on the ocean floor. Many of these stripes occur parallel to the great fissure or rift that runs down the middle of the Atlantic Ocean. The Theory of Plate Tectonics includes the idea that the ocean floor is gradually spreading away from the central fissure as new ocean floor material is produced and that the magnetic stripes on both sides indicate the state of the Earth's magnetic field during all of the millions of years that the spreading has been taking place. It is an idea that has gained a great amount of acceptance and currently represents one of the aspects of the overall idea that the Earth is very old.

The theory was developed to explain numerous observations, which had been made around the world including: the Mid-Atlantic ridge, submarine trenches, surface discontinuities and mountains as well as volcanoes and earthquakes.

The theory recognizes that large portions of continents are moving with respect to each other and seem to be sliding past each other. However, no force function is offered to explain how or why these movements are taking place. In order to move anything as large as a mountain or a portion of a continent, a very large force would be required. Supposedly an underlying fluid layer is moving in a manner which drags the continental portions of crust along as it goes but there is no explanation for why the fluid layer is moving in such an odd manner! A coastal portion of California is moving northward but immediately inland, the ground is not moving. What incredible force is causing this movement? There has never been any suggestion to account for this mysterious force or how or why it is acting. If a system of ideas (i.e. a theory) is offered to explain something which requires very great forces and incredibly high amounts of energy, then included among those ideas there should be a discussion of the source of both the force and the energy. However, no such discussion has been included.

The theory declares that the Atlantic Ocean is growing wider and has been growing wider ever since an original supercontinent cracked. Europe is separating from North America. (280) The source of this separation is the mid-Atlantic ridge, which is supplying new ocean floor material which could make the surface of the Earth larger. In particular, the Atlantic Ocean must be getting larger unless the extra material can be accounted for some other way. If an equivalent area of ocean floor was being consumed elsewhere around the Atlantic, it would not be getting any wider and consequently there would be no necessity to move the continents and no influence on the size of the Pacific Ocean. Subduction (i.e. the diving of ocean floor down into the interior of the Earth) could account for any new ocean floor being produced. This is thought to be occurring where there are deep ocean trenches near continental shores. However, the North American plate continues out across the Atlantic Ocean as far as the mid-Atlantic ridge, making the idea of subduction around the Atlantic Ocean untenable. Therefore since the floor of the Atlantic Ocean is not being consumed by diving into the interior, and since the mid-Atlantic Ridge is producing more ocean floor every day, the Americas must be moving and the Pacific Ocean floor must be getting smaller. It must be diving into the fluid interior of the Earth. Up until it started to dive, this material would have been floating on the molten interior material. It was floating supposedly because it was lighter or at least not heavier. The theory is therefore basically declaring that, an area of Pacific Ocean floor as large as the entire Atlantic Ocean floor has been unnaturally forced down into the interior of the Earth. However, this is in direct contradiction with well-known physics. It is very difficult to force an object that is floating, down into the material that it is floating on. A situation like this is an artificial construct. Nature is declared to be doing something extraordinary which it normally or naturally would be unable to do.

The rate of separation of the Americas from Europe and Africa has been declared to be one to two inches per year. (281) This is not a measured amount but rather a deduced amount, which is based on a calculation. The distance from the center of the mid-Atlantic ridge to certain magnetic anomalies (I.e. stripes) approximately 500 miles on either side of the ridge, were used for the calculation together with the 'known' ages of the rocks at these anomalies. Unfortunately the ages of the rocks are not known but only surmised from other theories. It is also unfortunate that 'older' rocks have been found closer to the ridge using these same theories, all of which places the separation notion in the dubious category.

Just to make matters worse, satellite measurements do not show any widening of the Atlantic Ocean at all (282) and neither do other observations. 'If the theory is correct, the motion of the continents should be observable at present; though Wegener claimed, on the basis of certain reports, that Greenland and an island near its western coast still move, repeated observations and triangulations do not support this claim. ... The land masses of today do not change their latitudes; the motive force claimed is insufficient by far. Coal beds in Antarctica and recent glaciations in temperate latitudes of the Southern Hemisphere all conspire to invalidate the theory of wandering continents.' (283)

Supposedly, the source of material for the new Atlantic Ocean floor is the mid-Atlantic ridge, which means that the mid-Atlantic ridge is somehow deforming from a mountain ridge down into ocean floor. Unfortunately, no evidence is offered to show that this is happening. Instead, the ridge material appears to have piled up on the ocean floor and is just sitting there. The theory is therefore in direct violation of the evidence.

In order to force the Americas away from Europe and Africa, a tremendous force would be required. This force must act outwards both ways from the central region and it must be directed horizontally without having any significant vertical component. Otherwise the central ridge would just get higher and higher and pretty soon it would stick up through the ocean surface. However there is no evidence that the central ridge is getting any higher at all. In order to have formed in the first place there was certainly a vertical force involved because the ridge material is piled up quite high. Alternatively, the vertical force was temporary and forced material up through the fissure where it just piled up on top of the ocean floor In this case the force that pushed up the ridge material somehow lost its vertical factor and is now only pushing horizontally. However, in order to achieve the stated objective of widening the ocean floor, it must still be pushing outwards in two directions at once. It would therefore have to be a type of wedge developing an outward push without piling the ridge material any higher and the ridge is supposed to be deforming into ocean floor. However, there is no evidence that any such actions are occurring.

Why isn't the ocean floor in a state of buckling failure? The theory requires that the outward force from the center of the ocean be strong enough to either force the continents apart or at least force ocean floor material down into the interior of the Earth. The magnitude of any such force would be astronomical. Why is it preferential for the continents to separate rather than for the ocean floor to buckle and crumble and just pile up? This is the type of failure we expect in the arctic with respect to sea ice. When the compressive forces become too great, sea ice crumbles and buckles up into great pressure ridges. The ocean floor is understood to be relatively thin (approximately three km. thick, (290)) in comparison with the distance from the mid-Atlantic Ridge to any shore. Long thin objects such as this cannot carry compressive stress. They would, like pack ice, buckle, crumble and just pile up. How then would the ocean floor, without buckling, be able to carry the forces required to separate the continents around the Atlantic Ocean or to force floor material to dive under the continents and into the interior of the Earth?

It is declared that the ocean floor has been forming for a considerable length of time and that as it formed, it has been moving towards the continents. Where the ocean floor meets the continents, it is understood to slide underneath, or subduct. Trenches adjacent to the continents are understood to be the evidence of subduction taking place. If this were the case, the trenches, should be full of the ocean floor sediment, which had accumulated over long eons of time and been dragged along on top of the ocean floor towards the continents for those many years. They are not full. (32) Actually many of them are empty with almost no material in them at all, which makes them quite easy to identify. In particular the Kermadee Trench north of New Zealand, the Chile Trench, the Middle America Trench and the Tonga Trench are rock bare. (291) Most ocean floor has a sediment layer but most of the trenches are virtually empty. Even though a lot of ocean floor is almost void of sediment in many places, if the trenches indicate where ocean floor is being subducted, they should be full of sediment. Being empty, they look like they formed quite recently by a rapid displacement followed by a slumping back. (292) A sudden one-time force could account for their appearance. Without being full of ocean floor sediment, the ocean trenches actually testify to an event that happened recently and they contradict the expectation that they have traveled one-half way across the ocean over long eons of time.

It is asserted that molten fluid magma is oozing up through a fault and forcing the ocean floor on either side to move away. The Mid-Atlantic Ridge is being built by magma oozing between two plates and forcing North America to separate from Africa. (293) As mentioned above, the force which would be required to cause the plates to separate would be very very large. How much force does it take to move a continent? While it would be difficult to specifically identify the magnitude of such a force, suggesting that the required pressure must be hundreds of thousands or even millions of pounds per square inch would be quite realistic. This high pressure magma is supposedly being contained in a fault between two slabs of the Earth's crust without blowing out into the ocean and causing continual volcanic activity. Further, the extreme pressure would hardly be confined to the fissure between the separating plates, as if they were in some kind of pressure vessel but it would also exist in all of the magma underneath. The pressure required to separate the ocean floor would be many orders of magnitude higher than the pressure that magma would normally develop due to its depth below the surface. If somehow the magma pressure became higher than normal, basic physics predicts that the entire ocean floor would be pushed up until it was well above sea level. It is inexplicable how the weight of the ocean floor is keeping such pressure from elevating the whole ocean. It is inexplicable how the fault is containing the extreme pressure without allowing it to escape at high speed up through the ocean and into the air, and it is also inexplicable how such enormous pressure is being generated. Therefore the above quotation explaining how ocean plates are being separated stands in total disregard for well-known basic physics. Consequently it is neither a statement of fact nor a statement with any validity. It is rather a tale that has just been made up.

A hotspot is a location on the Earth where volcanic activity is happening remote from faults where two crustal plates meet. One expert has defined a hotspot as persistent volcanism in a location that is moving very slowly relative to other hotspots. (294) It has been estimated that there are between 40 and 100 hotspots on the Earth. The wide spread in this estimate is due to disagreement among the experts as to exactly what constitutes a hotspot. However, there does seem to be agreement that the three best known are; Iceland, Yellowstone and Hawaii. Of particular interest at the present time is the fact that a new volcanic hotspot, named Loihi, is developing south of the main island of Hawaii.

One of the most important concepts in Plate Tectonics Theory is that hotspots provide a record of actual plate movement. By measuring the distance between rocks along the hotspot trail, as well as the ages of these rocks, a record of plate movement over many millions of years should be provided. The following quotation focuses on the Hawaiian Islands as evidence of plate movement. 'Each of the Hawaiian Islands formed over a hotspot. A hotspot is the result of a persistent region of molten or melted rock known as magma. There are at least 50 hotspots worldwide and they occur at several places beneath the moving plates of the Earth's crust. The Earth's crust is divided into a dozen or so moving plates, called tectonic plates. The plate that is over the Hawaiian hotspot is called the Pacific plate. These tectonic plates are adrift on the molten magma beneath. The Pacific Plate is currently moving northwest about 3.5 inches per year. Thus when a volcano forms over the Hawaiian hotspot, it is eventually pulled northwest and becomes extinct. In this way a chain of volcanoes is produced.' (295) The Hawaii Center for Volcanology offers a similar scenario. 'The Hawaiian Islands are volcanic in origin. Each island is made up of at least one primary volcano… In fact even beyond Kure, the Hawaiian chain continues as a series of now-submerged former islands known as the Emperor seamounts.' (296) The Kure Atoll is located about 1500 miles northwest of the big island of Hawaii and is the furthermost feature included as part of the Hawaiian Islands. While the Hawaiian Island chain runs generally northwesterly, the Emperor Seamount chain continues from the end of the Hawaiian Island chain and runs almost strait north for another 2500 miles to within 600 miles of the Kamchatka Peninsula of Russia.

The claim being made for the Plate Tectonics Theory is therefore that the entire Pacific Plate moved strait north over a hotspot for a distance of about 2500 miles. During that time, seafloor was being consumed by subduction zones to the north of the plate, as the plate was sliding along other plates on both sides. At the south end of the Pacific plate, new seafloor was being produced. Then the entire Pacific Plate changed direction. Instead of moving strait north, it moved northwest. A reason for the change in direction is not given. Subduction must then have commenced to the northwest of the plate while all along the trailing edge new seafloor was produced. However, the Pacific plate is completely surrounded by subduction zones which are referred to as the Pacific 'ring of fire'. (295) The theory is therefore internally inconsistent and does not agree with itself.

The force required to move the entire Pacific Plate would be incredibly high. A plate as extensive as the Pacific Plate could not be pushed because it would just buckle up. Further, whether the motion was north or northwest, the perimeters of the plate (which would be the sides in either case), are far from being strait. One therefore wonders how the edges of the plate slid in and out of the various recesses along the way. Then there is the problem of magma pressure. There is no pressure that could conceivably have been nearly great enough to move the plate without simply blowing out and no mechanism has ever been identified that could produce the required incredibly high pressures. In other words the entire idea is totally lacking of any realistic sense of basic physics and consequently isn't credible. Recently, more advanced study of the Hawaiian hotspot suggests that it might be moving and not the seafloor. If this is the case, the entire Plate Tectonics Theory would be rendered moot. (297)

A further problem is presented by the sulphides. A question that does not appear to have been asked by researchers is; if the ocean crust were formed by conveyor belt type spreading away from the ridge crest, why is not the ocean floor as enriched in sulphides as the ridge crests? If the sea floor spreading really has been occurring at a slow rate of a few cm/yr over millions of years, the surface of the ocean crust should be dotted with sulphide chimneys, sulphide mounds, and fossil vent communities, all under a blanket of sediment. (298) Since there is such a deficiency of sulphide chimneys, one is left wondering if the idea has any merit at all.

3.3.5.3 Magnetic Data Contradictory

The entire concept of plate tectonics relies heavily on the magnetic data from the ocean floor and without it, the Theory of Plate Tectonics would not have been developed in the first place. In fact, the so-called spreading zones are classified as either fast or slow as determined by calculations based on the magnetic data. (284) However the data are not convincing to everyone. 'The theories of continental drift and seafloor spreading are highly conjectural, but it is hard to stop anything as big as the floor of the ocean once it has been put into motion.' (285) Other commentators have similar reservations and caused one author to conclude that paleomagnetic data are so unreliable and contradictory that they cannot be used as evidence to support the idea of drifting of continents. (286)

3.3.5.4 Vertical Magnetic Reversals

Magnetic reversals in a horizontal direction are difficult enough to explain with any hypothesis but their occurrence in both vertical and horizontal directions completely contradicts the continental drift idea. '… these several vertically alternating layers of opposing magnetic polarization directions found in cored oceanic crust disproves one of the basic parameters of sea-floor spreading theory, namely that the oceanic crust was magnetized entirely as it spread laterally from the magnetic center.' (287)

3.3.5.5 Magnetism Too Strong

'An even more puzzling fact is that the rocks with inverted polarity are much more strongly magnetized than can be accounted for by the Earth's magnetic field. Lava or igneous rock, on cooling below the Curie point, acquires a magnetic charge stronger than the charge this rock would acquire in the same magnetic field at outdoor temperature, but only doubly so. The rocks with inverted polarity, however, are magnetically charged ten times and often up to one hundred times stronger than they could have been by terrestrial magnetism. "This is one of the most astonishing problems of paleomagnetism, and is not yet fully explained, although the facts are well attested."' (288)

3.3.5.6 Rapid Reversals Observed

The case for using magnetic reversals as indicators of time and hence of the rate of ocean floor formation, is weakened further by the recent observations that reversals can occur within a few days. A team of geoscientists investigated the Miocene lava flows at Steens Mountain, Oregon. They observed that the seven lava flows above B51 are of normal polarity and the ten below it are of reversed polarity. Numerous samples taken through the several-meter thickness of flow B51 show a bumpy but continuous transition from the reversed polarity below to the normal polarity above and this would have happened in about 15 days. The investigators thought that such a rapid change was unbelievable. The rapidity and amplitude of geomagnetic variation that we infer from the remanence directions in flow B51 truly strains the imagination. (289)

3.3.5.7   Erratic Magnetic Patterns

Magnetic reversals of the ocean floor have been used as evidence of seafloor spreading. However, this information is not as supportive as its proponents declare. It is not true that linear magnetic anomalies can be correlated from the North Atlantic via the Indian Ocean to the North-eastern Pacific. The magnetic signatures are very similar in limited areas, but are very different among different areas. Moreover, magnetic stripes need not be caused solely by alternate bands of 'normal' and 'reversed' polarization, differences in magnetic susceptibility values of adjacent rock types can produce the same. ... The so-called magnetic anomalies are not what they are purported to be – a 'taped record' of magnetic events during the creation of new ocean floor between continents. (299) Unfortunately, the rocks include erratic magnetic patterns. It is known that 'lightning can magnetize rocks. ... It is probable that much of the scatter observed is the result of lightning strikes. (144) (While it is clear that a lightning strike cannot affect the floor of the ocean, the fact that a surge of current can magnetize a rock leads one to wonder what other natural phenomena could do the same.)

3.3.5.8   Magnetostriction

It has occasionally been suggested that the only explanation for magnetic reversals in the ocean floor is Catastrophic Plate Tectonics. (i.e. a theory that Plate tectonics is a valid theory but that everything happened within a one year time frame instead of millions of years) However, there are other explanations for the magnetism found in rocks. As mentioned above, the magnetism of rocks is often many times stronger than the magnetism of the Earth. These stronger levels may be explained by magnetostriction. Another mechanism for changing the direction of magnetism in igneous rocks is magnetostriction. This is caused by distortions of the rock due to shock or stress. Since there has been considerable earth movement in the past, (i.e. particularly during asteroid impacts) magnetostriction could explain many reversed rocks. It is possible that some of the interesting magnetostriction of rocks stem from the effect of geologic stresses, rather than from changes in the geologic field. Shock induced magnetism from meteorite impact has been invoked to explain the magnetism in Precambrian rocks from Lake Superior. Shock induced magnetization has even been demonstrated in the laboratory by simulating meteorite bombardment. (301)

Magnetostriction could cause the magnetizations of rocks to be in directions that are not those of the fields acting when the magnetization was originally acquired. Some of the great upheavals and folds in the Earth beneath the oceans must certainly have yielded profound magnetostrictive effects would have altered the orientation of the magnetization in the rocks. (302)

There is no doubt that the Earth has experienced many asteroid impacts. The shock waves propagating from all of these impacts as well as from all of the major Earth faults (as they formed) would have been sufficient to cause magnetism of varying orientations at many places around the world. Such developments would have occurred whether the Earth had a magnetic field or not and the resulting magnetism left in the rocks would have no bearing on the strength of the Earth's magnetic field at that time!

3.3.5.9   Conclusion

If the magnetic field of the Earth actually reversed repeatedly, it must have died off completely in one direction before building up in the other direction. At some point as it died off it would have been too weak to offer protection from incoming radiation. All of the animal life on the Earth would have become exposed to incoming radiation and the effects would have been devastating. It would have been like living near a failed nuclear reactor. Almost every animal would have cancer. Most animals would have birth defects. Because of these two effects the Earth would have been rendered uninhabitable. The period of uninhabitability would have involved hundreds and possibly thousands of years as the field died down and then built up again. Then after an unknown time another period of uninhabitability would have happened and it would have happened every time the field reversed. It would have been totally impossible for animal life to have developed under such conditions. A strong and stable magnetic field is a necessary 'window of life' and without it, life on Earth would not be possible. This leads one to conclude that there haven't actually been magnetic reversals and that the Earth isn't really 4.5 billion years old at all.

3.3.6   The Period of Heavy Bombardment

There was a time in the history of the solar system that is referred to as; The Late Heavy Bombardment Period. The time of this period is commonly given as 4.0 billion years ago with a duration of 200 million years. (142) Other commentators agree with this and identify exactly the same period of time. About 4.0 to 3.8 billion years ago a period of heavy bombardment is thought to have peppered all of the planets including the Earth. (143) Other commentators have offered opinions based on radiometric dating. ' ... yielded dates for the heavy bombardment ranging from 4.1 to 3.4 (billion years ago) ... dating of rock derived from Earth's Moon defined the period of cataclysmic bombardment from 3.9 to 3,4 (billion years ago) The impact craters of Mars suggest an earlier period of between 4.25 and 4.1 billion years ago. (164) Numerous other commentators seem to concur with these statements which is suggestive that a widespread consensus exists except that in reality there is no consensus. Several hypotheses are now offered to explain the apparent spike in the flux of impactors (i.e. asteroids and comets) in the inner solar system but no consensus yet exists. (143) Just to add to the confusion, ... impact spherules collected around the Apollo 14, 16 and 17 landing sites indicate that the Moon was heavily bombarded as recently as 800 million years ago. One has to wonder just how many periods of 'heavy bombardment' have occurred in the past. (164)

From these comments it is clear that fairly-definitive dates have been assigned to these ancient events but it isn't clear how this could be determined with any certainty. Also, how does one know that they are holding a sample of material that was associated with an event that happened billions of years ago?

There does seem to be agreement on the severity of the event and that some major disruption in the solar system must have been the cause. The disruption that has been hypothesized is that Jupiter, Saturn and Neptune were all involved and interacted in a way that upset Neptune's orbit! One theory holds that the orbital interaction of Jupiter and Saturn sent Neptune careening into the ring of comets in the outer solar system. The disrupted comets were sent in all directions and collided with the planets. (143) This is an extreme hypothesis and there are several aspects of it that deserve some comment. What caused the disruption of the orbits of both Jupiter and Saturn to the degree that they could cause Neptune to basically leave the solar system and travel out to a 'ring of comets'. This hypothetical 'ring of comets' has not yet been discovered but is rather another hypothesis - which has its own set of fierce defenders - but there really isn't any evidence to support it. There might be a cache of comets away out there or there might not be any cache at all. The observation that comets are continually coming into the inner solar system only tells us that there are comets somewhere but whether or not they are congregated into a 'ring' or cloud or some other type of grouping is not known. There is virtually no doubt at all that there are objects well beyond Pluto and probably objects all throughout the universe among the stars. None of these objects would necessarily be associated with any particular star and could quite readily just be drifters without a home. This type of thinking has been confirmed quite recently when a 'planet' was found drifting in space without any apparent relationship to any star. Why wouldn't there be other objects like this in inter-stellar space? Who can say?

Neptune is a massive planet and to change its orbit would require a very large force. It would not happen easily. According to the offered hypothesis Neptune was caused to leave its orbit, drift beyond Pluto and then somehow return to its current location and then magically acquire a circular orbit. Circular orbits for planets are difficult to explain and if some mysterious disruption was involved it doesn't explain how the circularity of Neptune's orbit could have been re-established. Currently, Neptune is beyond Uranus by about a billion miles. Furthermore it is almost two billion miles farther away than Saturn. Was Uranus also involved in that major upset and how did it also recover a nearly-circular orbit? How could Neptune, which is so much further away, have been so severely affected without also upsetting Uranus?

The impetus which attempts to associate all objects in space with stars and in particular that all objects in space should be orbiting some star or another is derived from the now-debunked theory explaining how the solar system formed from a collapsing cloud of space dust. This was called the Laplace Hypothesis of Solar System Formation and held a seemingly solid position up until the very first exo-solar system was discovered. In that system three orbiting gas giants were identified orbiting 'too close' to their host star. (345) Other similar discoveries quickly followed and even at that time when only a few exo-solar systems had been identified several of them had huge planets in what we would call their inner solar system (i.e. where Mars, Earth, Venus and Mercury orbit). These discoveries all showed that a new theory was badly needed. Now consider the case of the star HD 114762, 90 light years from Earth. A wobble in the star's motion implies the presence of another body, 11 times as massive as Jupiter, orbiting ... closer to the star than Mercury is to the Sun. Then there is 70 Virginis ... (where) ... there is a body 6.8 times as massive as Jupiter orbiting at a distance about half that of the Earth to the Sun. Or take Tau Bootis; it has an orbital companion 3.7 times as massive as Jupiter with an orbital period of just three days and eight hours. ... The presence of such huge bodies so close to their stars challenged prevailing theory (i.e. The Laplace Hypothesis of Solar System Formation) How could gas giants form so close to their suns? ... The discovery of more than a dozen extrasolar planets has forced a serious rethinking of many details of the solar nebular theory. ... One cornerstone of the standard theory has been that the planets first formed at or near their present locations relative to the Sun. But the news from afar ... has suggested a more complicated ... scenario for planet formation. (144) In other words the old theory is dead and a replacement has not been offered. Therefore the associated expectation that every object in space should be associated with some particular star must also be abandoned.

Incidentally, a different theory was actually offered some time ago which discussed the possibility that the universe isn't driven by gravity at all but by electromagnetic forces. (145) The majority of scientists do not like this approach and they explain that gravity is the only known force that can act at great distances. Unfortunately for the majority opinion the gravity explanation doesn't work unless it is propped up by 'dark matter'. When the orbital motion of stars around the centers of their respective galaxies didn't agree with the predictions that were expected due to gravity – assuming that the amount of matter in a galaxy was known – 'dark matter' was invented. In fact, the gravitation-based understanding was so inadequate that galaxies were declared to have about 95% dark something – either matter or energy. A different approach might recognize that if a theory was 95% away from making predictions that matched observations there might be something wrong with it! Clearly it can never really be known how the universe actually operates but if the great majority of matter and energy must be invented to enable a theory to match-up with observation, a different approach should be considered.

If the major planets like Jupiter and Saturn were involved in the Late Heavy Bombardment event, it would certainly have been terrifying so the first thing that would have to be agreed upon is that there was no life of any significance on the Earth. No one has yet attempted to explain how any form of life survived even one major asteroid impact whereas the Late Heavy Bombardment Period is thought to have involved many impacts. 'Based on the impact history of other worlds, computer estimates suggest that during the 200-million-year period of the Late Heavy Bombardment, Earth suffered over 22,000 impact craters larger than 20 km, about 40 impact basins larger than 1000 km and several continent-sized 5,000 km basins.' (142) (We must note that the diameter of the Moon is about 3500 km (2159 miles). If the diameter of an impact basin is approximately the diameter of the impactor – or some modest multiple of it – it means that several moon-sized objects hit the Earth!)

The only way around the quandary of explaining how any life survived would be to declare that there wasn't any life here anyway and that life developed later. Unfortunately this has not been done. We must note at this time that the 'impact history of other planets' is not known but only surmised. Further the 'computer estimates' clearly identifies that the entire idea is really just a computer program anyway! (actually the Nice model (named for the French town where it was conceived) ((recall 'garbage in, garbage out')

Around 3.5 billion years ago something more emphatic became apparent. Wherever the seas were shallow, visible structures began to appear. ... 3.5 billion years ago when life was just getting going ... Stromatolites, a kind of bacterial rock, filled the shallows ... (however) ... without an atmosphere ultraviolet rays from the Sun, even a weak Sun, (i.e. the Sun was supposedly 25% weaker then) would have tended to break apart any incipient bonds made by molecules ... (yet) ... you have organisms almost at the surface. It's a puzzle. (155) Well it certainly is a puzzle but it is easily resolved when we realize that this declaration is not science at all but simply a statement by a scientist. However, it does recognize the problem in trying to have some form of life in the presence of ultra-violet light because ultraviolet light always breaks molecular bonds. This is done intentionally in sewage treatment plants and water treatment devices to destroy bacteria. It is also done in hospitals to sterilize instruments for surgery. The declared exception stated in this quote is therefore not credible! The further problem is the 'Shallow seas'. The dim Sun of that ancient time would not have been sufficient to bring the surface temperature of the Earth to above the freezing point of water, so the Earth would have been frozen solid. In fact the heat from the Sun is still insufficient to bring the surface temperature to the proper level if it were not for the heat retention characteristic of the greenhouse gases. The most influential greenhouse gas is water vapour. Without oxygen there would not have been any water and therefore no water vapour. Neither would there have been any clouds or carbon dioxide – two of the other most important greenhouse factors. ' ... the current average surface temperature is about +15C, whereas the radiative equilibrium temperature which would hold if the Earth had no atmosphere, is about -18C.' (51) In other words the Earth would have been frozen solid from this factor alone except that without water we cannot say frozen – just cold. This comment presumes that the heat from the Sun would be what it is at the present time but billions of years ago it would have been much less. It is only the greenhouse heat-retention factor in combination with the heat from the Sun that enables the surface temperature of the Earth to remain above freezing at the present time. If a world-wide cooling trend was ever introduced on the Earth, atmospheric water vapour would be lost and the surface temperature would spiral down until it was well below freezing. (51) In recognition of these difficulties either there could not have been any life on the Earth at that remote time or there simply wasn't any such remote time at all!

There are serious difficulties with every aspect of the Late Heavy Bombardment declaration. Firstly, evidence of these impacts is not currently recognized. None of them are listed in the Earth Impact Database where almost every major impact has a date assigned to it. If any entry had a date assigned to it that fell in the stated period of time between 4 billion and 3.8 billion years ago it might be associated with the Late Heavy Bombardment Period. However, no such assignment can be made! The oldest entries are; Acraman Australia 590 million, Beaverhead USA 600 million, Gardnos Norway 500 million, Holleford Canada 550 million, Janisjarvi Russia 700 million. Newporte USA 500 million, Paasselka Finland 1,800 million. Presquile Canada 500 million, Shoemaker Australia 1,630 million, Srangways Austalia 646 million, Sudbury Canada 1850 million, Suvasvesi Finland 1,000 million, Vredefort South Africa 2023 million, Yarrabubba Australia 2,000 million. From this comparison none of the 22,000 declared impacts have any current recognition.

The very large impacts come with a different set of questions simply because of their size. Using conventional reasoning the diameter of an asteroid would relate to the diameter of its crater by about a 1:10 ratio. Therefore a 5000 km (3100 m) crater would have been made by an object about 500 km across. This is a very conservative offering because as explained previously, if the impacted body has a crustal exterior and a molten interior, the crater size will more closely approximate the impactor size. This is simply because if the diameter of an impacting object was any appreciable fraction of the crust thickness, it would simply punch right through the crust and distribute its energy throughout the entire planet. This means that the energy of the asteroid would not have been dissipated at the impact site at all but would have been distributed throughout the entire body of the Earth making energy calculations based on crater diameter meaningless. It also means that crater diameter would be only slightly larger than asteroid diameter. An asteroid that was 500 km in diameter is more than terrifying – it is completely overwhelming. How could one that was almost 5000 km in diameter be absorbed by the Earth without it exploding? The diameter of the Earth is only about 8,000 miles or 12,800 km so an object that large would be more than one-quarter as large as the Earth and actually larger than the Moon. What would have happened if several objects as large as the Moon hit the Earth? If there were 'several 5,000 km monsters' and they had originated outside of the orbit of Mars, the energy level of every one of them would have been more than sufficient to cause the Earth to explode! At the very minimum any one of them would have displaced the Earth from its orbit by several percentage points. The problem to be recognized in this case is that the Earth currently has a 'Goldilock's orbit. The orbit is 'just right'. (186) If the Earth had ever been hammered by even one of those hypothesized monsters, its orbit would have been significantly changed. If this is what happened, how did the Earth subsequently and repeatedly acquire such an ideal orbit – exactly in the middle of the Sun's thermally-habitable zone? This seems just too coincidental to be credible. Asteroids orbit the Sun in the same direction as the Earth which means that large impacting asteroids would have hit the Earth from behind. The orbital effect on the Earth would have been to push it forward and into a higher orbit. The length of the year would have become longer. If any orbital change were to occur with the Earth, some major environmental factor of the Earth would also have had to change at the same time or the habitability of the Earth under the original circumstances as well as the present circumstances becomes very difficult to explain. (This was discussed in chapter 2.)

Comets can approach the Earth from almost any direction and they generally come with much higher speeds than asteroids – possibly 100,000 miles per hour. A comet could come from the direction of the Sun or from any other direction. Also they do not follow the general orbital plane of the planets. This means that they can come in from the 'side'. (i.e. any direction whatsoever) With any impact that was not in the direction that the Earth was moving, the Earth could have been knocked in closer to the Sun, or further away, or out of the orbital plane of the other planets altogether. An impact that came in from directly in front would slow the Earth's speed making it fall into an orbit that was closer to the Sun or even into the Sun. With any of these possibilities it is difficult to see how a nearly-circular orbit in the plane of the other planets could have resulted. Such large impacts would have pushed the Earth away from its ideal orbit at the midpoint of the Sun's thermally-habitable zone with the result that its habitability would have been reduced – most likely to zero.

Further, how did the Moon stay so closely associated with the Earth during all of this commotion? The Moon is in an orbit which is very close to circular around the Earth and without such near-circularity the Earth's angle of inclination – which is generated and actively maintained by the Moon – would have also been disrupted. Chaos, chaos everywhere!

The entire event is declared to have happened within 500 to 700 million years after the Earth became a sphere. Since the Earth would have been cooling continuously from its formation, a thin crust would be expected. This is recognized by the hypothesis because 'craters' are mentioned. If there wasn't any crust there could not have been any craters. However the magnitude and quantity of the hypothesized impacts would have been devastating to any crust - especially a thin one. The great mountainous circular tsunamis expanding from every impact site would have fractured a thin crust – or even a thick one – into innumerable pieces. Further devastation would have developed at every antipode on the far side of the Earth. The throbbing and pulsating action resulting would have kept the crust totally agitated for most of the period over which all of this mayhem was to have happened. While the crust of the Earth has certainly been fractured into many pieces one wonders what it would look like if it had received '22,000' hits including several which made 5,000 km. diameter craters!

The period of the Late Heavy Bombardment really is a dubious idea and should be set aside but doing so is suggestive that the idea that the Earth is 4.5 billion years old should be set aside as well.

3.3.7 The Coal Formation Theory and its Temporal Problems

3.3.7.1 The Basic Aspects of the Theory

The Swamp Theory of Coal Formation declares that coal has been formed from plants. This recognizes the observation that the outlines of numerous different species of plants are often observed in coal. It also recognizes that the structural component of all plants is carbon. Coal is carbon, which appears to have been pressed down and compacted into very tight and hard rock-like formations. Therefore to declare that coal is formed from plant material is in keeping with both well-known physics as well as certain direct observations.

Another aspect of the Swamp Theory of Coal Formation declares that the plants, which formed the coal, grew in situ where the coal is presently found. However evidence has not been offered to support this declaration. In fact a large boulder has been found in a seam of coal. Was the boulder brought in and placed among the dead trees or was the entire mass moved into position and just happened to include a large boulder? (253) Appeal to reason is not a substitute for evidence. This portion of the Swamp Theory of Coal Formation is therefore not substantiated and the available evidence contradicts the declaration because some coal beds are very thick and a lot of plants were required for their formation.

The theory also requires that the plants grew and died in such a way that they did not rot. New plants kept on growing and piling up on top of the old ones without rotting until a huge mass of dead, preserved plants had been accumulated. This aspect of the theory violates well-known physics. Current and readily repeatable observation indicates that when trees die and fall over, they rot. Much of the material of the dead tree is thereby oxidized into carbon dioxide and becomes part of the atmosphere. Any moisture, which was part of the tree, will simply evaporate. The rest of the tree will become part of the soil and in fact will form the soil. This is what is always observed. Therefore to declare that a tree can die and fall down without rotting violates observation. In order to circumvent this violation, it is declared that the plants, which would form the coal, grew and died in a swamp. When they fell into the water, the water kept them from rotting thereby keeping the carbon, from which they were formed, from leaving. Therefore, it is a necessary condition of the theory of coal formation that all of the plants, which formed the coal, grew in a swamp, died, fell over and were covered by the water of the swamp. This aspect of the theory violates neither well-known physics (water will keep a plant from rotting for a considerable time) nor observation (plants do die and fall over in swamps every day) as far as dying and falling over are concerned. However, the claim that the plants grew in the swamp is not valid.

The problem with the swamp origin of coal relates to the types of plants found in coal. Pine, spruce, hemlock, sequoia and other dry land conifers are found in European and North American lignites. Palms, birch, beech, magnolia, cinnamon and others are found in Cretaceous coals (254). None of these trees grow in swamps.

The next aspect of the theory declares that the plants lived and died in the swamp and were covered by the water of the swamp but the swamp never became filled up with plants. In order to produce even a modest thickness of coal, a great many plants are required. Some coal beds are several feet thick. It has been estimated that it requires ten feet of plant material to form one foot of peat and twelve feet of peat to form one foot of coal. (255) One of the thickest coal beds found to date is thirty feet deep. (256) If this estimate of plant volume is correct, 3600 feet of plants would have been required. (Of course coal is formed from trees as well as other types of plants. 3600 feet of compressed ferns is hard to imagine but 360 trees, ten feet in diameter is a little more comprehensible.) This means that many plants grew successively in the swamp and that the water level of the swamp kept rising at exactly the necessary rate to keep the area as a swamp but not get too deep to terminate the growth process. A great amount of time would be required to grow all of the plants required to form a seam of coal which was thirty feet deep. Therefore, the swamp water must have kept rising at just the right rate for thousands and thousands of years. This type of swamp has never been observed. In recognition of the difficulty of maintaining a swamp of this nature it was declared that the swamp must have sunk instead. The swamp sank at just the right rate to allow the water to keep the dead plants covered but still allowed the new plants to grow properly. Instead of rising for thousands of years the swamp sank for thousands of years. A swamp of this nature has never been observed either.

3.3.7.2 Extended Times Not Involved

The final aspect of the Swamp Theory of Coal Formation declares that the plants went through their normal life cycle of living and dying for an extended period of time, which involved millions of years. According to the theory, succeeding generations of plants grew and died and added to the accumulation of material, which was forming a future coal bed. New plants grew on top of old plant material and the entire mass of material kept building up. Entire forests of plants grew and died thereby accumulating the carbon for the future coal deposits. This extended period of time has been called the Carboniferous Period and is declared to have lasted 60,000,000 years. (257) However, the idea that a vast amount of time was required to form the coal-producing plants requires artificial construct as will be shown in the following discussion.

Certain coal beds are observed in layers. There are layers of coal interspersed with layers of rock. This necessitates that after the swamp had sunk at just the right rate for an extended period of time, it became covered with a layer of material which would later turn into rock. Then a new swamp formed on top of the layer of non-swamp material. Then the whole thing sank at just the right rate and another layer of coal-forming material was deposited. Next, another layer of rock-forming material was deposited and the sequence repeated.

Of course, a lot of time would be required. Geologists have suggested that it would require 1000 years to form one inch of coal. (256) In the process, any portion of any tree, which lived and died and did not fall down properly, would be subject to rot and disappear. However, the theory is now in direct violation of observation. It has been repeatedly observed that the fossil remains of trees project right up through several layers of coal as well as the intervening rock material, which separates them.(254) How could this be? During the extensive times, which are required by the theory, the trees would have been exposed to the atmosphere and would have rotted. In fact they would have had time to rot a thousand times. This aspect of the theory is therefore not valid because it violates well-known physics.

If the trees and other plants, which were the source of material for the future coal beds, were not covered and isolated from the atmosphere quickly, they would have rotted and hence become unavailable for coal formation. In particular, the trees, which extended up through several layers of material must have been buried quickly or they also would have rotted. It may therefore be concluded that the material, which would become coal was gathered, placed and covered within a short amount of time (i.e. at least well within the 'rot' time).

Meteorite material is never found in coal. Well over 50 billion tons of coal, have been mined and never once has there been a report of meteorite material being found. This has even led to the speculation that meteorites did not fall during the 'millions' of years that the coal was being deposited. (258) Alternately, since meteorites are observed to be continually impacting the Earth (259), the period of time involved in forming the coal and covering it with overburden, was simply less than the interval between impacts.

There is another possible explanation. The material for the coal beds was transported and placed during a time of world-wide chaos as a shower of asteroids hit the Earth. Great tsunamis were sweeping back and forth across the Earth and arranging and rearranging the material for the coal beds as well as the material for the numerous sedimentary layers – some of which would soon be folded into mountains. It was a scene of utter chaos and catastrophe. The plant material for the coal beds was semi-fluid and was being rolled and shoved into place and then pushed and shoved back and forth until the massive wave action died out. During all of that time it was a mass of vegetation in a fluid-like state and if a meteorite did fall into it, it would just pass right through. There wasn't enough shear resistance in the massive assembly of plants to stop a speeding meteorite and hold it from going right through to the bottom. Meteorites of any size have an incredible amount of energy and stopping them with solid rock is difficult enough. The coal-forming plant assemblies could not have stopped them because they would not have had enough structural integrity to do it. Even much of the future sedimentary rock material that was being placed would have had a difficult time stopping a meteorite (or asteroid). Soon the asteroid shower petered out and the wave action quieted down but it would have continued for several days or even weeks after the last asteroid hit. During much of that time the massive assembly of plants that would become coal were still being agitated and would not have been able to retain an incoming meteorite but would have let it pass right through.

On the other hand Erratic Boulders are commonly found in coal. (415) However with respect to time, Erratic Boulders are thought to have been placed during the Great Ice Age which was only a few thousand years ago compared to the millions of years declared for the Carboniferous Period when coal was formed. At the very minimum this means that coal would have been formed very recently in keeping with the other evidences that indicate the same thing.

However Erratic Boulders would have been caused to fly through the air as one result of impacting asteroids which would also have caused the continent-crossing waves that placed the coal material. An explanation is needed for Erratic Boulders in coal and recognizing that they were scattered about from impact sites is consistent with the idea that the upheaval also caused water movement which would have been sufficient to place the coal material. The presence of Erratic Boulders in coal casts a long shadow of doubt on the idea that coal was formed many millions of years ago and by association that the Earth is billions of years old.

There is another type of evidence which indicates that the coal-forming process only occurred a few thousand years ago and there are three elements to it. Firstly, coal has a carbon14 count, indicating that it is only about 50,000 years old. (260) (If the carbon14 production level was only a fraction of its present level at the time that the coal-forming trees were growing the low carbon14 count would indicate that coal is only a few thousand years old.) If it is really as old as the carboniferous period, which is claimed to have been more than 100,000,000 years ago, there would not be any carbon14 count at all!
Secondly, the uranium238/lead206 ratios identified in the inclusions found in coal are very high. If the process had been ongoing for hundreds of thousands of years, this ratio would not be high. The high ratio indicates that the uranium has only been decaying for a few thousand years and that coal has only been formed recently. (261,262)

Thirdly, the trees which formed the coal were formed almost instantly. This evidence has been recovered from the study of very tiny coloured circles found in both rock and coalified wood. At the centers of these circles is a very tiny inclusion of polonium and there is no evidence for any decay product before polonium which suggests that polonium was present and decayed but that it did not have any parent material. Hence it is referred to as 'parentless polonium'. (263) ... The circles that are usually found are indicative of polonium218, which has a half-life of 3 minutes, polonium214, which has a half-life of 164 microseconds and polonium210, which has a half-life of 138 days. When these circles are found in rock they require nearly instantaneous crystallization of the rocks with the creation of the polonium atoms. (264) The first two of these rings are not found in coal. This indicates that the wood that would become coal, did not form around the polonium at the same time that the polonium formed. In fact it indicates that it did not form for at least five half-lives (or about 15 minutes) after the polonium218 started to decay. However, polonium210 is included indicating that the wood had enclosed the polonium within days of its formation. (i.e. in between about 15 min, and 690 days (138 x 5)) (Five half-lives are mentioned because after this much time has passed, there is very little radiation activity taking place.) (We could note at this point that Holy Scripture states that trees were formed on the third day of Creation.)

A further complication regarding time relates to finding a fossilized forest in the ceiling of a coal mine. Mangrove-like plants were found but this type of plant was not supposed to appear until much later than the carboniferous period. (i.e. at least 200,000,000 years later (265))

All of these examples indicate that a great deal of time was not involved in the formation of coal. Therefore, the declaration that a great amount of time passed, during the formation of coal, is not valid.

3.3.7.3   Extended Times are not Needed

If a great amount of time was required for coal formation it would be because time was needed for the volatiles to leave the plants leaving only the carbon. Such a declaration contradicts the following three repeatable observations.

Charcoal is made from hardwood using a process which only requires a few hours. The wood is placed in a furnace and set on fire. The furnace has a restricted oxygen supply. A small portion of the wood burns and heats the rest of the wood. The heat drives away the volatiles from the unburned wood leaving only the carbon.

There is a second way that wood has been reduced to carbon in just a few minutes. Construction procedures have occasionally required that piles be driven into the ground around the perimeter of a site. The soil was thereby stabilized and construction could proceed in safety. When these wooden soldier piles were driven into the ground, they would sometimes overheat. It has been observed that when a recently-driven pile was cut, only carbon remained in the interior of the pile. Since the pile had been driven into the ground within the space of a few minutes, very little time was required for the interior of the pile to be changed from wood to carbon .

Another example illustrating how wood can be reduced to carbon in a short time also involves heating the wood and driving away the volatiles. This process is referred to as pyrolysis (i.e. chemical decomposition by heat action) and occurs if wood is slowly heated. As this happens, the wood will become more and more simply carbon and its ignition temperature will be reduced. This may occur accidentally if wood is enclosed in a wall behind a hot stove. After a period of time, which may involve several years, the wood, (which in this case is completely out of site), might catch on fire. The gradual loss of the volatiles (or parts of the wood, which can evaporate), leaves only the carbon which could then be properly called charcoal and charcoal will ignite at a much lower temperature than green or moisture-containing wood. The second way that this same drying process can happen involves wood, which is near an open fire. If a green log is placed near a fire, it will gradually dry out. Then it might ignite even though it is several feet from the fire. The time required for this to happen is measurable in hours - not years.

3.3.7.4   Layers Indicate Marine Environment

In some locations, numerous layers of coal are found interspersed with layers of rocky material containing sea shells. (266) 'The plants that went into the formation of ancient (coal) beds include chiefly ferns and cycads; layers of later ages are composed of sassafras, laurel, tulip tree, magnolia, cinnamon, sequoia, poplar, willow, maple, birch, chestnut, alder, beech, elm, palm, fig tree, cypress, oak, rose, plum, almond, myrtle, acacia, and many other species. … It is said that the plants fall, but before they decompose in the air they are covered by the

138

water of the swamps. A layer of sand is deposited over them, forming the soil for new plants, and thus the process repeats itself. In order that the sand be deposited, it is necessary that these marshy areas be covered by water in motion. Since almost regularly marine shells and fossils are found on top of coal beds, the sea must have covered the swamps at one time; then, for new plants to grow there, the sea must have retreated. There are places where sixty, eighty, and a hundred and more successive beds of coal have formed; ... many times the sea trespassed ... and as many times retreated. Fossils of marine clams, snails ... are abundant in the shales just above each seam of coal. Later, with fluctuating sea level, the salt water withdrew and another freshwater marsh came into being, giving rise to another bed of coal. ... Ohio displays more than forty such cycles and in Wales more than a hundred separate seams of coal have been discovered.' (267) This type of evidence is not supportive of the notion that sea and land rose and fell in unison over the extended times required to grow multiple forests on top of each other. The swamp theory of coal formation is stressed further by the split seams of coal. '... a coal bed, undivided on one side, sometimes splits on the other side into numerous beds, with layers of limestone or other formations between.' (267, 268) We must recall that limestone is a type of sedimentary rock and that the material for this type of rock would have been placed by water. All of this evidence places the 'large amounts of time' question in the dubious category.

3.3.7.5   The Carbon Cycle Theory

The carbon cycle theory is a system of ideas, which identify the several ways in which carbon is transferred in nature from one place to another. The carbon cycle theory may be validly thought of as being well established. It is supported by a host of observations. The following discussion of the carbon cycle theory will explain how carbon is transferred from plants to animals and back again to plants. Also, the way in which carbon is removed from the cycle and trapped away so it cannot be recycled will be outlined.

The structural component of all plants is carbon, and there are three ways by which this structural carbon is introduced into the atmosphere as carbon dioxide. First, as animals eat plants, their digestive systems convert part of the plant material into sugar. Sugar is a molecule, which is an assembly of carbon atoms. The carbon, which was the structural component of the plants, is converted, by the digestive system of the animal, into sugar, which is able to circulate through the circulation system of the animal. In each cell of the animal, the sugar molecule is brought into close contact with oxygen, which was also brought to each cell by the circulation system. As the carbon combines with the oxygen, heat is produced. This process of combining the carbon and the oxygen is a chemical reaction and the heat, which is produced by this reaction, keeps the animal warm. It follows that if an animal or a person does not have enough food, the heat-producing reaction cannot occur and the animal could chill and die. For example, if a person exercises or works to excess, the amount of sugar in the bloodstream will diminish. The resulting inability to produce heat may cause hypothermia (or chilling). It is unfortunate that people have died from hypothermia even when the temperature was well above freezing. A sign that hypothermia is developing is excessive convulsive shaking. The cure is warming and supplying food, which will resupply the circulation system with sugar. The carbon has therefore served a useful purpose and it is absolutely essential that an animal bring in food to keep warm and continue living. As the carbon and oxygen are combined, carbon dioxide is produced and released into the atmosphere as the animal breathes out. This is one way by which carbon is circulated from plants back into the atmosphere.

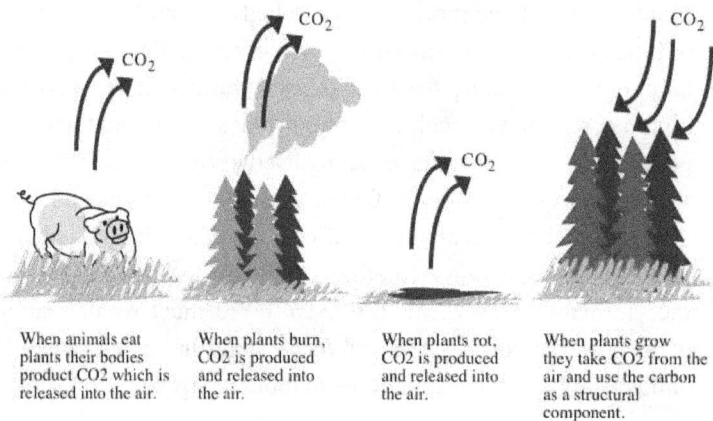

When animals eat plants their bodies product CO2 which is released into the air.

When plants burn, CO2 is produced and released into the air.

When plants rot, CO2 is produced and released into the air.

When plants grow they take CO2 from the air and use the carbon as a structural component.

The Carbon Cycle

Carbon is continually cycled through plants and back into the air. Additionally, it cycles through plants, then animals, and then back into the air.

There are two other ways in which the carbon in plants is converted into carbon dioxide and released into the air. These two ways are rotting and burning. When plants either rot or are burned, the carbon, which is in these plants, combines with oxygen, which is in the air. Carbon dioxide is thereby created and becomes part of the atmosphere.

In summary, there are three ways for plant carbon to enter the atmosphere. The plant may be eaten by an animal which will subsequently exhale the carbon as carbon dioxide. Secondly, the plant may rot during which process the plant's carbon combines with atmospheric oxygen to form carbon dioxide. Thirdly, the plant may be burned which is a heat-producing process combining the plant carbon with atmospheric oxygen.

While there are three ways to return carbon from plants to the air, there is only one way to get the carbon from the air to the plants. The plants must grow. As plants grow, they take carbon dioxide from the air and form their respective structures. The carbon therefore becomes locked up as part of the plant and will remain as the structural portion of the plant until the plant is eaten, burned or simply rots.

As shown in the diagram, The Carbon Cycle, the carbon cycle has two main branches, both of which are required to form the complete cycle. As plants grow, carbon from the atmosphere enters the plant and becomes its structural component. Then when the plant is either eaten, burned or rots, its carbon re-enters the atmosphere and the carbon cycle is complete.

3.3.7.6   The Interrupted Carbon Cycle

As coal beds are formed, (or the great peat bogs, i.e. peat is also carbon and may be on its way to becoming coal - if there is enough of it and it is properly packed down) the carbon cycle is interrupted. We understand that all of the plants, which became part of the coal beds, were formed from carbon obtained from atmospheric carbon dioxide. However, as the coal-bed carbon accumulates, it is effectively trapped and is no longer available to circulate as part of the carbon cycle. It is being trapped off into a great carbon storehouse. In fact, if such a situation were allowed to continue, more and more carbon would become unavailable and the great carbon cycle would have less and less carbon in circulation. Since both plants and animals need carbon to keep circulating, life would consequently become less and less viable. This process could have led to a carbon-starvation death for the Earth if the Industrial Revolution had not taken place. In recognition of how much carbon is presently stored in the coal beds of the Earth, in comparison to the amount in the biosphere, it is a wonder that this has not already happened. The problem of losing carbon is worsened by the carbon which enters the ocean, combines with other material and then sinks to the bottom and stays there. Indeed carbon-starvation might have happened, except that the industrial revolution reintroduced great quantities of carbon back into the atmosphere. If there had not been an industrial revolution and vast quantities of coal had not been burned, carbon dioxide levels today would probably be much lower than they were prior to the industrial revolution and life in general would be less viable.

Now we are in a position to recognize the great problem which exists in trying to explain the coal beds. The carbon from the atmosphere must have formed the plants for the coal beds, but the carbon in these particular plants has not been allowed to circulate back into the atmosphere. It is still in the coal. Therefore, the carbon, which is in the coal beds, has been diverted from the carbon cycle, and has become trapped out of circulation. It is therefore appropriate to ask where the carbon in the coal came from in the first place. It is obvious that it was not exhaled by any animals, which had eaten plants, because the carbon from the plants, which formed the coalis still in the coal, which hasn't been eaten at all. Neither did it come from any plants, which were burned, because if they had been burned, their carbon would have combined with atmospheric oxygen to form CO2 and would consequently not be in the coal beds either. The same carbon cannot be in two places at the same time.

Once coal is formed, its carbon is tucked away in the coal storehouse and it is out of circulation. Therefore, all of the carbon in these coal formations has become completely unavailable for carbon dioxide formation and the possible production of more plants. Hence it is appropriate to seek an explanation for the source of the carbon dioxide, which supplied the carbon for the plants which formed the great coal deposits of the world.

The coal beds do contain a very great amount of carbon. Various estimates have been made and compared to the amount which exists in the biosphere (the total of all living things). The coal beds might contain 50 times as much carbon as the biosphere. (269) All of this carbon has been trapped away from the carbon cycle and has not been available to recirculate since it was trapped and because of this trapping, the carbon, which formed the coal-bed plants must have come from some source other than the metabolization process of animals. Neither did it come from the burning or rotting of plants. Of course, it came from the air because that is the source of all plant-forming carbon, but how did it get into the air? Where did it come from?

Three possibilities present themselves.

1. The first possibility is that the required CO_2 was formed by a burning process, which used primeval (i.e. virgin) carbon as a source. This burning process introduced the virgin carbon into the carbon cycle at just the right rate to enable the trees and other plants, which would form the coal beds, to grow. In order to be internally consistent, it must be recognized that a vast amount of virgin carbon was required. In fact exactly the same amount of ancient virgin carbon was required as is presently found in the coal beds. This type of arrangement is recognized as an artificial construct because nature has been conveniently arranged to bring about a result, which isn't otherwise credible.

2. Is it possible that there could have been enough carbon dioxide in the air at some ancient time to enable the coal-forming plants to develop simply by depleting this CO_2? The amount of carbon dioxide which is in the atmosphere at the present time is about 395 parts per million (270) which means that 395 out of every million molecules in our atmosphere at the present time are carbon dioxide molecules. If all of the carbon in this carbon dioxide were assembled together to make coal, about $6 \times 10(11)$ metric tons of coal would result. Current estimates of the world's coal reserves are $15 \times 10(12)$ metric tons. (269) Atmospheric carbon is therefore equivalent to $6/150 \times 100$ or about 4% of the world's coal reserves. Therefore, if, prior to formation of the coal beds, the amount of atmospheric CO_2 had been about 25 times as great as it is now, the coal beds could have been formed by growing plants and depleting that higher level of CO_2 down to its present level. Therefore, one possibility for coal formation is that an ancient CO_2-rich atmosphere could have been depleted to an atmosphere with much less CO_2.

3. Prior to the formation of the coal beds, the ancient biosphere was 50 times more extensive than it is at the present time. Forests, swamps and meadowlands were filled with an abundance of all kinds of plants. In addition a greater area of the Earth was involved including the high arctic lands, Antarctica and areas now below sea level. Then suddenly this ancient biosphere was annihilated and its carbon is now found in coal. This explanation basically shoves the question back because now we must ask where the carbon came from to form this massive assembly of ancient plants. Did they just suddenly appear? Were they created?

In summary, there appear to be three possibilities for the formation of the coal-forming plants.

1. The ancient biosphere was formed from CO_2 which was produced by an unknown, virgin-carbon burning process.
2. The ancient biosphere was formed by depleting even more ancient atmospheric CO_2 from a level approximately 25 times as high as the present level of 395 ppm.
3. A massive ancient biosphere was created virtually instantly and the plants from it were available to form coal.

However, all of these possibilities come with attachments. If the ancient atmosphere had 25 times as much CO_2, the average temperature of the world would expectedly have been much higher and the trees would not have been able to grow properly because they would have sweat too much and dried out – unless the humidity was excessively high (but there is a definite upper limit to humidity level). Also, it would have been above the body temperature for most types of animal life. In other words the world would have been much less habitable and possibly uninhabitable.

While the third possibility is totally unacceptable to many people, with both the first and second possibilities, it would have been necessary that none of the plants which grew during the extended times required, were burned, eaten or decayed. There were no forest fires caused by lightning (which currently strikes the Earth several thousand times every day). Also it was a rot-free forest wherein no significant quantity of material was consumed or depleted in any way.

As a corollary to the dilemma involved in explaining the origin of the coal beds, one is caused to wonder why there was an abundance of plants during the Carboniferous Period but not at other times! Why was there so many plants growing during one small portion of the great antiquity of the Earth but not at other times. Alternately, perhaps the Earth has never had a history involving hundreds of millions of years!

3.3.7.7   Coal Formation Conclusion

The Swamp Theory of Coal Formation contradicts both available evidence and well-known physics. Neither is it comprehensive. There is no explanation why certain swamps sank at a rate which kept the water level just right and one is prompted to ask why the coal-forming plants lived and died only during the Carboniferous Period. Weren't there any trees growing during the extended time periods involved with other periods? Was it a treeless, swampless world for most of the history of the Earth?

The great difficulty in explaining the existence of coal leads one to consider the creation option for the coal-forming trees. However if that is the way that it happened, vast periods of time are not the least bit necessary.

The presence of carbon14 in coal indicates that the entire coal-forming procedure must have happened only a short time ago (i.e. well within a few tens of thousands of years) and contradicts the idea that there was a Carboniferous Period lasting for millions of years and ending 100 million years ago. This contradiction jeopardizes the assertion that the Earth is 4.5 billion years old but it is suggestive that the entire idea of a very ancient Earth should be set aside.

3.3.8 The Recency of Carbon14 Formation

3.3.8.1 Introduction

Carbon14 came to the attention of the public during the mid-nineteen hundreds through the work of a man named Willard F. Libby. He identified a procedure, which was supported by solid theoretical work, to determine how long ago an object had been made. This procedure was primarily applicable to objects, which were made of wood and assumed that the manufacture of the item had taken place soon after the wood had been harvested. Measurements and calculations would determine the time which had expired since the wood stopped living because when a plant dies it stops taking in carbon. Other previously-living plants could also be used but since wood was very common, wood was most often used.

As long as a plant is living, it will interact with the atmosphere. The interaction of interest in this case concerns the respiration of the plant - in particular the take-up of carbon dioxide from the air. All plants take up carbon dioxide and use the carbon as a structural component. All plants, including trees, are built of carbon. The carbon is taken in by the plant and used for construction purposes. Therefore, when the plant dies and is harvested, no more carbon will be taken in. The carbon, which is already part of the plant, is expected to remain in place indefinitely. This expectation, on its own, was not of much use in determining how long it had been since the plant died until it was realized that there are three different types of carbon and that one of them could be a time indicator.

Most of the carbon, which is found in nature, is carbon12. There is some carbon13 and a very small amount of carbon14. Carbon13 is a little heavier than carbon12 and carbon14 is a little heavier than carbon13. However, carbon is still carbon and any interactions, which carbon has with the rest of nature, will be similar, no matter which type of carbon is involved. Therefore, when carbon reacts with oxygen and forms carbon dioxide (CO_2), all three types of carbon will be involved expectedly in the same ratios in which they would normally occur in nature. While this doesn't always happen, the expectation that it will happen is a necessary part of the idea that carbon14 could be useful as an indicator of time.

Neither carbon12 nor carbon13 change with time. They always stay the same no matter how much time is involved. However, carbon14 does not stay the same but gradually changes into nitrogen. Therefore, there will be less and less carbon14 as time goes by and the diminishing amount will provide us with an indicator of how long it has been since the plant (i.e. usually wood) was alive.

Carbon14 is radioactive. As it changes from carbon back to nitrogen, a small part of the carbon atom is lost resulting in nitrogen being formed. Radioactive changes are predictable and fortunately, all radioactive changes follow a similar pattern of change. This pattern of change is called the half-life law. No matter how much material there was at the beginning of the period of interest, there will only be one-half of it left after one half-life has passed. Of course, different radioactive materials have different half-lives. In the carbon case, the half-life of carbon14 is of particular interest and it has been determined to be approximately 5730 years. (271) Since

95% of a sample will be gone after about five half-lives, a half-life of 5730 years is of very great interest as a dating technique for artifacts, which might be up to a few thousand years old. If the half-life were only 100 years, the active ingredient would be virtually all gone after 1000 years and so would not hold the potential to be useful for dating something that was several thousand years old. Since the history of humanity has occurred in the last few thousand years, the development of the carbon14 dating technique aroused a great deal of interest in the scientific community. Since the middle of the twentieth century, this technique has been applied to many samples to determine how old they were. However the use of carbon14 as a time indicator has a serious limitation.

3.3.8.2   Nature's Way

In nature, responses to change often follow the half-life relationship. For example, with processes such as heating a kettle, filling a tank or charging a capacitor, the parameter of interest will very likely follow the half-life response curve. In some of these cases, a half-life parameter will not be measured but another idea will be introduced, which is called a time constant.

The time constant relationship is shown in the diagram, Time Constant Curve and Half-Life Curve, which in this case is showing how the voltage across a capacitor varies as the capacitor is charged. After the amount of time, which is called the time constant, has passed, the capacitor will be approximately 63% charged. When two time constants have passed, the capacitor will be approximately 86% charged. After three time constants have passed, the capacitor will be 95% charged.

A very similar situation develops when a leaky bucket is being filled with water. As shown in the diagram, Leaky Bucket, water is being poured into the top of the bucket. In this example, the bucket has a lot of small holes in it and so as the water is being poured in the top, some of the water will leak out through the holes. After a while when the water level has risen, there will be so many holes which are leaking that the leak rate will equal the fill rate. When this occurs, the water level in the bucket will not rise any higher. The parameter, which is of particular interest in this example, is the water level change with time. As time passes, the water level gets higher and higher but as it gets higher, the leak rate increases and the rate of water level increase slows down. The increase in the water level will keep getting slower until it actually stops. This occurs when the leak rate is the same as the fill rate. The change in water level will follow the same pattern of change as the voltage on the charging capacitor. The curve, which describes the charging capacitor, could be placed right on top of the curve for the filling bucket and it would be found that the two curves had exactly the same shape. It is very curious that this would be the case but nature has many other examples of changes, which follow this very same pattern.

Another example of this pattern of change occurs when a kettle is being heated on a stove. As it is heated, the temperature of the water will increase. The temperature will not jump up suddenly. As the water gets hotter, further increases in temperature will take longer because as the kettle gets hotter, it loses heat at a faster rate. A time will come when the kettle will not get any hotter. At that time, the heat, which is being put into the kettle, will be exactly the same as the heat, which is leaving the kettle. When the heat, which is going in, is equal to the heat, which is leaving, the temperature will just stay the same. If the change in temperature were measured up to the time that the kettle did not get any hotter, a pattern of change would be found, which would be similar to both the charging capacitor and the leaking bucket.

Time Constant Curve and Half-Life Curve

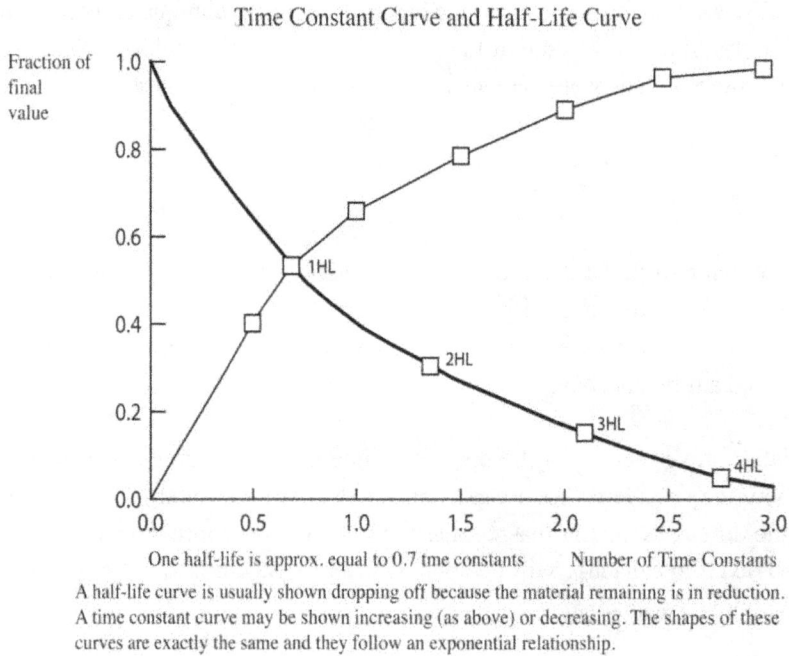

One half-life is approx. equal to 0.7 tme constants   Number of Time Constants

A half-life curve is usually shown dropping off because the material remaining is in reduction.
A time constant curve may be shown increasing (as above) or decreasing. The shapes of these
curves are exactly the same and they follow an exponential relationship.

This type of change has a very particular mathematical description which is called an exponential. This type of pattern of change is found a great many times in nature. When either the half-life of a radioactive material is known or the time constant of a capacitor circuit is known as well as either the starting point or the finishing point, both the future and the past of the particular parameter of interest can be determined. These well-known physical relationships will be particularly helpful in determining the validity of the carbon14 dating procedure.

3.3.8.3   Carbon14 Production

The atmosphere of the Earth is approximately 80% nitrogen and 20% oxygen. In the upper atmosphere, a very small amount of the nitrogen is continually being changed into carbon14. Energy is required to make this change and is supplied by incoming cosmic radiation, which liberates neutrons, which become captured by nitrogen making the nitrogen heavier than usual. Soon, this heavy nitrogen loses a proton and carbon14 results. The production rate of carbon14 was determined by Libby, the scientist who did the initial research, to be 18.8 SPR (Specific Production Rate). Others disagreed with this value and placed it at 27. (272)

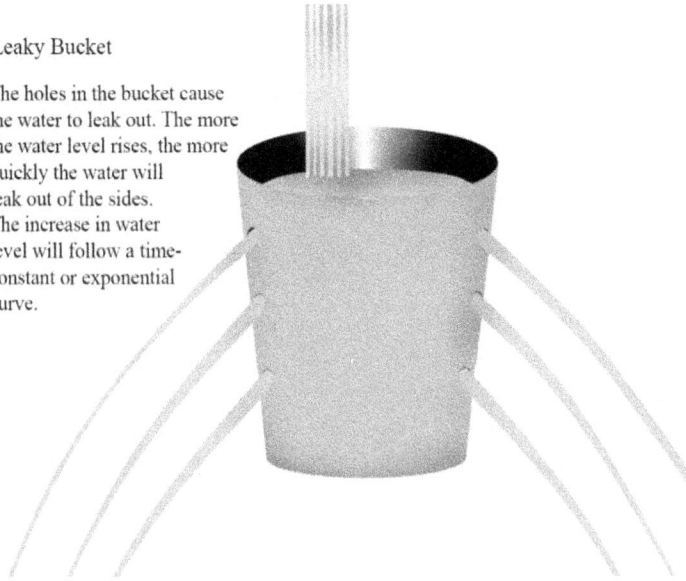

Leaky Bucket

The holes in the bucket cause the water to leak out. The more the water level rises, the more quickly the water will leak out of the sides. The increase in water level will follow a time-constant or exponential curve.

3.3.8.4 Carbon14 Decay

Carbon14 is radioactive. Therefore it is continually disappearing. Actually it does not disappear but simply changes back into nitrogen. The disagreement concerning production rate means that the decay rate, which is usually given as 16.1 SDR (Specific Decay Rate) is between 20% and 40% less than the production rate. (273, 274)

3.3.8.5 Carbon14 Buildup

The production and decay rates of carbon14 are not the same. They should be the same. They should be the same if the process has been in operation for more than eight or ten half-lives. This puzzled Libby, because if the world (was) millions of years old, the SDR should have long since come into equilibrium with the SPR . (272) This situation can be understood from the chart entitled Carbon14 Buildup. The decay rate should gradually build up until it is the same as the production rate. The increase in decay would be the same as the increase in voltage of a capacitor, which was being charged. As shown above, when a capacitor is being charged, the voltage gradually increases. It will follow a pattern of increase, which is called an exponential. The shape of the exponential curve is the same as the shape of the half-life curve, which indicates that the two different processes will have similar patterns of change. As discussed above, this pattern of change will also be similar to the change in water level, which occurs with the leaky-bucket example. When a kettle is heated on a stove, the temperature of the water will change in a similar way. This type of change is very common in nature and is in fact the way nature usually responds to a change in circumstances.

With the leaky-bucket example, it is clear that as long as the quantity of water which is leaking out of the bucket is not as much as the quantity which is being put in, the water level in the bucket will keep on rising. The example of the charging capacitor is very similar. As long as the voltage across the capacitor is less than the source voltage, it will keep right on rising until it matches the source voltage. Voltage measurements would determine if the capacitor was fully charged. It would be expected to keep on charging until the source voltage had been reached and it could not be charged any further.

The situation with radioactive material is very similar. Once the process has begun, the decay rate should build up and keep right on building up until it matches the production rate. Therefore if a situation is found where the decay rate is less than the production rate, it is clear that the decay rate is still building up. This means that the process has only recently begun. (i.e. It would only have begun within a few half-lives ago.)

3.3.8.6   The Present Time

The chart entitled, Carbon14 Buildup, shows how the decay rate of carbon will relate to the production rate. This chart represents a considerable amount of time involving several thousand years. In the situation, which is shown by the chart, the decay rate starts at zero and then gradually builds up. At the right side of the chart the decay rate almost matches the production rate which is constant at the 100% level. This chart would be very useful to us if we knew where on the chart, the 'present time' was located. What location on the chart represents the carbon14 situation in nature at the 'present time' (i.e early in the third millenium)? What location on the curve represents the 'present time'? From the 'present time' position, the part of the chart to the right would represent the future and the part to the left would represent the past.

The 'present time' location is determined by noting on the chart where the production and decay rates differ by 20% for the 20% case and where they differ by 40% for the 40% case. The present time position for the 40% case, on the chart is therefore a little further to the left of the 20% case. (Recall that Libby recognized the 20% difference and others recognized a 40% difference.)

3.3.8.7   Carbon14 Startup

The time since the process began, can be determined from two factors. The half-life of the material must be known and the difference in production and decay rates must be known. When these two factors are known, the time since the process began, can be calculated. On the Carbon14 Buildup diagram, we would then simply follow back down the buildup curve until the zero line is reached. First we locate the 20% difference location on the chart. (See location A) Then by following the radioactive half-life curve back down to zero, the time since start-up can be determined. On the diagram this is approximately 2+1/4 half-lives or about 2+1/4 x 5730 years (from 10.1.1 above) = 12,000 years. Secondly, the 40% case will be located (see location B) and once again the curve will be followed back to zero. In this case, the time since startup is approximately 1+1/3 half-lives or about 1+1/3 x 5730 years = 7,000 years. From these determinations, the time since carbon14 started to be produced in the atmosphere of the Earth, is between 7,000 and 12,000 years ago.

3.3.8.8   Partial Startup

The situation is modified if it is assumed that the process was always in operation at a low level and then suddenly increased to a higher level. If, for example the process was formerly operating at about one-half of the present level, the time since the increase, would be much shorter. Suppose that the process was formerly operating at the 50% level and then suddenly increased to the present level. In this case we only follow the curve back down to 50% of the present level. When this line of reasoning is followed, the time since startup is reduced to 6,000 years for the 20% case and to about 1,000 years for the 40% case. Similarly, if the process were operating at 25% prior to a sudden increase, the time since that increase would be approximately 9,000 years for the 20% case and 5,000 years for the 40% case. All of these determinations can be made directly from the Carbon14 Buildup diagram.

3.3.8.9   The Half-Life Curve

It is a curious reality of the half-life curve that no matter where we are on the curve, the shape of the curve is exactly the same. This feature is what makes the above discussion valid. If we are in the middle of the curve and only chose to use 10% of it, the shape of that 10% would be the same as the whole curve. Even if we are at one end and extended the curve further, the shape would be the same as the original curve. It might be that the beginning, (of whatever portion it is that we want to study), is part way up from the beginnings, which are shown in the above diagrams. The discussion can still proceed because the shape of part of the curve is the same as the whole curve. In fact the half-life curve can be extended both ways indefinitely and this whole new curve would still have the same shape as the original section.

3.3.8.10   Catastrophic Implications

The time since the process either began or suddenly increased is not nearly as important as the fact that the process did begin and the fact that it had a definite beginning in the very recent past. If it began, something caused it to begin. The production of carbon14 is understood to occur in the upper atmosphere when incoming radiation collides with nitrogen atoms and causes them to change into carbon14 atoms. Therefore, since the process seems to have had a definite and recent beginning, either the incoming radiation suddenly began or the nitrogen in the atmosphere was suddenly exposed to it. Prior to such an event, the nitrogen in the air it must have been shielded from the incoming cosmic radiation. If this were the case, the catastrophic event was the removal of the protective shield. A layer of atmosphere, which consisted of water vapour, could have provided a partial shield to incoming carbon14-forming radiation. The water vapour layer could have affected the C14 dating method because it would have lowered cosmic ray incidence causing less C14 to be formed. (319) Therefore, the catastrophic event could have been the removal of this water vapour layer. As soon as that happened, carbon14 production would have increased to the present level and the decay rate would start to build up towards the present level.

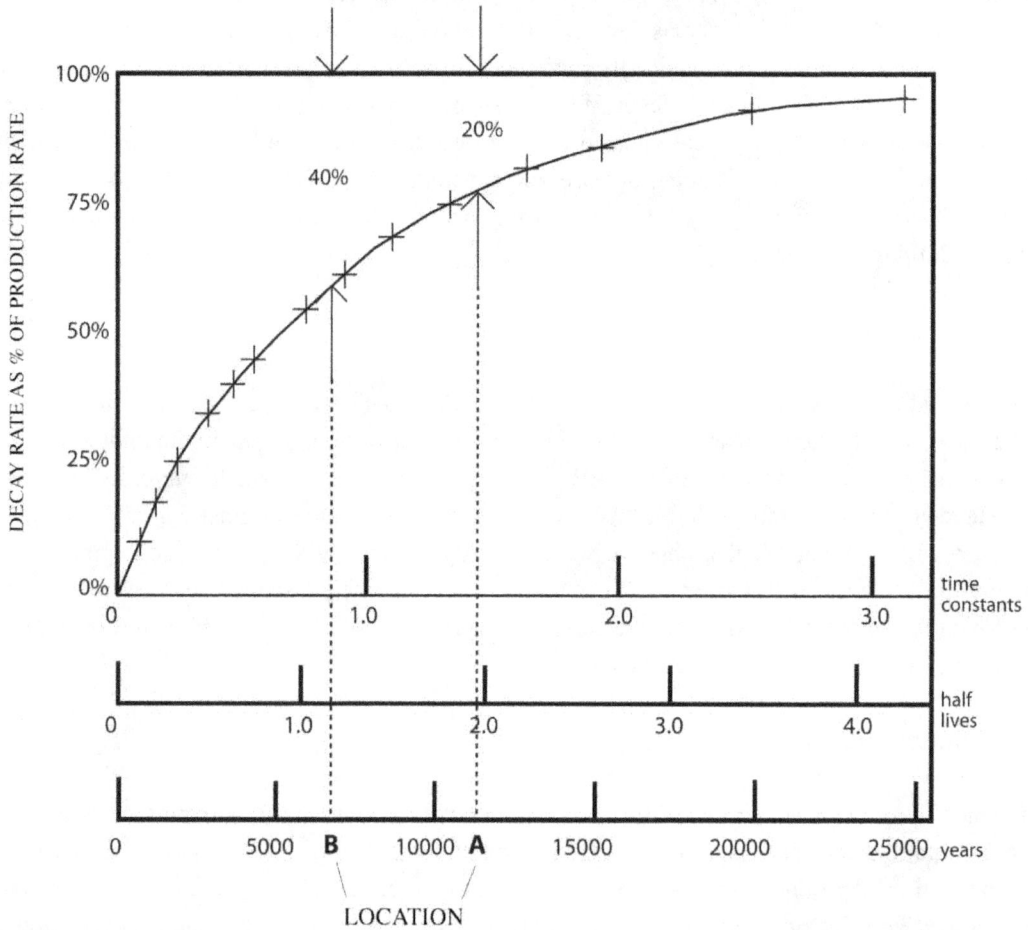

Carbon 14 Buildup
C14 HALF-LIFE = 5730 YRS
How long ago did the Carbon 14 production process begin?

3.3.8.11 Theory Correlation

The previous chapter entitled; 'The 360-Day Year', identified several factors, which indicated that there might have been a water vapour layer in the upper atmosphere at one time. This idea is supported by the deduction that the carbon14 process had a definite and not-too-far-in-the-past beginning. If there was a protective water vapour layer and it was suddenly lost, the carbon14 production process would have started. Alternately, if the process was already underway at a low level, it would have increased to a higher level. With either of these possibilities, the idea that a pre-existing water vapour layer was suddenly destroyed and the independent conclusion that there was a definite beginning to carbon14 production, correlate each other.

150

The conclusion that the carbon14 process had a beginning, or at least a significant increase in activity quite recently, also agrees with the observation that coal has a low level of carbon14 activity. (i.e. 50,000 yrs. (275)) Whereas if coal is as old as claimed for the Carboniferous Period – 100 million years – it should not have any carbon14 count at all! If coal consists of the plants, which existed prior to the impact of a large asteroid, these plants could have been growing in an atmosphere, which was at least partially protected from incoming cosmic radiation and one which would have had very little carbon14 production happening. In that case, the carbon14 count would indicate that coal is older than it is in reality. (i.e. 50,000 years instead of a few thousand years)

Another scientific measurement that reinforces the conclusion that the production of carbon14 has only recently begun is the decaying magnetic field. The magnetic field of the Earth very significantly affects the amount of incoming cosmic radiation. ... The magnetic field forms a barrier to cosmic radiation. ... If the magnetic field was stronger in the past, it would be certain that incoming cosmic radiation would have been less than today. This means that the production of neutrons in the atmosphere would have been less in the past. It further means that the collision of the neutrons with the nitrogen atoms would have been less and the production of carbon14 would have been less. (318)

3.3.8.12   Carbon14 Dating Upset

Unfortunately, if the carbon14 process had a beginning or even a significant increase in activity within the last few half-lives of carbon14, the use of carbon14 for dating ancient artifacts is compromised. Any dates, which are determined will be in error and will indicate more time than is actually involved. A radioactive process such as this could only be expected to give valid results if the process was in equilibrium and the production and decay rates were the same. When they are not the same, it means that the process is still ramping up and is not yet well established throughout nature. Even though it appears that several thousand years are involved, it is clear that steady state has not yet been realized and the process will not really be very useful until that happens.

An upgrade to the use of carbon14 as a dating procedure might be realized if a time-dependant correlation or adjustment factor were included. Even though the originator of the carbon14 theory recognized that production and decay were not in balance, (276) no adjustment factor was ever introduced.

3.3.8.13   Catastrophic Atmospheric Change

In order for carbon14 production to either have been initiated or to have undergone a major increase in activity, there must have been a catastrophic irreversible change in the atmosphere of the Earth. If the previous atmosphere included a water vapour layer, this is exactly what would be expected if a large asteroid crashed into the Earth.

3.3.8.14   Carbon14 Startup Conclusion

Carbon14 production is not in balance with carbon14 decay and the discrepancy is being more widely recognized than it was shortly after carbon14 was first recognized as a time indicator. Statements now appear in textbooks indicating that acknowledge the C14 decay rate in living organisms is about 30% less than the production rate in the atmosphere and that. from this difference in rates, it can be concluded that the atmosphere is less than 20,000 years old. (331) The discrepancy is substantial and means that the production of carbon14 in the atmosphere has only recently begun. This evidence from science has catastrophic implications and necessitates a drastic departure from the ideas involved with a world that is said to be several billion years old. What was happening with the atmosphere and the plant life of the Earth for all of those millions and millions of years when there wasn't any carbon14 being produced? Or did those millions and millions of years actually exist? Clearly if coal has a carbon14 count and the carbon14 production has only just begun, the Carboniferous Period could not have been over 100 million years ago. Further, why was plant life so prodigious for the span of 60 million years of the Period and then just disappeared? Thereafter it doesn't show up in the geological record at all! Weren't there any living plants taking up carbon at other times throughout the vast ages? The recency of carbon14 startup, the correlation with the decaying magnetic field, and the fact that coal has a carbon14 count all cast a long shadow of doubt over the entire idea that the Earth is 4.5 billion years old.

3.3.9   The Ice Cores

On both Greenland and Antarctica the glaciers are about two miles thick. This is approximately 10,000 feet of ice. It is understood that it requires about twelve inches of snow to make one inch of ice. Glaciers that are two miles thick would therefore require about 120,000 feet of snow to have fallen and been retained. This would certainly have been a great quantity of snow. In the Greenland case it has been declared that '... there is snow that fell on central Greenland before the start of the last ice age, more than a hundred thousand years ago.' (341) This is, of course, a very bold declaration because there aren't really any credible time indicators embedded in the ice to enable any temporal cross-checking to take place. The situation on Antarctica is declared to be similar.

With respect to the Greenland situation and using rough approximations it seems that 120,000 feet of snow fell over a time span of (roughly) 120,000 years. It is immediately obvious that this represents about one foot of snow per year for the extended time of more than 100,000 years. Does this type of declaration have any credibility? One foot of snow might present an impediment to exiting one's driveway but it really is not very much snow. This small amount of snow must not only have fallen every year for that long period of time but it was also retained for all of those thousands of years! What type of situation could have been in place to enable such a small amount of snow to fall repeatedly and then be entirely retained? Why would such a situation continue year after year for thousands of years?

The situation that is declared to have happened on Antarctica is even worse. The time period in that case is declared to have been about four times as long. The Antarctica cores indicate a history of more than four hundred thousand years. (343) The average snowfall would therefore have been between three and four inches – every year for 400,000 years. This type of declaration has no credibility whatsoever!

Now when we switch from declarations to actual measurements we have The Lost Squadron report to investigate. In 1942, in support of the war effort, the United States of America sent planes to Europe. Gas tanks were not large enough to enable direct flight so planes had to go across the North Atlantic via Newfoundland, Greenland, Iceland and Scotland. The group of planes of particular interest could not land when they got to Greenland so pushed on towards Iceland. However they could not land there either so returned to Greenland and landed on the glacier. There was only one casualty of a broken arm and otherwise everybody returned home safely. The location of the landing was noted and some pictures were taken. In 1992 a group of people equipped with instruments for looking down through the ice identified an object that appeared suspicious. A shaft was opened down to the object which turned out to be one of the lost planes. This particular plane was dismantled and brought to the surface. Other planes were identified but deemed too damaged to be worth recovering. In 2000 the recovered plane was flown again. (342) Locating the plane buried deep in a sea of ice was a remarkable achievement but the main point with respect to the time question is that the plane was covered by about 250 feet of blue ice and all of this material had to have accumulated in just 50 years. This represents an average of five feet of ice per year which, as before, represents about 60 feet of snow per year. How do we compare this amount with the four inches on Antarctica or the twelve inches on Greenland from the previous declarations?

The Lost Squadron Case documents that the snow on Greenland is still accumulating and accumulating at a very significant rate even though we are not even having an ice age! One would naively expect that during an ice age, snow accumulation would be much more rapid than when an ice age was not in progress. Furthermore, the ongoing accumulation is occurring (high on the glacier) even though the glacier overall is shrinking! (346) The minimum conclusion from all of this evidence is that the declarations that vast amounts of time have been involved is seriously in doubt which in turn casts a doubt on the entire idea that the Earth is 4.5 billion years old!

3.3.10   River Formation

3.3.10.1   The Hydrologic Cycle

Water constitutes a major portion of the material at the surface of the Earth. There is water in the air in vapour form. There is water in the soil. There is water in underground streams and reservoirs and there is a great amount of water in rivers, ponds, lakes and oceans. Water in its solid form, ice, covers most of Antarctica as well as Greenland. With the exception of some of the ice, all of this water is in motion. The source of energy to move the water is the Sun. As the Sun heats the Earth, water evaporates from everything that the Sun shines on including trees, grass and open ground. Lakes, rivers and oceans evaporate as the Sun heats the surface layers of water. It requires a lot of energy to evaporate water but as the Sun shines on it, the temperature increases and evaporation results. In particular, as the Sun shines on the oceans in the tropical regions, the ocean water evaporates into the air leaving behind any salts or minerals, which were in the water.

This evaporated water then becomes part of the atmosphere, mixes with the other components of the atmosphere and is carried along by the moving air. As the weather patterns of the Earth shift and change, the temperature of the atmosphere will occasionally drop. Rain will result. Some rain falls right back on the ocean but a lot of rain will fall on the land. Some of this rain will supply the trees and crops with needed moisture and some will either fall directly on the lakes and rivers or percolate through the soil before either evaporating again or trickling into streams and lakes.

The water of the Earth is always on the move. This great pattern of movement is referred to as the Hydrologic Cycle. The Hydrologic Cycle involves a pattern of water transfer from the oceans to the land and then back to the oceans again. The heat from the Sun evaporates the water in the ocean. Clouds are formed. The water vapour in the clouds is transported by the great wind patterns of the world. When the clouds pass over the land, rain falls on the land. Some of this water soaks into the ground. Some of it runs off right away into streams and ditches. Lakes and rivers become filled with the water, which runs off the land. Most rivers and lakes empty into the ocean. A few do not reach the ocean but depend directly on evaporation for the release of water. All of the oceans evaporate and a few bodies of water, like the Dead Sea, which do not empty into the ocean, also evaporate. Either way, the water, which previously fell as rain, is recycled into the air before travelling back over the land to fall again as rain.

This cycle of recirculation is quite predictable. Most rivers do not undergo a significant change in level year after year. In fact, the water levels are so predictable that bridges are built and boats operate on regular schedules. It is also true that some rivers go into excessive flood mode quite often. The Amazon River is a good example of this characteristic. Water levels in the Amazon River can change dramatically overnight. This happens when there is a heavy rainfall in the mountainous regions where the river originates. This unpredictability means that bridges cannot be built and any activity along the shores must recognize that serious flooding can occur at any time. The general cycle of activity, however, is the same as with the more predictable rivers. Rainfall is the source of water for the river, which returns the water to the ocean and the cycle just keeps on going. The evidence shows, however, that this was not always the case and that the entire process had a recent beginning.

3.3.10.2   River Deltas

Many of the major rivers of the world, which flow either into the ocean or one of its adjacent bays or estuaries, have a buildup of material at their mouths. These buildups are called deltas. Deltas are formed from material, which the river carried from locations upstream. When the mouth of the river is reached, the silt drops out of the water because the water slows down and cannot carry the silt any further. As time goes by, the silt accumulates and forms an extension of the land. Fortunately, the rate of buildup of all of the major deltas of the world, have been calculated. For example the Mississippi River is pushing its delta into the sea at the rate of a mile every 16 years. This represents an advance of 250 miles in only a few thousand years, so the river cannot be older than that. This determination assumes that the flow rate has always been the same (as it was up until quite recently when some of the flow was diverted to the west). Such a short time frame is quite alarming but even more alarming is the notion that the entire process had a beginning, which beginning was when the silt of the delta started to be deposited.

The time since the rivers started to flow seems to be quite short. From another viewpoint, it can be deduced that if these rivers had been flowing for several hundred thousand years, their deltas would be much more extensive by now. In the case of the Mississippi River delta, at the current rate of 2 million tons per day, it would require only 10 million years to fill the entire Gulf of Mexico. Similarly, if the Amazon had been flowing for a long time, the Amazon River delta would extend much further out into the Atlantic Ocean than it does now.

3.3.10.3 River Startup

The river deltas of the world have a very definite and measurable size. It must therefore be concluded that they had a definite and relatively-recent beginning. If the deltas had a beginning, the rivers must have had a beginning.

3.3.10.4 Niagara Falls

The Niagara River flows between Lake Erie and Lake Ontario. The elevation of Lake Erie is much higher than Lake Ontario and most of the elevation change occurs at the Niagara escarpment. At this location there is a sudden drop from the higher elevation region which includes Lake Erie to the lower level which includes Lake Ontario. Both Niagara Falls and the American Falls occur where the Niagara River plunges over the escarpment.

Every year the Niagara River works its way back further and further into the rock and the falls move a little further upstream continually. Observations have been made concerning the rate at which the river erodes the rock where it plunges over the edge. Reports are also available from the early observers to the area. These data along with the variations in the thickness of the various rocks along the Niagara gorge enable estimates to be made concerning how long the river has been flowing. The town of Queenston is located right at the edge of the escarpment and is actually partly on top and partly at the bottom. Important historical battles were fought in this area and this is the place where the Niagara River started to work its way back into the rock. Queenston is now several miles downstream from the falls. This distance is therefore a time indicator. When this distance, the observed and reported erosion rates and the variations in the type of rock along the way are considered, it can be determined that the water started to flow over the edge of the escarpment approximately 7000 years ago. Further, since flow was greater at the end of the Ice Age, the time might have been even less. It was concluded that seven thousand years might be the maximum length of time since the birth of the falls. When immense masses of water were released by the retreat of the continental glacier, the rate of movement of Niagara Falls must have been much more rapid so the time estimate is sometimes lowered to five thousand years. (26)

Whether the time was five thousand years or seven thousand years is secondary to the conclusion that Niagara Falls was formed in the very recent past. It had a definite and recent beginning.

3.3.10.5 Weather Pattern Startup

Rivers are fed from rain and rain results from the various weather mechanisms around the world. If there is too much rain, rivers will flood. If there is not enough rain, rivers diminish and dry up. The conclusion is therefore that our present weather mechanisms, including rainfall, started in the recent past at the time that the rivers started to flow, Niagara Falls was formed and the river deltas started to accumulate.

3.3.10.6 Catastrophic Geological Change

If there were no flowing rivers prior to the beginning of delta formation or the beginning of the migration of Niagara Falls, the geology of the Earth must have been dramatically different from the way it is at the present time. Therefore it may further be concluded that there was a world-wide catastrophic change in the geology of the Earth. However, this is what would be expected if a large asteroid crashed into the Earth. There would be major changes in the geology of the Earth, in the weather patterns of the Earth and in the location and indeed the actual existence of the great river systems of the Earth. Since this entire scenario is a recent development, another shadow of doubt is cast on the notion that the Earth is very very old.

3.3.11 Earth's Mass Stability

In order for the Earth to have a long-term stable orbit and hence have long-term habitability its mass must remain constant. The magnitude of the pull of the Sun on the Earth is dependent on both the mass of the Earth and the mass of the Sun. If the mass of either one increases, the gravitational pull of the Sun on the Earth will increase. This means that the Earth would spiral inward closer to the Sun. We recall that the distance of the Earth from the Sun is a very important parameter with respect to the habitability of the Earth. The orbit of the Earth is just right at the present time and in fact is referred to as a 'Goldilock's orbit'. Deviation either way is not acceptable and in fact the allowable deviation is not more than 5% either way.

It has been observed that material is being added to the Earth every day and that it amounts to about 3000 metric tons. This is not really very surprising because the Earth is quite large and has a very strong gravitational attraction for space material. However it is quite surprising that the Earth is exactly in the middle of the habitable zone of the Sun at the present time and one wonders how this could be the case if the Earth has been here for extended periods of time involving millions of years. Clearly we cannot have millions of years and and an optimum 'Goldilock's orbit' at the same time while material continually arriving. Why would the Earth just happen to be at the optimum distance at the present time? However, the orbit is an observation while the 'millions of years' is a declaration. There obviously cannot be millions of years.

3.4   Near Space

3.4.1   Mercury

The Earth and the other planets in the inner solar system have been impacted by some very large asteroids. The craters that remain testify to this and now that all of the inner planets have been photographed these large impact sites are available for study. On Mercury there is a very large crater called the Caloris Basin. 'Caloris, a behemoth 1300 km in diameter, is the largest of these craters. (i.e. the craters on Mercury) The impact that created it established a flat basin ... on which a fresh record of impacts has built up. ... The Caloris impact probably occurred about 3.6 billion years ago.' (277) It must be noted that this declaration of time was made by observing various photographs which were taken by a camera on a satellite near Mercury. Nobody has ever been to Mercury and no rocks have ever been recovered to study. The problem with respect to very long time periods is that there are very few impacts in the Caloris Basin. 'The impact which created the Caloris Basin must have occurred after most of the heavy bombardment had finished because fewer impact craters are seen on its floor than exist on comparatively-sized regions outside the crater.' (277) In a few areas inside the crater rim impact marks are dense enough to overlap. There are also many areas that show no impacts at all! How can this be explained if the crater is 3.6 billion years old? There are still many hundreds of thousands of asteroids orbiting the Sun and impacts with the major planets are not unusual. 3.6 billion years is a very long time and one would expect the Caloris Basin to be completely littered by impact marks after such a long time has passed.

3.4.2   Mars

Mars also has several very large impact features. The largest one is called Hellas and it is approximately 2100 kilometers (1304 m) across. 'The basin is thought to have formed during the heavy bombardment period of the solar system about 3.9 billion years ago when a large asteroid impacted the surface.' (154) Isidis is slightly smaller with a diameter of 1500 km. (939 m) 'The crater was probably created three or four billion years ago when a comet or big asteroid slammed into Mars. (157) 'There are hundreds of thousands of craters on Mars'. (158) The floors of all of these large craters show impact marks. However in light of the 'hundreds of thousands' and the declared ages of several billion years they seem to be almost void of craters. After several billion years one would expect all of the large craters to be completely covered by impact marks by now. The situation is more suggestive that the impactors came as part of one major event which was comparatively recent and that there has been very little activity since then.

3.4.3   The Martian Moons

'Mars has two very small satellites named Deimos and Phobos. Both are pock-marked or pitted as if they had encountered a considerable quantity of smaller asteroidal debris.' (159) It is the impact marks that are of interest with respect to the previous declarations of billions of years being involved in the history of the solar system. Deimos is the smaller of the two Martian satellites (or moons) at approximately '6 x 7 ½ x 10 miles'. (159) The largest impact marks (or pock marks) on Deimos have shapes which suggest the state of Deimos when these marks were formed. The craters themselves are reasonably regular in appearance but the crater rims are not. The rims are rounded and smooth looking. This indicates that these marks were not formed in the usual way when an asteroid crashes into a solid body. These impact marks could only have been formed in semi-molten material which allowed the impacting object to form a more-or-less regular appearing crater but caused some of the material that was pushed aside to just well up and form a smooth rounded rim. This was more like dropping a stone into a slurry of mud which was thick enough to allow a crater to form but a little too thick to slump back down. The shape of the craters on Deimos indicates that Deimos was in a semi-fluid state when the impacting objects arrived.

The gravity of Deimos is very small. However it could have been sufficient to attract small nearby objects with enough energy to enable them to make a pockmark on impact if the material of Deimos was soft and unsolidified at that time. The material of Deimos would have had to have been very close to solidifying completely which is exactly what we would expect if Deimos had been cast into space in a semi-fluid state. All of the impact marks on Deimos indicate similar formation conditions. Why isn't Deimos riddled with regular-looking hard-surface impact marks? If it has been around for a billion years this is what we would expect.

Deimos therefore provides evidence of recent asteroid formation from an exploding planet that had a fluid interior but a very viscous or solid shell. If the original planet had been a completely-solid object just before it was destroyed, Deimos would not now appear as an irregular blob. It would be jagged looking instead. On the other hand if the original object was completely low-viscosity fluid, Deimos would now have a spherical shape. An exploding planet that had a fluid interior and either a semi-fluid or solid shell would generate all types of objects including broken sharp-edged fragments, semi-fluid blobs and perfectly spherical balls and this is how asteroids appear.

The other satellite of Mars is named Phobos and is larger than Deimos at 12 x 14 x 17 miles. Phobos also has a significant pit mark which is really quite large in comparison to the size of Phobos. (161) As with Deimos, this mark appears to have been formed in semi-fluid material just before Phobos hardened. If Phobos had been solid at that time, an incoming object would not have been able to penetrate so deep into the surface. Neither would the rim be smooth and rounded. However semi-molten material would have allowed an impacting object to penetrate quite deep and then retain the shape of the crater that was formed. In such a condition, the self-gravity of Phobos would not have been sufficient to cause the shape of the original surface to be recovered before solidification. Phobos must have been just ready to freeze when the impactor arrived. The smoothness of the rim supports this conclusion.

Both of the moons of Mars testify to recency. If they were ancient, their surfaces would show the types of marks that happen when an impacting object hits a solid surface. Instead they only show marks that could have been formed if they were in a semi-fluid state. Over millions and billions of years there should be marks of the more generally-expected nature but neither of these satellites have any such marks. Their appearance is suggestive that both of these moons were cast into space along with a large number of much smaller fragments and that their limited gravity attracted some of the smaller objects which hit them before either of them solidified.

Further, they must have been cast into space in a semi-fluid state quite recently because the type of pit-marks that form in solid material have not had time to accumulate. The recent formation of the moons of Mars casts a serious doubt on the idea that the solar system is 4.5 billion years old.

3.4.4   The Asteroids

Further supporting evidence for the partial fluidity of an asteroid-originating planet is provided by the 'asteroid 2005 YU55, a round mini-world that is about 1,300 feet (400 meters) in diameter.' (160) This asteroid passed the Earth in November of 2011 closer than the Moon which enabled amateur astronomers to photograph it. An object this small could not have achieved a spherical shape on its own if it had been solid or composed of small solid pieces. It had to be fluid from the very beginning in order that it's own self-gravity could cause it to become spherical. The shape of this asteroid confirms that the origin of the asteroids had to have been an object that was at least partially fluid.

There is another asteroid that tells a very similar story. In this case asteroid Hektor appears as two basically-spherical balls being pushed together. It has sometimes been described as two potatoes stuck together. The pressed-together appearance of the asteroid is very suggestive that both spheres were not quite solid when they came together. As with other asteroids they have very little gravity so the impact energy would have been very low. Therefore they could only have achieved their squeezed-together shape if neither of them was solid. If they had already cooled down and become solid, coming together would have resulted in an object that appeared like two balls touching rather than being significantly deformed at the contact area.

Other examples confirm the semi-fluid assertion. Asteroid Eros has several very large impact marks in comparison to its size. All of them have a soft rounded appearance. Some of the impacting objects appear to have been quite large and have thrust themselves very deep into the asteroid's material. A solid object hitting a larger solid object of this size would not have entered it at all! It would have made some mark and just careened away. (522)

Asteroid Ida provides similar evidence of having had a historically-fluid state. It is oblong with the usual soft impact marks and rounded features. Even more compelling is Ida's moon Dactyl which is described as being egg-shaped with approximant dimensions of 1.0 x 0.87 x 0.75 miles (i.e. almost perfectly spherical) (445) and in some photographs it appears as a sphere. (446) This means that it must have been fluid when it was cast into space. In fact, the heating factor is suggested by the following; ' ... it has been altered by strong heating-evidence so far suggests that Ida is a piece of a larger object that has been severely heated.' (445)

Numerous other objects in the solar system have the appearance of having been liquid when they were formed. 'Recently we have become aware that many objects in the Solar System have a dumbbell shape. If we go to a lava field from an active volcano, we find volcanic 'bombs' which are shaped like dumbbells. These are produced from lava which has been ejected from the volcano.' (351) (Volcanic 'bombs' are blobs of material that fly through the air when a volcano explodes. They are called 'bombs' because they explode when they crash into the Earth.) This type of observation leaves very little doubt that the asteroids were formed from liquid material and the deep soft impact marks reinforces this conclusion. Further, the minimal, virtually-zero evidence of hard impacts indicates that the asteroid-forming and crater-forming event was quite recent.

With respect to the asteroids location, it is understood that the vast bulk of them orbit the Sun between Mars and Jupiter. However we are also instructed that Jupiter has a disturbing influence on the orbits of the asteroids causing them to relocate. At the same time we are told that they formed very long ago. 'The asteroids themselves formed about 4.6 billion years ago. ... Over time, the gravitational influence of Jupiter gradually modifies their orbits, occasionally shifting them into trajectories that carry them inward to where the Earth circles the Sun.' (29) These two claims are incompatible. If their orbits are gradually being shifted, why is it that the vast bulk of them still orbit between Mars and Jupiter after 4.6 billion years? Perhaps there hasn't been any 4.6 billion years!

3.4.5   Ceres

Ceres is the largest of the asteroids and because it was recently orbited by a spacecraft (named Dawn) a great amount of information has become available. Several of the features of Ceres indicate recency. There are a significant number of large crater basins which are quite shallow. Many of them have a rebound mound at the center and most of them have a smooth flat appearance. While the roundness and the smoothness indicate a molten state at the time of impact, the smoothness also indicates recency. If these marks had been formed long ago we would expect numerous additional impact marks and they would be impact marks as made in a hard surface. The lack of such marks and indeed the lack of hard impact marks on the asteroid in general indicates recency. There is no shortage of asteroid debris in the solar system and no shortage of objects crashing into other objects so if these features are ancient they should show more impact marks and in particular more hard impact marks. The lack of such features tells us that the formation of Ceres was quite recent and by association tells us that the age of the Solar System might not be as old as promoters of the 4.5 billion year old solar system idea would like us to believe.

There are two other features of Ceres that support this contention. Ceres has several very bright lights. They were detected while the Dawn spacecraft was still thousands of miles away. This immediately means that a lot of power is being expended to keep these lights operating. Even the most powerful searchlight would be difficult to detect if the observer was 20,000 miles away. However instead of being a few feet in diameter like a searchlight, the lights of Ceres are miles in diameter. This means that a very great amount of energy is being expended just to keep them operating. If Ceres was 4.5 billion years old, one would expect that it would be very quiet by now and that these lights would have gone out long ago.

The second feature which is similarly curious is the massive plume that is being emitted from the surface of Ceres. This plume extends into space and occupies a volume which is comparable to the volume of Ceres itself. A gaseous structure such as this would immediately dissipate in space and indeed it is dissipating. However, this plume is fresh and being constantly replenished. Has this been going on for 4.5 billion years? If Ceres was close to a large planet, gravitational interaction might be credited with keeping Ceres active. However Ceres is millions of miles from any other planet and hundreds of millions of miles from the Sun. Why is it still so active after all of this time?

3.4.6 The Moon

The Moon has been observed by large telescopes on Earth and in near-Earth orbit as well as lunar orbiters and lunar landers and a very large number of photos have been taken. There seem to be two types of impact marks. On the lunar maria and other places where lava appears to have flowed the impact marks are smoothly shaped with rounded rims. This is particularly evident at the location where the Apollo17 mission landed. (347) Numerous impact marks are obvious and all of them have the soft rounded appearance as do impact marks in other locations where it is even more obvious that the surface was once molten. (348) The Apollo17 landing site is on the lunar maria which are understood to be of volcanic origin. (349)

The Tycho crater is a large (85 km across, 53m) located in the south central part of the visible surface of the Moon. 'The freshness of the crater and the rays of material radiating from it suggest that this is a young crater; there has been little time to erode it. The circular crater is surrounded by a bright ejecta blanket. Rays of ejecta extend across the lunar surface.' (350)

3.4.7   Near Space Summary

Both Mercury and Mars have very large craters that are declared to have been formed between three and four billion years ago. Mercury has numerous other impact marks but Mars has 'hundreds of thousands'. Similarly the Moon has a very large number of crater marks – mostly on the far side. The giant craters of Mars and Mercury have very few impact marks but in the Argyre Crater on Mars there are quite a few. But are there enough to support the idea that these craters were formed so long ago? Four billion years is a lot of time to accumulate impact marks! The situation on the Moon is even more mysterious. The maria cover more than half of the near side but the entire area is virtually free of crater marks. The maria on the Moon are thought to have been formed at about the same time as the giant craters on Mars and Mercury several billion years ago. Where then, are the random asteroid impact marks that would have been forming every few years as they are declared to be doing now? We recall that every listing of impact craters for the Earth includes a column showing the believed date of the impact. From these listings it is clear that major impacts are thought to have occurred on the Earth at least every few millions of years. How did other heavenly bodies avoid such activity for so long? The declared extreme ages for these features is not supported by the evidence which raises a doubt about the certainty of the great age declarations in general. The apparent youth of the Lunar Tycho Crater only further adds to the doubt. We recall from above in a previous discussion that most of the asteroids are still concentrated between Mars and Jupiter. Why would this be the case if the solar system is billions of years old?

3.5   Review and Summary

The Earth (and by association the Sun, Moon asteroids and planets) has been declared to be 4.5 billion years old. Supporting evidence includes the idea that the Sun is powered by nuclear fusion as well as the dating of certain rocks here on Earth and from the Moon. Unfortunately the way that the Sun is powered cannot be known with any degree of certainty because there is no way to tell from which direction any detected neutrinos might be coming. Also the conventional dating procedures for rocks are dubious due to contradictory measurements. Further there is no way to obtain independent time co-relating evidence. However the evidence for recency is over-whelming in its diversity and includes;

A.   From the Moon

1.   The Moon is receding from the Earth at a rate that will make the Earth uninhabitable within another 100 million years and would have made it uninhabitable about 100 million years ago.

2.   If the Moon is really very old it should be inactive by now but it is continually showing lights and emitting clouds of gas.

3.   If the Moon is really very old it should be cold by now but instead it is very hot inside.

4.   If the Moon is really very old, the repeated impacts of small rocks from space should have pulverized the surface into dust so that there should be a thick layer of dust over the entire surface whereas there really isn't any appreciable amount of dust at all. The Moon doesn't have an atmosphere so there is no wind or rain to wash dust away so it should just stay where it formed – but where is it?

5.   The Moon has been pummelled by more than 200,000 objects able to create impact marks more than one mile in diameter. While it has been struck numerous times, if it is really billions of years old, why is more than one half of the Earth-facing surface virtually free of impact marks?

6.   Why does the lunar Tycho crater appear so young?

B.   From the Sun

1. If the Sun is really very old its mass will have increased because it is continually drawing large amounts of material into itself. This means that all of the planets, including Earth, would have been spiraling inward for 4.5 billion years and the Earth could only be in the thermally-habitable zone temporarily. Why does the Earth just happen to be in the thermally-habitable zone at the present time and how long will it stay there?

2.   Is it possible that the Sun is being powered by gravitational collapse? If it was, it wouldn't necessarily be billions of years old at all.

C.   From the Earth

1.   If the Sun is really very old and is powered by nuclear fusion it would have been considerably cooler 4.5 billion years ago and the Earth would have been frozen solid. If this had ever been the case, the primary greenhouse gas, water vapor, would not have been available to permeate the atmosphere and provide the

greenhouse-gas warming effect which is so necessary to keep the average temperature on the surface of the Earth in the habitable zone. If the Earth had ever been in a frozen state, water vapour would never have been liberated and without the greenhouse gas effect that water vapour provides, the Earth would still not be receiving enough heat from the Sun to bring it into a thermally-habitable state - even up to the present time. Therefore it would still be frozen solid.

2. Why doesn't the ocean contain a much greater amount of silt? If the Earth is really very old, erosion from the exposed continents and material falling through the atmosphere from space would have accumulated on the ocean floor in much greater amounts than is currently present and would have been sufficient to displace enough water to completely flood the land. However some areas of ocean floor have no silt at all.

3. Trying to date asteroid impact times is dubious because the dates cannot be cross-checked by any other procedure. How, with any certainty, can anyone know that some particular rock or other material can actually yield a date for a particular impact?

4. Why haven't the continents, which would barely have stuck out of the ocean several billion years ago (due to the differential density of continental and oceanic material) have been eroded away after all of this time?

5. Why is there a difference in density between the two types crust (i.e. continental and oceanic) and why has the difference been maintained for so long?

6. How could animal life have ever survived more than 100 major mass-extinctions with one occurring every few million years?

7. How will animal life survive when the magnetic field totally collapses in a few thousand years because when it does, there will not be any protection from incoming cancer-causing radiation? Why is it just collapsing now if it has been around for several billion years?

8. Why doesn't the dating of any of the currently-recognized asteroid-impact sites coincide with the supposed time of the period of the Late Heavy Bombardment? How did the writer of the computer program (i.e. the Late heavy Bombardment is a computer program) decide when the bombardment happened and why are the supposed dates for the bombardment so very different for the various places where it took place?

9. Since the Late Heavy Bombardment involved the impact of several very large objects with the Earth (i.e capable of forming craters 5,000 km in diameter) why haven't any of these monster impact sites been identified?

10. Why wasn't the Earth driven out of its very precise, necessary, narrow and endlessly-repeatable nearly-circular Goldilock's orbit by one of those monster objects – any one of which would have modified the orbit resulting in an uninhabitable Earth?

11. If coal is 100 million years old, why does it have a carbon14 count?

12. If coal is 100 million years old, why does it include Erratic Boulders which were supposedly placed by the moving ice of the Great Ice Age only a few thousand years ago?

13. If the carbon14 production process has been going on for millions of years, why aren't the production and decay rates in balance?

14. Why did carbon14 production only just begin?

15. Why is it that all of the rivers of the Earth only started to flow a few thousand years ago? Weren't there any rivers on the Earth for the declared long eons of time?

16. .Neglecting friction energy loss, the recession of the Moon will cause the Earth to stop rotating within another few hundred million years. A daylight period will then be about three weeks long followed by a period of dark about three weeks long. This will cause both over-heating and over-chilling resulting in an Earth that will be uninhabitable. Why is it that the day-night period is so ideal at the present time?

17. Tidal friction energy loss is causing the rotation rate of the Earth to slow down so rapidly that a day will be twice as long in about 130,000 years. This means that the equatorial bulge will have almost disappeared and the pull of the Moon on an enlarged tidal bulge will cause the axial tilt to increase and the Earth will tip over. In fact tip-over will probably happen within another 20,000 to 30,000 years because the axial-tilt stability zone is not very wide. When that happens, the axial tilt will increase dramatically making the Earth totally uninhabitable. These two factors extended into the past by a similar amount of time would also have made the Earth uninhabitable only a few tens of thousands of years ago! How then could it be billions of years old? Why is it that the equatorial bulge and the tidal bulge offset each other so ideally at the present time?

18. Why is it that the decreasing rotation of the Earth, the collapsing magnetic field and the recent start-up of carbon14 production all indicate an Earth that could only be a few tens of thousands of years old?

19. Why did the carboniferous period only last for a small fraction of the Earth's history? Weren't there any trees growing throughout the rest of the Earth's history?

20. Why didn't the Earth stop rotating long ago?

D. From Near Space

1. Why is the Caloris Basin on Mercury so free of impact marks?
2. Why are the huge Martian craters so free of impact marks?
3. Why do the Martian moons show evidence of being recently molten?
4. Why do the asteroids show evidence of being recently molten?
5. Why are most of the asteroids still located between Mars and Jupiter?

E.   Conclusion

We understand that the gravity of every one of the planets is very strong and that the gravity of the Sun is enormous. It is also clearly evident that the Solar System and beyond is totally littered with all kinds of objects that travel around in every conceivable direction. It is also evident from what we observe here on the Earth, that material from space is continually impacting the Sun, the Moon, and the planets. Since it is understood that it is the mass of a planet(and the mass of its moon) that determines the orbit that any moon will have around it, any increase in the planet's mass or the moon's mass would result in steadily-decreasing orbits for all of its moons. Similarly, a steadily increasing mass for the Sun or any of its planets would result in gradually-decreasing orbits for the planets. After several billion years one would expect that all of the stray inter-planetary material would have been swept up by now but more importantly that the steadily-increasing mass of the planets (and their moons) would have caused all of the moons to have spiralled into their respective planets and the steadily-increasing mass of the Sun (and the planets) would have caused all of the planets to have spiralled into the Sun. Why are the orbits of the planets and their moons so close to circular at the present time?

Of more immediate concern is the recession of the Moon and the associated tidal-friction loss that is slowing the rotation of the Earth so quickly that it will only be rotating about half as fast in another 130,000 years!. By then the stabilizing factor provided by the equatorial bulge on the angle of inclination will have been lost and the Earth will have tipped over making it uninhabitable! It would similarly have been uninhabitable within about the same amount of time into the past. Why does it just happen to be habitable right now?

Collectively, these as well as, all of the other evidences from science discussed above, indicate recency and totally contradict the idea that the Earth (and the rest of the Solar System) is 4.5 billion years old. The conclusion must therefore be reached that the old-age idea is unsupportable from science and should be abandoned.

4.0   The Astronomical Theory of the Ice Age

4.1   Introduction

The Astronomical theory of the Ice Age declares that; For a period of about two million years, a period of several great ice ages created huge sheets of ice that built up at both poles of the Earth and moved outward under their own weight. Glaciers more than a mile thick spread over the Northern Hemisphere gouging valleys and shaving hills. (247). 'The concept of an ice age, when climates were colder and glaciers were far more extensive than they are today … The building of vast glaciers where none existed before requires that vast quantities of atmospheric moisture be precipitated in the form of snow. As this moisture must come from the sea, the result is lowering of sea level.' (248)

The requirement therefore, is that the climate over the continents be colder but that atmospheric moisture be much more available. Unfortunately, these two criteria are counterproductive. In order for ice to accumulate on the ground, it must snow. Atmospheric air currents pass over open water, which evaporates and fills the air with moisture. When this moisture-laden air passes over land, which is cooler, snow falls. However, if the climate becomes too cold, open water will crust over with ice thereby inhibiting evaporation and subsequent precipitation. There is an optimal temperature range where snow will fall. If it is too warm (i.e. just above freezing), rain falls. If it is too cold, everything freezes over and there won't be any precipitation at all. The maximum-possible-snowfall temperature occurs just below freezing. Well-known physics teaches that the warmer the air is, the more moisture it can hold. Since in this case snow is required, when the temperature is just cold enough to enable snow to fall instead of rain, maximum snowfall will occur. When it gets colder, and the lakes freeze over, snowfall will drop off dramatically because there won't be any source of moisture to form the snow. It is therefore logical to ask how the ice-age snow accumulated if the oceans were frozen and the air was deprived of the necessary moisture. If the entire world had been chilled, large areas of ocean would have been covered with pack ice. Hence they would have been unable to contribute moisture for the ice build-up. Further the water that wasn't covered by icewould have been cold and not a good source of moisture. Therefore, the assertion that ice could build up under the constraints of a cold climate, is in direct contradiction with well-known physics.

At the present time, major snowfalls only occur when the open-water sources of moisture are 'warm'. In this case, 'warm' is only a few degrees above freezing but the difference in moisture availability is tremendous. In a recent major snowstorm in Southern Ontario, more than five feet (153 cm) of snow fell within just a few days and was enabled to fall because Lake Huron was still 'warm'. (521)

Over the years several explanations for the Ice Age have been proposed. At one count over 60 theories had been proposed. (249) One would expect that this would mean that there were more than 60 different ideas advanced to explain the Ice Age but this is far from the case. Rather, they are all based on the same premise or assumption namely; 'a cold Earth is an Earth having an ice age'. Based on this common element they all simply offer slightly different explanations for how the Earth chilled. These ideas are still current and still being advanced to explain 'mini ice ages' of very recent time when the glaciers of Norway and Greenland advanced. The advance and retreat of these glaciers seems to be reasonably well documented although with the general scarcity of good world-wide data recovery over the last few centuries, supportive information such as ocean temperature, salinity, circulation and surface temperature of the Earth as a whole, it is difficult to be conclusive concerning why these glaciers advanced. Was the incoming heat from the Sun reduced or was there some other temperature modifying factor at play. It does seem quite logical that if a glacier is advancing there must be something like an ice age happening but it wouldn't necessarily mean that the Earth was cooling down.

However, the basic idea does have a certain element of logic to it and gained credibility from the work of a man named Milutin Milankovitch. This was before the age of computers so - to his credit - he did all of his calculations by hand. These calculations involved the Earth's orbit and showed that it would change slightly over long periods of time causing the Earth to periodically chill slightly. Since the modifications would have been caused by planetary motion, the idea became known as the Astronomical Theory of the Ice Age. Further, since the predicted orbital changes would result in the Earth repeatedly drifting back and forth slightly from the

Sun as it orbited around it, there must have been repeated ice ages - possibly one every few thousand years. It has consequently become a well-established belief that there have been repeated ice ages.

A variation on having the orbit of the Earth change slightly would be to have the entire solar system drift across its arm of the galaxy and into regions where there would be more inter-stellar dust. This dust would permeate the solar system and result in a little less heat reaching the Earth. We might call this a 'cold' region of space. Based on the idea that 'a little less heat would cause an ice age', whenever such drifting occurred there supposedly would be another ice age. The weakness with the idea is that it isn't really known if a little more dust would have this result and neither is it known how long the solar system might remain in such a location. The idea therefore appears whimsical and as a result is somewhat lacking in credibility. Never-the-less it fits in quite well with the over-riding notion that 'a cold Earth is an Earth having an ice age'.

The basic idea intrinsic in the Astronomical Theory of the Ice Age does have a certain sense of logic about it because cold is necessary to make snow. While the ideas of Milankovitch have received considerable criticism they still dominate explanations for the Ice Age!

4.2   Problems with the Theory

4.2.1   Insufficient Heat Reduction

The theory did not gain universal acceptance among scientists however because the variations in solar heat that were postulated were really quite small and would not have really had very much influence. This would partly be due to the temperature equalizing mechanisms that continuously operate throughout the Earth to stabilize our temperatures and keep everything the same. The changes in summer sunshine at high latitudes postulated by the theory are actually quite small and would not be enough for an ice age. In addition heating at high latitudes is only partially dependent on sunshine. Transport of heat by the atmosphere and oceans is also important. (48) In spite an abundance of contradictory evidence such as this, numerous scientists have persisted in arguing that the theory is valid. It is recognized that such a scenario does fit in quite well with the over-riding conviction that the Earth is 4.5 billion years old and that major developments such as an ice age would only happen over periods involving many thousands of years. However, if the Earth isn't really that old, it would cast even more doubt on the Astronomical Theory of the Ice Age.

There are other reasons that the Earth could experience a chill. The receipt of a constant supply of heat from the Sun is crucial to maintaining a steady temperature here on the Earth so if the output of the Sun diminished slightly there would be a chilling effect. Solar flares diminish the Sun's output because they appear as slightly cooler than the rest of the solar surface. If there was an increase in solar flare activity the output of the Sun would be reduced. The Earth would therefore chill. There is a possibility that this is what happened during the relatively recent advance of the Scandinavian glaciers. Also, it could be happening even at the present time because a reduction in the output of the Sun has been observed during the last few years. As quoted in a previous chapter 'A meeting of the US Geophysical Union in 1986 brought together a number of independent studies which all agreed that the Sun has steadily declined since 1979. Satellites in Earth orbit, rockets, balloons and measurements at ground level all showed that the Sun was fading.' (35) With the basic premise in mind (i.e. a cold Earth is an Earth having an ice age') does this mean that the Earth is actually in the early phase of another ice age even as it seems to be over-heating?

As discussed in 4,4,2 Volcanic Winter below, volcanic activity can and has reduced the heat input from the Sun by significant amounts resulting in the entire Earth being chilled. An unusual amount of cloud cover caused the temperature drops associated with both Tambora and Krakatoa but the Earth was far from totally covered by cloud. However, there was enough to partially and temporarily interrupt the incoming flow of heat from the Sun by several percentage points. The immediate difficulty that the Astronomical Theory of the Ice Age faces in the light of these events is that the chill caused by the cloud cover from the Tambora and Krakatora events was several times as great as the chill that is thought to have caused the Ice Age. In fact instead of the 1% or 2% anticipated by the Earth drifting a little in its orbit, the heat reduction caused by the cloud cover produced by Krakatoa was estimated at 12% and Tambora was even worse! An ice age did not result from either of those developments which causes one to wonder how it could have happened with the much lesser chill of 1% or 2%.

4.2.2   Basic Physics Violation

The basic and unspoken premise underlying the Astronomical Theory is that the whole event developed gradually and required several thousand years to take place. It was a slow process and caused entirely by the Earth being very slightly chillier than before the process began with the reduction in solar energy being in the one or two percent range. (49) The Sun was still shining of course and clouds were still forming except that the ocean was getting colder and becoming more and more covered with ice. The great northern ice pack extended much further south and completely covered the North Atlantic and North Pacific. The ice shelves around Antarctica extended for hundreds of miles and covered much of the Antarctic Ocean. It was an ice age after all and the Earth was colder than usual. Even the surface temperature of the water near the equator was reduced.

With the accompanying reduction in evaporation that these developments would bring, the great torrential rains of the tropics would not have occurred because the water near the equator would not have been evaporating very much. As a result of the drop in evaporation it would have been much drier than usual in the tropics and when the air entrained in the Hadley atmospheric circulation cell returned to the surface of the Earth at the usual descent regions 30 degrees north and south of the equator it would have been so dry that the deserts that currently occur in these areas would have been much more extensive and would have occupied the southern half of the USA. Colder always means drier and with one half of the ocean covered by ice and the other half cooler than usual it certainly would have been a drier Earth. Consequently, one is urged to ask how thousands of feet of snow could have accumulated across both North America and Eurasia under such conditions. The reality of the situation would have been exactly the opposite of what the theory predicts. Cold water and dry air are counterproductive as conditions to produce an ice age and extending the time element will not solve this dilemma.

Losing money on a small transaction cannot be compensated for by carrying out more similar transactions. The other old saying that captures the situation goes something like 'you cannot make a silk purse from a cow's ear.' The necessary ingredients are not present. The theory fails on the basic facts of science that 'frozen water does not evaporate at all, cold water evaporates very little and cold air cannot carry very much moisture anyway'. The idea that it would be possible to have an ice age where the snow and ice accumulated over several thousand years caused by such conditions is therefore not valid and must be set aside.

The next point is closely related to the last. As stated, as air gets colder it cannot hold as much water vapour. However water vapour is a greenhouse gas and a certain amount of it is essential to help keep the surface temperature of the Earth at the proper level. A general world-wide chill would reduce the amount of water vapour in the air and therefore the ability of the atmosphere to retain some of its heat. This would cause a further reduction in temperature with a further loss of water vapour. If such a chilling scenario ever got started the temperature of the Earth would spiral down until any further reduction in temperature did not result in any further significant loss of water vapour. This terminal situation would only develop when the surface temperature was well below freezing. (55) However as soon as the temperature drops just below freezing, water surfaces would crust over with ice and there would not be any moisture available to form an ice field. Consequently there would not be any ice age. The Ice Age required that an enormous quantity of water leave the ocean to form the great ice fields but this would not happen if the ocean became chilled down to near the freezing point or if some significant portion of it became crusted over with ice.

4.2.3 Greenhouse Gas Violation

While the atmosphere of the Earth consists mostly of oxygen and nitrogen, it also includes relatively small quantities of several other gases, some of which are referred to as greenhouse gases. These particular gases have the peculiar characteristic of being able to reflect heat energy back to the surface of the Earth. Of course, all of our energy comes from the Sun but when some of it contacts the surface of the Earth its frequency is changed and it becomes heat energy instead of light energy. Then as it radiates from the surface of the Earth, it cannot easily get back through the atmosphere because the greenhouse gases reflect it right back to Earth. Consequently heat is basically trapped in between the lower atmosphere and the surface of the Earth. This is very much in our favour and without this heat retention factor the surface of the Earth would seriously chill at night. This actually happens over the Sahara Desert because the greenhouse gas inventory (i.e. water vapour and clouds) in the atmosphere over the Sahara is depleted.

The average surface temperature of the Earth at the present time is about +15C (+59F) (51). There are two sources of heat that cause the temperature to be exactly where it is. The first is the direct heat from the Sun. The second is the characteristic of the greenhouse gas inventory to retain a certain portion of this heat and to prevent it from being radiated back into space. It is necessary to retain some of this heat to keep the average surface temperature at a reasonable level. Over most of the Earth this works quite well but there are a few exceptions. As mentioned, one of these exceptions occurs over the Sahara Desert. The Sahara desert is reasonably close to the equator and becomes quite hot during the day especially during the summer when the Sun is directly overhead. The surface material of the Sahara is mostly sand, stone and bare rock. Such materials are easily heated. Therefore during the day they heat up. Since there is a lack of water reservoirs on the Sahara there is very little water to produce cloud to shade the surface. All of the heat of the Sun therefore reaches the surface and heats it up. The opposite happens at night. At night it cools off and the temperature often goes right down to freezing and even below. The main reason for this is the diminished greenhouse gas factors in the atmosphere so very little of the heat that radiates up from the surface is reflected back to the surface to help keep it a little warmer. The missing factors are water vapour and clouds. Both of these contribute to the greenhouse effect because they will reflect heat back to the surface. Heat that would otherwise be lost is thereby retained and this is not only curious, on a world-wide basis it is absolutely necessary for our survival.

There are several greenhouse gases in addition to water vapour and clouds. On the average water vapour provides about 50% of the overall greenhouse effect and clouds provide about 25%. (165) Carbon dioxide is the third most influential at about 20%. (165) There are several others, including methane, which collectively only make up a small portion of the total effect. CO_2 is currently getting considerable attention because it has been observed to be increasing continually since the industrial revolution and has now reached a level that is causing alarm. The historically high level is being blamed for the increase in the average surface temperature. While the expected increase during the coming years will really be quite small and only in the range of a few degrees it is thought that such an increase will cause great havoc around the entire Earth. Ocean water will warm up and hence sea level will rise because of expansion. The great glaciers on Antarctica and Greenland will melt causing an even greater rise in sea level. The warmer water will release some of its CO_2 inventory which will only aggravate the situation even further. But possibly the most serious result would be an increase in the amount of water vapour in the atmosphere because the amount is solely dependent on temperature. Therefore an increase in temperature will result in an increase in the amount of water vapour. A situation like this can be self-perpetuating and only make the changes that occur, even greater. (166)

The test to confirm if this is happening is to observe if further increases in temperature are occurring faster than the ongoing increase in CO_2. If this is the case the positive feedback aspect of the situation will probably already be in operation and the intensity of the alarm bells will increase. There will be calls to stop burning carbon-based fuels altogether but the conviction of many is that it is already too late so there isn't really anything that can be done anyway. The release of multi-megatons of Methane and Carbon Dioxide from the melting permafrost in Arctic Canada, Alaska and Siberia overwhelms human emission cuts so the process is beyond human control. (52) The greenhouse gas factor has a direct bearing on the validity of the Astronomical Theory of the Ice Age so is of interest from that perspective as well.

The Astronomical Theory of the Ice Age is built on the assumption that the Earth suffered an unusual and prolonged chill. There is absolutely no mention of any unusual heating factor at all. It was just chilly. Everything was chilly. While the problem of getting cold water to evaporate in unusually high volumes during chilly conditions would have been difficult enough to overcome, the problem presented by the cold on the greenhouse gas inventory is insurmountable. We absolutely must have the greenhouse gas inventory in place as a heat retention and control factor or the temperature at the surface of the Earth will drop below the survivable range. However there is a problem with changing the level of water vapour in the air which can be self-aggravating and self-propagating. If the temperature of the atmosphere drops, some water vapour will be lost. This could cause the temperature to drop some more. This could cause more water vapour to be lost. Once this starts to happen, the temperature would only stop dropping when any further reduction in water vapour would not result in any further reduction in temperature. By then, the average surface temperature would be about -25C and most of the surface of the Earth would be frozen solid and we would have a 'Snowball Earth'. (50)

If a 'snowball Earth' ever developed it would no longer matter if there were any greenhouse gases because the incoming solar energy would be simply reflected. No portion of the visible light energy would be converted to heat energy so there would not be any heat energy to trap and reflect back to Earth. In fact, if only an appreciable fraction of the Earth's surface appeared as a 'snowball' a similar fraction of the incoming solar energy would be immediately reflected. The remainder might slightly warm the snow but not enough to heat it so there would be some heat energy to trap in the lower atmosphere. This means that any greenhouse gases that existed would be totally ineffective and might as well not even exist! It is difficult to see how such a situation could be overcome with very little hope of ever thawing the Earth out again. (166) Since this has not happened it is clear that the ideas which form the Astronomical Theory are not valid and it should be set aside.

The other factor that only makes the situation worse is the reality that cold water holds more CO2 than warm water. The amount of CO2 dissolved in the surface water of the ocean increases as the temperature drops. Therefore cooler water absorbs more CO2 than warmer water. (54) Therefore if the ocean cooled it would take up more CO2. However since CO2 is a major greenhouse gas, removing some of it from the atmosphere would aggravate the developing chill and contribute to a drop in surface temperature. There is a conviction that this actually happened during the ice age and that the amount of carbon dioxide in the atmosphere fell to 180 parts per million at ice age maximum. (53)

Based on the above recognition of the greenhouse gas effect, it will be appreciated that any introduction of prolonged cold absolutely must be offset by prolonged heat. In order to have an ice age, which will already be recognized as a rare and unusual event, there must be a balance between heating and cooling. An ice age would require as much heat as it would cold – not only to make it happen but to preserve the greenhouse effect as well. Otherwise the loss of the greenhouse gases, due to the chill, would result in the Earth just locking-up cold and staying there. Since that has not happened, any commentator offering an explanation for the ice age must offer a suggestion of how the required heat was supplied. Otherwise any such offering will be incomplete.

4.2.4   Faulty Time Assumption

The Astronomical Theory necessarily includes the idea that everything would happen slowly and require thousands of years. If an ice age was starting it would require many hundreds of years to accumulate the ice and this would supposedly happen as the Earth became chillier and chillier. The entire Earth would gradually cool down and this would give time for plants and animals to adapt. This would enable them to live right through the entire cycle and thrive once again when the Earth warmed up. The idea that things might have developed quickly is totally foreign to the whole concept – partly because plants and animals would not have had time to adapt to the changing conditions and partly because a source of heat of the magnitude suggested above is completely out of the range of the imagination of the proponents of the theory. However the possibility that things proceeded much more quickly is supported by the very recent saga of The Lost Squadron (Note; This report has been discussed elsewhere but because it is relevant to the time question it will be repeated here.) as well as by our own collective and recent experience. (i.e. The average temperature of the Earth is rising. (417))

During World War II a group of planes were being delivered to Europe from North America by way of Greenland. At that time fuel tanks were not large enough to enable direct flight so several stops were planed. One of these stops was Greenland. Then the planes left Greenland and flew to Iceland for another scheduled stop. However they could not land at Iceland and thought that they did not have enough fuel to continue to Europe so went back to Greenland. However by then the landing strip at Greenland could not be used either so they decided to land on the Greenland glacier because their fuel supply was low. The landing went well with only one casualty of a broken arm caused by one plane flipping over. Everyone made it safely back home. The location of the landing was recorded and 50 years later in 1992 a group went back to the site to try to find the planes. They did find them with the help of an instrument that could look down through the ice. One plane was recovered and returned to North America where it was restored and flew again in 2000. It really is an interesting story but the point of interest with respect to ice ages is that the plane that was recovered was buried under 250 feet of snow and blue ice. (about 2/3 was blue ice and the rest was snow). (41) Since we understand that it requires about 12 inches of snow to make one inch of ice this means that the planes were covered by the equivalent of more than 2000 feet of snow that fell during those 50 years. There are two possibilities that could have reduced this amount. The planes could have melted down into the previously existing snow after they landed. Also some snow could have blown in from other areas as well. In any event this quantity of material had accumulated in just 50 years. If we allow that one-half of the snow either blew in or was the result of melting there would have still have been about 1000 feet of snow that fell in the usual manner. This reduces the snow requirement to only about 20 feet of snow per year. Otherwise we need an average of about 40 feet of snow per year for the 50 years that the planes were covered. Either way it seems like a lot of snow but it is a very realistic amount based on current observations. By extrapolation the 10,000 feet of blue ice that currently covers Greenland could have accumulated within a period of between 3000 and 6000 years. Since snow continues to accumulate on Greenland if it has been 5000 years since the end of the ice age the post ice age accumulation could readily be in the 8000 foot rage which only leaves 2000 feet to have accumulated during the ice age. Subsequently, it will be argued that the ice age did not last very long but that since ice accumulation conditions were ideal during that time the required accumulation of several thousand feet of snow(possibly 50,000) could realistically have happened within decades and not millenniums.

Recent experience supports these ideas. Abundant snowfalls happen when there is an area of chilly land adjacent to an area of open 'warm' water. Warm in this case would only need to be a few degrees above freezing. The accompanying; Table 1 Water Vapour Pressure, clearly shows that as temperature increases the vapour pressure of water (i.e. its ability to evaporate) increases rapidly. Also as the Vapor Capacity of Air diagram shows, (par. 7,3,5) the ability of air to hold moisture also increases dramatically as temperature goes up. Therefore the ideal conditions for snow production include a stream of 'mild' air flowing across a body of 'warm' water and then coming to an area of land that is below freezing. The damp air will not be able to hold the moisture that it has picked up over the water and so it will precipitate out. The system will act like a conveyor belt and just keep on bringing snow until the wind changes direction or until the water cools down. With this mechanism, a great deal of snow can accumulate within a few days and even several feet of it would not be surprising. This type of development occurs every winter at the eastern end of both Lake Erie (Buffalo, New York) and Lake Ontario (Watertown, New York). It also occurs along the eastern seaboard of North America (Boston, Mass. and Halifax, Nova Scotia).

Table 1	Water Vapour Pressure
(reference; College Physics by Weber, White & Manning)	
Temperature degrees C	**ressure mm of mercury**
5	.5
10	.2
12	0.5
14	12.0
16	13.6

With this particular snow-producing mechanism, snowfall can be substantial as long as the water is several degrees above freezing. In the case of the lakes, no further heat is likely to be added to the water so it will cool down as the snow-producing evaporation continues. The period of intense snowfall will be brief because as the water evaporates to release the moisture for the snowfall, it will cool down. It will be recalled that in order to evaporate a pound of water 570 BTU of heat are required. The heat in the lake water will supply this heat but in the process the water will cool down. This confirms that there is a limited amount of snow that a relatively warm lake can provide. As a snow-producing mechanism a 'warm' lake is limited. The situation would dramatically change if additional heat was added to the water.

Let us suppose for a minute that both Lake Ontario and Lake Erie were heated. In this case the heat would be provided by a long fissure on the bottom of each lake and up through these fissures molten magma from the interior would ooze. It would be red hot and it would well up from the interior and cover almost the entire bottom of both lakes. The water would warm up. It would warm up until it was well above temperatures that would be suitable for swimming and in fact probably above the comfort zone altogether. Under such conditions the lakes would evaporate at unusually high rates forming massive moisture clouds that would drift downwind until they were over the land. If the land was chilly, it would snow. The snow would keep coming. The streets would soon be filled and driving would be impossible. The people would stay inside but pretty soon they would not be able to see out of their windows. Next the houses would be buried. Then, when the snow had accumulated to more than 50 feet the houses would collapse. There would be no escape. Suppose that the molten material just kept oozing out for more than one month. Would the snow then be 100 feet deep or would it be several hundred feet deep? With everything buried in several hundred feet of snow there would be no escape and all of the people would perish. Something like this happened during the ice age.

With the idea in mind that an expanding glacier is evidence of an ice age happening, one would be justified in suggesting that the glacier on Greenland was expanding during the 50 years that the planes were buried and that Greenland was therefore having an ice age. Also, if snow is accumulating at the rate of several feet every day is an ice age starting to happen? Such suggestions would not seem to correlate very well with the current over-riding conviction that the world is over-heating. But it might suggest that we require heat to have an ice age - as well as cold. Beyond that idea however the entire event shows that sometimes events happen much more quickly than we might otherwise expect. The notion that an ice age could happen within a few hundred years is in this category and if such a development was possible it would discredit the Astronomical Theory altogether.

4.2.5 Structural Characteristics of Ice Ignored

Ice, as a solid material has certain structural characteristics - as has wood - as has steel. The best that anyone can do is to recognize what these characteristics are and proceed accordingly. As an example, in the far north during the winter, trucks are used to transport supplies to various settlements. A road is prepared which runs right across lakes and rivers simply because there isn't any way to go using non-water surfaces. Forest, uneven ground and bogs all work together to prohibit even temporary roads from being laid out across country. Bogs, for example, are a mixture of water and a tangle of vegetation and the frozen assembly is not structurally reliable. Solid ice on the other hand has known structural characteristics which do not significantly vary from year to year. Ice is reasonably predictable. However not just any ice will do. It must have reached a certain thickness before it would be safe for a truck to use – as safe, that is, as can be reasonably assessed. Ice is not as predictable as steel and does not go through a controlled manufacturing process. While a sample of ice could be tested for compression, bending and shear strength it would never be expected to develop some super-characteristic where it could double or triple any one of these basic characteristics. The plan for winter roads therefore includes doing assessments of ice thickness during the fall before allowing the over-ice winter 'roads' to be used. After all, driving across a frozen lake with a truck-load of supplies is dangerous enough anyway. In fact a significant element of danger is always present with ice-roads - even when the most conservative decisions are taken. The fact that ice has certain structural characteristics must always be recognized where winter roads are involved.

The ice that formed during the Great Ice Age was expectedly no different structurally from the ice that forms winter roads. It would also have been the same as the ice that forms on lakes and rivers at the present time. The ice on lakes is of particular interest with respect to ice-age ice. Ice forms on lakes when the temperature remains below freezing long enough. The colder it gets the faster the ice forms and the thicker it gets. Pretty soon even with a large lake like Lake Superior, ice will cover the lake. Then the wind will blow. It will blow across the surface of the ice and press the ice against some shoreline. The ice will become jammed against the shore and will not be able to move but the wind will keep right on blowing. Since ice is basically smooth there would not appear to be very much force transmitted from the wind to the ice but over long expanses the aggregate force can become quite large. At some point the ice will not be able to withstand the compression forces that develop and it will buckle and pill up into pressure ridges. The pressure in this case has been supplied by the wind as it blew across the surface. The failure is with respect to both compression and bending. Ice cannot be bent. If you force it to bend, it will fracture. If you compress it too much it will crush.

We are given to understand that the ice of the great Ice Age was pushed for long distances by some unexplained force in the north. It was pushed down through lake basins, scouring them out as it went. Then it was pushed up out of the lake basins and over any escarpments that were in the way and on across the countryside. Force was required to get through the lake basins and up the other side. Force was required to slide across the land - particularly where the land was uneven. Much of the land would have provided serious resistance to a moving slab of ice. In fact land is mostly uneven with cliffs, escarpments, valleys and hills. Even the great western prairies have hills, river valleys and ridges. Ontario is much worse. The broken unevenness of the Laurentian Plateau would have necessitated that extreme force be applied before the ice would move over the numerous ridges and escarpments that cover much of the province. How could the ice have withstood such force? Some of the valleys run east-west and some run north-south. Sometimes converging escarpments are encountered and the ice would have had to squeeze through before expanding again downstream. Of course if the ice was two miles thick only the bottom layer would have encountered these obstacles and the forward movement of the upper layers would have been unobstructed. Therefore the ice would have been stressed towards shearing off whenever such solid formations were encountered. As the ice scraped along the surface, some of the surface material of the Earth would have been removed and shoved ahead of the ice as long as it didn't offer too much resistance whereupon the ice would have slid up over it like it had to do with so many other obstacles. When the forward movement halted, this material would have formed terminal moraines and these moraines would have consisted of whatever material the ice had been able to push along in front of itself. In fact anything that was basically loose and not rigidly restrained would have been pushed along. The terminal moraines would therefore consist of all of that loose material. Vegetation, not even large trees, would ever have been able to resist the force of the moving ice and neither would loose soil or gravel.

Terminal moraines are supposedly an assembly of material that was pushed ahead of a moving glacier. When the glacier stopped moving forward all of the accumulated material would have also stopped moving and just stayed in place as the glacier melted. The assembly of material would therefore consist of everything that the glacier encountered as it moved for, in some cases, more than 1000 miles. The great Oak Ridges Moraine in Ontario would be typical of this type of situation and supposedly it must be situated where the great glacier came to rest. Actually we are instructed that the glacier also moved across the border into the northern USA so the Oak Ridges terminal moraine must have been from the forward movement of a later glacier because the one that continued into the USA would have scraped it all away. Since the earlier one would also have scraped across the country gathering material on its way, one is caused to ask why the earlier glacier left behind so much material for the next glacier to gather? The first glaciers must have missed all of the material that is included in the Oak Ridges Moraine!

The situation could be compared to a very large bulldozer crawling forward and gathering material as it moved. In this type of situation it can readily be seen that loose material like gravel would be readily gathered and would pile up in front of the blade. As more material was gathered the pile would get bigger and pretty soon the gathered material would spill over on top of itself. It would be a type of crude mixing process where the highest material was continually falling forward and covering the most recently gathered material. Unfortunately the evidence doesn't quite support this basic idea. The Oak Ridges Moraine includes gravel, large pockets of sand and streaks of clay and the entire assembly is several hundred miles long, several miles wide and several hundred feet thick. The streaks of clay occur randomly throughout the assembly. Why the entire assembly would wind up being several miles wide if it had been pushed into place by a moving slab of ice is hard to imagine but the streaks of clay introduce a serious element of incredibility. The material should have all been mixed by the time it had been pushed for hundreds of miles. It isn't the least bit mixed. The clay streaks occur randomly throughout the assembly and they should not occur at all! The clay material should have been well mixed into the entire assembly long before the terminus was reached. While this is impossible to explain the absence of any vegetable matter is even harder to explain. Why are no trees found in the gravel? Why isn't there any soil or leaves? Why is all of the gravel so free of vegetable matter that it can basically be scooped up, crushed and placed on roads?

One also wonders how the drumlins survived because they mostly consist of soft unconsolidated material. It is far from clear how the drumlins formed under such crushing circumstances. (66) How did the ice pass over them without scraping them away? The entire scenario of scraping, sliding, crushing and relentlessly, forward-moving ice would have been most devastating to everything that was either movable or structurally fragile and the soft material that formed the clay would have been dispersed almost without trace. There should also be a few thousand crushed trees throughout the deposit but there isn't even one!

In order to grasp the structural implications of such a situation consider a mind experiment where a slab of ice two miles thick is being pushed by some incredible, without-limit force. Let us consider a slab of ice about one hundred miles long and ten miles wide. At one end there is a force engine – like a giant bulldozer. The other end is free to move. A slab of ice with these dimensions has a slenderness ratio of 2/100 or 2%. The pushing force required is very large. Will the ice move forward? Would the ice on a lake move forward? The ice on a lake would only move forward by fracturing and pilling up into pressure ridges. Even on Lake Superior the pressure ridges can be several meters high and in the high arctic they can be much larger. Similarly a slab of ice two miles thick that was being relentlessly pressed forward would fail in both compression and bending and only move forward as a great pile of fractured pressure ridges.

Also the question of the force engine cannot be ignored. What was supplying the incredible force needed to move the ice of the Ice Age and how was it able to do this without crushing the ice? The ice is declared to have been two miles thick but our atmosphere can barely form ice that thick because at 10,000 feet up, there is very little moisture in the air. Even if it could have formed four miles thick, would a great pile of ice even that high, have been able to spread out for one thousand miles? Would it even spread out for ten miles?

4.2.6   Erratic Boulders

Erratic Boulders are boulders that appear completely out of place. There isn't any 'parent' rock nearby from which they might have broken off. In most cases the 'parent' rock is found at some remote location far from where the boulder is found. Why are these rocks so far from home? The usual explanation offered is that the moving ice of the Ice Age did it. The ice moved all of these boulders until they were far from their original locations. It is likely that the reason this particular explanation developed was from observing that occasionally the wind will push the ice on a lake up onto a shore. As it is pushed up it might pick up a few stones which then stay on the ice when it is blown the other way. People would see these stones sitting on the ice and if the wind blows them to a different location they might then be considered 'erratic' or out of place. There is no doubt that this kind of thing happens and that stones are occasionally scooped up by the ice on a lake and wind up in a different location from where they started. They would be restrained to that particular lake however which means that they would never be found at a higher elevation than when they started but more likely at the bottom of the lake when the ice melted. The other obvious fact is that they had to be individual stones to begin with and they must have been 'free' enough so that the ice could scoop them up quite readily. At the same time we understand that ice which is able to slide up onto a beach cannot be very thick. In fact it has to be quite thin or it would just push anything on the beach farther inland. Then of course there is the structural ability of the ice to carry them. Rocks are really quite concentrated matter much denser and heavier than ice. While small boulders might be picked up it is difficult to see how a large boulder could be carried without breaking through the ice. Further, it is most unlikely that a slab of ice several meters thick would ever ride up onto a beach and if it did, how would the thickness enable boulders to be picked up at all? The whole idea therefore seems limited to rocks that are quite small whereas Erratic Boulders come in a great variety of sizes and some are very large. 'Some of them are thousands of cubic feet in size, and Pierre a Martin is over 10,000 cubic feet. ... The block near Conway, New Hampshire is 90 x 40 x 18 feet and weighs about 10,000 tons. Equally large is Mohegan Rock which towers over the town of Montville in Connecticut. The great flat erratic in Warren County, Ohio, weighs approximately 13,500 tons and covers three-quarters of an acre. The Okotoks, thirty miles south of Calgary, Alberta consists of two pieces of quartzite ... with a calculated weight of 18,000 tons (and is shown in the photograph; Okotoks Erratic Boulder.) Blocks of 250 to 300 feet in circumference however are small when compared with a mass of chalk stone near Malmo in southern Sweden, which is three miles long, one thousand feet wide and from one hundred to two hundred feet in thickness. ... A similar transported slab of chalk is found on the eastern coast of England "upon which a village had unwittingly been built."' (121) In recognition of the size of these monsters, to call them Glacial Erratics seems very dubious.

Supposedly, if a slab of ice was pushing a boulder along in front of itself there would be some evidence that the boulder had been rolled and pushed into its final location. The great Okotoks Boulder does not show any sign of having been pushed! In fact, it appears to have plunked down right where it is currently resting and that the landing has broken it in two! None of the other monster rocks show any sign of having been pushed either and one is left to wonder how they, particularly the flat ones, could have been pushed anyway.

The Okotoks Erratic Boulder located south of Calgary Alberta sw of the town of Okotoks

Many erratic boulders are found in locations that make the ice-carrier explanation totally inadequate because of both the elevation and the distance from their supposed source. 'The loose rocks lying on the Jura Mountains (west of the Alps) were torn from the Alps. ... These stone blocks lie on the Jura Mountains at an elevation of 2000 feet above Lake Geneva. Some of them are thousands of cubic feet in size and Pierre a Martin is over 10,000 cubic feet. They must have been carried across the space now occupied by the lake and lifted to the height where they are found. ... In the British Isles, on the shore and in the highlands, are enormous quantities of them, transported there across the North Sea from the mountains of Norway. Some force wrested them from those massifs, bore them over the entire expanse that separates Scandinavia from the British Isles, and set them down on the coast and on the hills. From Scandinavia boulders were also carried to Germany and spread over that country, in some places so thickly that that it seems as though they were brought there by masons to build cities. Also high in the Harz Mountains, in central Germany, lie stones that originated in Norway. From Finland blocks of stone were swept to the Baltic regions and over Poland and lifted onto the Carpathians. Another train of boulders was fanned out from Finland, over the Valdai Hills, over the site of Moscow, and as far as the Don. In North America Erratic Blocks, broken from the granite of Canada and Labrador, were spread over Maine, New Hampshire, Vermont, Massachusetts, Connecticut, New York, New Jersey, Michigan, Wisconsin and Ohio; they perch on top of ridges, (Please refer to the photograph of an erratic boulder in upper New York State – in the Adirondack Mountains sitting literally within a few feet of the very top of Mount Colden.) and lie on slopes and deep in the valleys. They lie on the coastal plain and on the White mountains and the Berkshires, sometimes in an unbroken chain.' (121) (This quotation is taken from Earth in Upheaval by Immanuel Velikovsky and used here with the kind permission of his grandson Rafi Sharon.) Some of these boulders have been moved for hundreds of miles across terrain that is very difficult to cross and then elevated for thousands of feet higher than their origin. By comparison, the great Okotoks Erratic of Alberta has only been moved about 30 miles from the rocky Mountains to the west, which does not seem very far when compared to the hundreds of miles involved with the boulders that went all the way from Canada into many of the northern states. How is any of this to be explained by a moving block of ice? How is it to be explained by a moving block of anything? The currently popular theories of the Ice Age are inadequate to explain these vast movements and it is totally impossible to imagine how they could explain Erratic Boulders that are even further from their place of origin.

'In innumerable places on the surface of the Earth, as well as on isolated islands in the Atlantic and Pacific and in Antarctica, lie rocks of foreign origin, brought from afar by some great force.' (122).

Adirondack Erratic Boulder

Other commentators have noticed Erratic Boulders in seemingly strange locations. Erratic Boulders are often found in coal far from any similar stone. ... Fragments of rock are the foreign bodies that are the most perplexing. Several were found in a coal bed in Upper Silesia. Since they were a type of crystalline rock they were unlike anything known in Silesia. In 1874 piece of talcose slate was found in coal at Mineral Ridge, Ohio, and it was thought that it might have come from the Canadian Highlands. (123) While distance is an issue with these examples, the idea that they have been mixed in with coal is even more perplexing. This means that the Erratic Rock fragments must have been transported and placed during the time that the material for the coal beds was being assembled. Did the Ice Age and the Carboniferous Period occur at the same time? We recall that the Ice Age is understood to have happened a few thousand years ago whereas the Carboniferous Period is understood to have happened millions of years ago. Something is obviously wrong with these assertions!

Any explanation for Erratic Boulders must therefore account for; placement in coal, transport for hundreds of miles, elevation for thousands of feet and lifting across the North Sea. It does seem incredible that moving ice could have done any of this. However there is an explanation that fully recognizes the realities of the geography involved as well as all of the other difficulties mentioned. Impacting asteroids could have placed the Erratic Boulders. 'One of the best documented ... (impacts) ... formed the Ries Crater in the southern plains of Bavaria. (Germany) The diameter of this crater is given as 24 kilometers.' (15 m) (134) The projectile in this case was a rocky asteroid with a diameter of nearly a mile and a mass of more than a billion tons. ... Around the basin rim are broken and upended rock strata, and huge boulders are found in the surrounding countryside to distances of

35 miles or more. (135) Clearly, when the Ries asteroid struck the ground, the rock was broken and pieces of it were thrown all around the countryside. The radius of this crater is about 7 ½ miles so from the center of the crater, rocks have been hurled about 35/7 ½ = 4.7 times the crater radius. By comparison, the great Sudbury Crater in Canada is about 250 kilometers across. (156 m) (418) If two different asteroids hit the Earth at the same speed we would expect that their respective energies would simply be related by their diameters but when a diameter is doubled the volume increases eight times. This is called a cube law and means that the energy level of the Sudbury asteroid in comparison the Ries asteroid would have been about 5,000 times as great. This assumes, of course, that the size of the Sudbury asteroid relates to the size of it's crater the same as the Ries asteroid relates to it's. This is very unlikely. The Ries asteroid struck the Earth and blasted out a crater as if the Earth was made of the same material all the way through. This, however, was not the case with the Sudbury crater because the Earth has a crust. Before we go very deep into the Earth, the viscosity of the rock diminishes. The crust or rocky shell is very thin in comparison to the size of the Earth and means that if an asteroid was energetic enough to blast right through the crust, the size of the crater that would be formed would not relate to the size of the asteroid in the same way that it would if the asteroid encountered solid material. Once the diameter of the asteroid was comparable to the thickness of the crust, it would simply punch right through the crust and dive deep into the interior of the Earth. The reason that the size of a crater would not relate to the size of the asteroid that formed it for large asteroids is because most of the energy of a large asteroid would be dissipated far from the impact site and actually include the entire Earth. This means that the size of the large asteroids that hit the Earth would have been much closer to the size of their craters than the relatively small ones that are totally stopped by the crust before they even penetrate a few hundred feet. In the Sudbury case, the asteroid would have been much larger than 10% of its crater's diameter and could quite readily have been more than 50% of it. How far could rocks be hurled by an asteroid that was 80 or 90 miles or more in diameter? Could they have been hurled from 'Canada and Labrador ... over Maine, New Hampshire, Vermont, Massachusetts, Connecticut, New York, New Jersey, Michigan, Wisconsin, and Ohio?' (121)

A thick slab of ice could not move boulders from Canada and spread them all over the northern USA but large impacting asteroids could!

4.2.7   Drumlins

Drumlins are features on the surface of the Earth that are attributed to the moving ice of the Ice Age. They are understood to indicate the direction of movement as well. However the directions indicated by many of the ones in Ontario directly contradict the general notion embodied in Ice Age theory that insists that the ice moved down from the north. It was to have scraped along with a leading edge that has been suggested to have been hundreds of feet high. Some commentators even suggest that it was more than a mile high. This would have been a most spectacular site - a great moving wall of ice crawling down across Canada and continuing right into the northern USA. Included in the idea is that it would have been pushing a great load of material in front of itself and this material would become a terminal moraine when the ice stopped moving forward. Such an arrangement would seem to agree with what is observed in mountains. When a glacier slides down a mountain it melts and leaves a pile of loose material at the location where it stopped moving. In confirmation of this idea there is a great streak of sand and gravel right across Southern Ontario and it is referred to as a terminal moraine (i.e. the Oak Ridges Moraine which is hundreds of miles long and several miles wide). Apparently this is where at least one great sheet of moving ice came to a halt otherwise there would appear to be a contradiction because the ice is also declared to have moved farther than this right down into the northern USA. Then it 'retreated'. However it did not actually retreat, it just melted. 'Retreat' suggests that the ice stopped moving forward and then moved backward instead. It is difficult to understand how it moved forward. It is impossible to understand how it could have moved backward. In spite of these difficulties the forward motion somehow caused drumlins to form at ground level underneath the ice. When the ice disappeared, the drumlins became visible.

Commentators on these matters are quite adamant that; a. The ice moved down from the north, b. The ice scraped a lot of material along in front of itself, c. The material became a terminal moraine when the ice 'retreated', d. Drumlins formed under the ice and indicate the direction of ice movement. All of this does sound pretty straight forward and easy to understand except that there is great inconsistency throughout the entire report.

Drumlins are mostly very soft structures. They consist of silt. They do have a 'head' and a 'tail' seemingly to indicate direction of movement. They commonly include other material within their structures and one of the most common items is a large boulder. However, they are never a great mound of solid rock. If they were solid rock one might understand that a great slab of moving ice would somehow ride over them and possibly smooth out any projections. However if they consist of unconsolidated material like silt or soil it is more difficult to understand why such loose material would not have been scraped away by the moving ice. It should have been scraped away by the leading edge of the massive ice formation but if that did not work they should have been scraped away by the incredibly-thick and heavy slab of ice as it slid over top of them. The ice is thought to have continued well into the USA which makes one wonder why the moving ice didn't simply erode all of the drumlins away like the eraser on a pencil erases pencil marks. It wasn't that the ice only moved for a few feet. After passing over Southern Ontario it is understood to have moved well down into the northern USA before it stopped. So the first question that requires an answer is 'why didn't the drumlin material get scraped away by the moving ice?' since they are easily erodible. (66)

However even before that question is tackled there is another one of even greater consequence. How did the drumlins form in the first place? No one has offered an explanation in spite of the fact that there has been a great deal of discussion on the matter. Apparently the question of drumlin formation is brought up quite often when attendees at conferences about the Ice Age discuss such matters. In spite of there being great insistence that the moving ice formed the drumlins there has never been any suggestion concerning how it happened. How a glacier forms drumlins is a great mystery. There isn't any agreement among the 'experts' so they remain as mysterious as ever. In fact the origin of drumlins is one of the great unsolved problems of the Ice Age.' (66) Clearly, a serious assertion is being made without any explanation to support it!

The news gets worse from here. The declared direction of movement of the ice does not line up with the direction that the drumlins indicate! The ice was thought to have moved south across Ontario but the direction indicated by the drumlins is not south at all. '... the orientation of drumlins indicates they were formed by a flow towards the northwest. This is an enigma in the glacial theory ... the cause of the northwesterly flow is a mystery ... the ice sheet must have flowed uphill, out of Lake Ontario over the cliffs of the Niagara Escarpment to form these drumlins.' (67)

There is a possibility that the drumlins were not formed by ice movement at all. There is a possibility that the great glaciers of the Ice Age had nothing to do with drumlin formation. Flowing water could have formed the drumlins. Similarly-shaped formations occur at the present time when a silt-laden flow of water encounters an obstacle in its pathway. The obstacle will cause little eddies of turbulence to develop around and down-stream of the obstacle. Some of the silt will drop out of the flowing water and pill up downstream and on top of the obstacle. A mound will form. This mound will have a shape which will indicate which way the water was flowing with a trailing edge stretching out down-stream of the original obstacle. The obstacle itself might become completely buried as the silt accumulates until there will be no indication that it is there at all.

A second development that has similar characteristics is the mound of snow that forms down-wind of a snow fence. At the air moves through a snow fence, eddies of turbulence form and some of the wind-carried snow settles out. As the mound of snow accumulates, the downwind side of the mound provides the eddy formation function and the original snow-fence could actually be removed and a mound of snow would still continue to accumulate. In fact whenever a material-loaded mass of flowing fluid, whether it is wind or water, encounters some means to create a little turbulence or cause the flow to slow down, some of the entrained load will settle out and form an accumulation of material where previously there wasn't anything. This also happens when a silt-loaded river enters a deeper body of water. The silt settles out and a delta is formed. The Mississippi River delta (as well as all of the other deltas of the world) was formed this way. Similarly, certain terminal moraines declared to have been formed by ice-age glaciers were more credibly formed by moving water. The great continent-crossing tsunamis generated by the impacting asteroids could quite readily have formed the terminal moraines as well as the drumlins. The further piece of evidence supporting this assertion is that the gravel of the 'terminal moraines' consists of rounded stones. 'Round' implies water movement! In fact gravel always implies water movement. All of the gravel formations that more or less follow along the southern edge of the Laurentian Plateau were placed by water movement as were the drumlins!

Drumlins are relatively soft structures and would have been destroyed by an over-riding slab of ice but they would have continued to accumulate material as long as an over-riding mass of silt-loaded water continued to flow past.

As pointed out above, the Astronomical Theory of the Ice Age has other serious shortcomings and the drumlin-formation uncertainty only adds further doubt to the entire scenario. Fortunately, there is a more evidence-recognizing, logical explanation for the Great Ice Age which will be presented in the following sections.

4.2.8 Rapid Freeze-up Ignored

One of the characteristics of the Astronomical Theory of the Ice Age is that the Earth drifted very slowly out farther from the Sun as it orbited around it. In fact it took several years for it to move outward during which time the Earth cooled down slightly. Unfortunately the evidence does not support this assertion.

Included in the material that became frozen are the tusks of Woolly Mammoths. These tusks are made of ivory the same as the tusks of present-day elephants. Ivory has always been of interest because it is readily carvable into useful shapes for both decorative as well as practical purposes. Piano keys were once covered with a thin layer of ivory. Small statues as well as abstract forms are still popular. In order to be carvable however the ivory must be 'fresh'. If the ivory isn't harvested until long after the animal is deceased then it will be considered 'stale' and useless for carving. Freshness can be retained for many years however if the ivory is frozen soon after the creature dies.

Woolly Mammoth tusks have been recovered from the great Russian bogs for hundreds of years and continue to be recovered right up to the present time. (440) In most cases the recovered material is 'fresh' and hence has value which prompts the harvesters to continue harvesting. The only way that this type of situation could have developed is if the tusks had become frozen very soon after they became separated from the animal. Of course, if they are still attached to the animal, then the entire animal must have become frozen immediately after dying. This could never have happened if the temperature had just drifted down slowly over time-spans of hundreds of years as the Astronomical Theory declares.

4.3 Summary and Conclusion

According to the Astronomical Theory of the Ice Age, the cause of the ice age was a cooling trend introduced by the Earth drifting slightly further from the Sun due to long term variations in its orbit. However the amount of cooling this would have caused would only have been 1% or 2%. This would have even been a much smaller amount of cooling than the Earth has recently experienced due to events right here on the surface of the Earth. During the eighteen hundreds the Earth went through two serious cooling episodes each of which resulted in more than a 10% drop in the heat from the Sun reaching the Earth. At the very minimum these events involved more than five times the reduction in solar heat than that which is predicted by the Astronomical Theory of the Ice Age as a result of slight orbital changes. Therefore if such predictions had any validity the Earth should be well into another ice age at the present time. However it isn't. Instead it seems to be going into a heating episode instead. There isn't any sign of an ice age at all. The Astronomical Theory of the Ice Age should therefore be set aside.

The chill predicted by the Astronomical Theory of the Ice Age would have caused all of the bodies of water on the Earth to chill. Chilly water (or frozen water) does not evaporate very well whereas the moisture required for the Ice Age involved several hundred feet of ocean water – just to form the ice. All of this water had to evaporate from the ocean and it would have required a tremendous amount of heat to do this. If the water was chilly it would not have been able to evaporate so where did all of the heat come from to enable all of that water to evaporate? Even if it was warm, as it is understood to have been in ancient time, there would not have been enough heat in it to evaporate more than a few feet. Therefore the Astronomical Theory is predicting a setup which is exactly the opposite of what basic physics demands.

Cooling the Earth on a global basis would result in a reduction of the two main greenhouse gases, CO_2 and water vapour. This cannot be done without irreversible consequences. The Earth would simply freeze up and remain that way. It is a wonder that this didn't happen during the cooling periods caused by the excessive cloud cover produced by the Tambora eruption and the Krakatoa explosion during the nineteenth century. If a drop in received heat of one or two percent could cause an ice age to happen as declared by the Astronomical Theory then a drop of more than ten percent (as occurred following both the Tambora explosion and the Krakatoa explosion) would have been more than enough to do the job. However in neither case did an ice age start to develop.

To summarize; The Astronomical Theory of the Ice Age fails on numerous basic points and therefore must be set aside as invalid and relegated to the 'Fairytales for Adults' category.

5.0　Beringia, an Ice-Age Fairytale

5.1　The Setting

The north-western portion of North America including Alaska and part of Canada as well as the eastern portion of Russia together with land which would have been above sea level at Ice Age maximum, has been given the name, Beringia. The name is in honour of an explorer named Vitus Bering who came from Denmark and surveyed the area in 1725 and showed that Asia and North America were not connected above sea level. (441). The name is still in current use as the straight between Alaska and Siberia is called the Bering Straight and the portion of the Pacific Ocean enclosed on the north by Alaska and Siberia and on the south by the Aleutian Islands is called the Bering Sea. The Aleutian Islands extend from mainland Alaska almost all the way to the Kamchatka Peninsula of Russia.

The water across most of the Bering Sea, the Bering Straight and the adjacent part of the Arctic Ocean is not more than 400 feet deep. While the reduction in ocean level at ice age maximum is not exactly known it is generally thought to be in the range of 300 to 500 feet. This means that all of this area would have been above sea level as the accumulation of ice and snow during the Great Ice Age reached its maximum level. Accompanying the belief that the ocean floor was exposed at that time is the idea of a 'land bridge' – a means for animals and people to get from one continent to the other without having to deal with the ocean. However getting there by ocean would actually have been much easier because when this land was exposed, the Great Ice Age was in progress!

There is a widespread conviction that several species of very large mammals lived right through the Great Ice Age and only died out after it was over. This conviction is so deeply ingrained that some animals are referred to as 'Ice Age Animals'. Harrington, one of the world's experts on the fauna of the last glaciations, has hauled thousands of ice-age oddities from the northern mud. … an ice-age elephant – better known as a woolly mammoth … ground sloths … beavers … bears … moose ... etc. (555) Other commentators concur with the basic assumption. 'This "muck" (i.e. the muck of Alaska) contains enormous numbers of frozen bones of extinct animals such as the mammoth, mastodon, super-bison and horse. These animals perished … at the end of the Ice Age or in early post-glacial times.' (556) 'Few controversies rage more fiercely in paleontology than why the megafauna vanished – not just in Australia but in North America, where mammoths, horses, camels, and dozens of other large Ice Age mammals all vanished … the giant mammals of the Ice Age.' (557) In summary, the overall declaration being made is that thousands of animals lived right through the Great Ice Age in a northern region at the same time that glacial ice was accumulating.

5.2　The African Serengeti

The secondary name given to Beringia is "An Ice-Age Serengeti". (149) The African Serengeti is a large area of Central Africa where great numbers of animals live. Numerous rivers cross the Serengeti. It covers several thousand square miles and includes portions of several countries. The herd of Blue Wildebeest (or Gnu – a species of large antelope) alone numbers many thousands. Other species included are giraffes, hyenas and lions. These animals live on the Serengeti Plain all year and migrate from place to place to take advantage of the best grass. Generally speaking, the area was considered to be an ideal setting for a great diversity of species which infers that the food supply is at least adequate and usually much better than adequate.

5.3   Normal Alaska Winter

However that's Africa where the weather is always warm and the Sun always shines. Alaska, on the other hand, is in the far north where winters are long and cold and where the Sun does not shine all of the time. Blizzards sometimes rage for several days in a row and all of the rivers are frozen solid all winter. Animals do live in Alaska. There are several very hardy species such as grizzly bears, moose and wolverine. Most of the bears hibernate all winter. Moose live on vegetation which is available through all of the seasons. Wolverines are not fussy about what they eat and will accept almost any kind of food. Even with such a basic approach they sometimes go without eating for several days in a row. No one would suggest that life in Alaska during the winter is the least bit easy – either for animals or people. In fact it is quite the opposite and animals commonly perish during the winter due to starvation or cold. While animals do live in Alaska through the winter, they are certainly not in abundance to the extent that a person could snowshoe through the land for days and never see one single animal. This provides immediate cause for wonder. How could a great diversity of thousands of animals have not only survived in the area but actually thrived when many of them were dependant on a good supply of grass and other vegetative nutrient throughout the entire Ice Age?

5.4   The Great Ice Age Reality

During the declared "Ice Age Serengeti" period, the Great Ice Age was underway in earnest. The land was chilled by the ever-present and abundant cloud cover to the extent that it was dark all of the time. Even at the present time, a thick layer of cloud can cause darkness well before sundown. One example of particular interest is when a massive storm cloud wedged up against the Rocky Mountains and caused darkness before sundown. When storm clouds build up over the Rockies, the prevailing westerly winds usually carry them away from the heights and out across the plains to the east. On this Saturday in July however, the strong westward movement of the Canadian air mass held the growing thunderheads pinned against the eastern slopes. Between 6:00 and 6:30 p.m. several large thunderstorms developed over Boulder and Larimer counties, northwest of Denver. The moisture-laden air from the south began to mushroom upwards until within an hour, the tops of the thunderheads had reached an altitude of 60,000 feet in the evening sky, darkening the land before sunset. (454) While this example is quite unusual it is not unusual to observe the dark underside of a storm cloud. The darkness is simply caused by the thickness of the cloud which prohibits sunlight from getting through. During the Great Ice Age, darkness was much more over-whelming because there would have been two layers of cloud involved. The lower moisture-laden snow-carrying layer would have been effective at sun-blocking on its own but there was also a second layer of cloud – a dust cloud produced by the impacting asteroids and the volcanoes which followed. This second type of darkness occurs occasionally at the present time when a volcano explodes and erupts. It can be quite alarming because it can be so prevalent that nothing whatever can be seen. This type of total darkness developed when Mount Katmai erupted in Alaska in 1912. (423) While it cannot get any darker than dark with two layers of cloud present during the Great Ice Age virtually no sunlight would have reached the surface of the Earth at all.

The Great Ice Age brought both kinds of cloud. The moisture clouds brought the snow which would have fallen at prodigious rates because the ocean was still warm and well able to supply the moisture. In addition to accumulating at the rate of several feet every day it would have been piled into endless drifts. It would have been a virtually continuous blizzard all across the regions where glacial ice accumulated. In between these regions there would have been less accumulation but even in such places the snow could easily have been 100 feet deep because it requires at least 80 feet of snow before it will weigh enough to press down and form any ice at all. Of course prior to the temperature dropping to below freezing and staying there the snow could have been mixed with rain or ice pellets - none of which would have allowed any animal activities to carry on. It was a time to take shelter and stay there until the storms subsided. Unfortunately the great storms would have raged on for years and any shelter would soon have become well buried - entombing its inhabitants.

5.5   The Food Supply

While it has been declared that the animals survived by accessing 'nutrient-rich grass' one wonders how this would have been possible under the admitted circumstances of cold and dark. '(He) ... believes that these species thrived in great numbers despite bitter winter winds and long months of darkness. The secret to their survival and the key to the giant size of so many of Beringia's animals, he explains, lay in the rich nutrients contained in the ecosystem's grass.' (558) Indeed this would have been a very special type of grass because it enabled animals to become exceptionally large and was abundant enough to support thousands and thousands of animals. This type of grass is unknown at the present time! 'Nutrient rich grass' has never before been credited with enabling an animal to reach exceptional size – in particular since only the grass-eaters would have eaten it while the other animals would have eaten the grass-eaters. However most of the animals were exceptionally large even though many of them would never have eaten the special grass at all.

5.6   Climate, Dry or Damp

The Great Ice Age was underway and the snow was falling in unprecedented amounts. In fact in some areas it has been declared that the glacial ice built up until it was hundreds of feet deep. This in turn would have necessitated that ten to twelve times this amount of snow would have actually fallen because it takes that much more snow to make ice. This in turn indicates that the air was not dry because dry air cannot hold very much moisture and the air was loaded with snow-making moisture. Therefore the declaration that; 'In the exceptionally "dry" air of Beringia, native grasses were likely cured to perfection in fall, preserving high nutrient levels' (413) is dubious.

This "damp-air" reality is recognized by certain other commentators discussing the Great Ice Age. The warm ocean would have provided a lot of heat for the atmosphere. (501) A large source of heat means a large amount of moisture. This is not the case at the present time when there is no 'large heat source' but where snowfalls are still occasionally quite large. Massive quantities of snow cannot accumulate when the air is dry because dry air does not include any significant amount of moisture. However massive quantities are required or we are not having an ice age. Since glacial ice is recognized as having existed across various parts of Beringia, it must be recognized that the 'ice-free' areas within the general region also experienced great amounts of snow accumulation because it would not be realistic to expect enough snow for glacial development in one county while in nearby areas there was virtually no accumulation at all.

5.7   Snow or Grass

The glacial ice of the Great Ice Age was distributed throughout Beringia. In particular it is understood to have covered the southern third of Alaska as well as numerous other locations throughout Alaska and Eastern Siberia. These glaciers would have been typical of glaciers of the Ice Age wherein the ice accumulated until it was hundreds or even thousands of feet deep. This in turn means that thousands and tens of thousands of feet of snow must have fallen.

Therefore the situation which is presented is that grass and other vegetation grew in abundance while interspersed throughout the area the ice was accumulating until it was thousands of feet deep. The snow was falling in the next field but it wasn't falling in this field. It was cold enough in the next field to keep the snow from melting but in this field it was warm enough to enable abundant vegetation to grow every summer so a diversity of fauna could thrive. It was warm here but cold over there. Over there the snow was piling up to incredible depths but over here there wasn't any snow at all. Summer came and the grass grew. Then winter came and the animals accessed this grass through a few inches of snow. With this arrangement it is declared that the great diversity of Ice Age animals thrived and grew to be very large.

5.8   The Darkness

The darkness is deserving of further comment. During the great Ice Age as mentioned above, the land was covered by two layers of cloud. This assured that total darkness was perpetual and that the snow did not melt and that it remained as snow for the entire time that the glacial ice was accumulating. This was a necessary condition. Otherwise the snow could not have accumulated and continue to accumulate for years – possibly several hundred years. It had to remain frozen right up until ice-age maximum had been reached. This would have required several decades at the minimum so the darkness had to have persisted for that entire time. Once ice-age maximum had been reached the ice proceeded to melt and apparently it melted quite rapidly.

If the snow and ice had been allowed to melt every summer (allowing nutrient-rich grass to grow) there would not have been any accumulation. Therefore it would not have been an ice age at all but only a series of snowy winters.

There is no doubt that the Great Ice Age was an unusual event and that the circumstances that caused it to happen will not readily recur. However suggesting that thousands of large animals survived the prolonged darkness, the cold and the accumulating snow is totally absurd.

5.9   Gross Contradiction

Unfortunately the declarations associated with Beringia are gross contradictions. It cannot be snowing profusely a few miles away and not snowing here at all. Wasn't there any wind blowing? If not, what air movement brought the snow-laden storm clouds from the ocean? It is contradictory to assert that it was warm enough to enable grass to grow in apparent abundance throughout the area while at the same time it was cold enough for snow to fall and remain frozen. We cannot have major build-ups of snow right next to areas that are snow-free. It cannot be warm and cold at the same time in virtually the same place. If it is snowing so heavily only a few miles away in every direction, it must surely be snowing here as well.

Arctic Islands and continental areas in both Siberia and North America are declared to have been glacial-free during the Ice Age. (152) While this may have been the case it does not mean that these areas were snow-free. It is more likely that there wasn't enough accumulation of snow to form significant thicknesses of ice. There still could have been an overwhelming abundance of snow. After all, it requires at least eighty feet of snow to just start to form ice. If two hundred feet of snow accumulated, 120 feet of it would possibly compress into about ten or twelve feet of blue ice. The remaining eighty feet would remain as snow. Years later when the area was investigated for glacial evidence, would ten feet of ice have left any sign of its existence? Glacial ice in a flat area would not have moved. There would not be any signs of movement. How then could it be determined that it had or had not been there? It is even likely that several hundred feet of ice, which would only have been on the ground for a few years, would not leave behind any particular indication that it had ever been there. But two hundred feet of ice represents at least two thousand feet of snow. An animal has never existed that could thrive in two thousand feet of snow. In fact, an animal has ever existed that could thrive in one hundred feet of snow – especially if it never melted!

However, all of the Ice-Age animals are presumed to have survived the climatic extremes of the Great Ice Age only to die when it was all over. 'The megafauna species … survived each glacial and interglacial period. But at the end … (they) … became extinct … What caused the virtually simultaneous demise of mammoths, mastodons and saber-toothed cats, not to mention native horses, ground sloths, native camels, armadillo-like glyptodonts, giant peccaries, mountain deer, giant beavers, four-pronged antelopes, dire wolves, native lions, and giant short-faced bears? Scientists have grappled with this question for nearly two centuries … We have accumulated facts on the nature of ancient floras and faunas, on past climates, on human prehistory, and on the chronology of it all. These are precisely the kinds of facts that scientists have assumed all along are needed to provide an adequate explanation of late Pleistocene (Ice Age) extinctions. Nonetheless, from a historical perspective one of the most interesting lessons to be learned from this volume is that we are apparently no closer to that adequate explanation …' (559)

Either the snow was there or it wasn't there. Either the animals were there or they were not there. If there was fifty feet of snow there could not have been any animals. If there was a great diversity of animals, there could not have been any snow. To assert that thousands of animals lived in the snow fields during the Great Ice Age is not a true story. Beringia, an Ice-Age Serengeti, is a fairy tale.

6.0   Asteroid Shower or Separate Impacts?

6.1   Introduction

The most popular and widely-held understanding concerning the arrival of the asteroids that have hit the Earth is that they came at widely time-spaced intervals. Further, large asteroid impacts are always associated with mass extinctions. It is understood that whenever a large asteroid hit the Earth a mass extinction of animal life would surely follow. In fact certain impacts have been directly connected with very specific mass extinctions. This is most evident concerning the Chicxulub impact on the Yucatan Peninsula in Mexico. That event has been credited with the extinction of the dinosaurs – every last one of them, large and small. (75) Of course other creatures would have died as well but in the case of the dinosaurs it is thought that they were wiped out completely. Other mass extinctions are also recognized and even assigned impact times. For example, one commentator states that there have been three mass extinctions in addition to the Chicuzulub impact which is recognized as having happened '65 million years' ago. The other three mass extinction episodes occurred 230, 365, 445 million years ago. (18) Further, when commenting on the Chiczulub impact it was suggested that an asteroid 10 miles in diameter would have enough energy to 'wipe out a hemisphere'. (20) It would be reasonable to equate 'wiping out a hemisphere' with a mass extinction. In support of the argument that there have been numerous mass extinctions, we note that the Earth Impact Database (EID) lists several that would meet the 10 mile-diameter criteria as well as several more 'suspected' sites that would do the same. Even an asteroid that was only four or five miles in diameter would be responsible for the death of large numbers of animals over very widespread areas and any such event would properly be called a mass extinction.

The same list (i.e. the EID) identifies ages for almost every impact and they are spread out over vast periods of time involving hundreds of millions of years. (2) Ideas such as this are intimately connected to the basic idea that the Earth has been here for a very long time (i.e. for 4.5 billion years as discussed in chapter 3) and that evolution explains how life has developed down through the ages. In fact, a very old Earth is a necessary condition if asteroid impacts are to have occurred at widely time-spaced intervals. However, as will be pointed out in the following discussion there is a serious problem with major impacts being separated by millions of years. While dealing with an asteroid shower would have been a most difficult undertaking, dealing with major impacts one at a time spaced apart by millions of years would have been totally impossible. The Earth would simply cease being habitable and life on Earth would have never again been possible. Ironically, while the Earth could not survive a single major impact, it could possibly survive a shower of impacts that happened within a short period of time. On the other hand, if the Earth has only been here for a few thousand years, the arrival of the asteroids could not have been spread over many millions of years anyway. Ideas such as these cannot be approached separately and independently but must be understood with respect to all other relevant factors and conditions. The following discussion will be dealing with the major problem associated with the 'one at a time, single impact, widely time-spaced scenario'.

6.2   The Greenhouse Gases

Very tight temperature regulation is absolutely necessary in order for life to exist on the Earth. The temperature must be held within a very narrow range which obviously must be between the freezing point of water and the temperature at which seeds can germinate. If the average surface temperature of the Earth (or any other planet) was below the freezing point, most of the water would be confined to ice-covered reservoirs and just stay there all the time. However in order for water to be useful it must be mobile and able to leave the reservoirs and easily evaporate into the air which can then carry it for use elsewhere. In this manner plants could be watered and animals would have a source of drinking water. These criteria basically rule out Mars as a place for life because the surface temperature of Mars is below freezing virtually all of the time. (21) Similarly if the temperature was too high, seeds could not germinate and plants would not be able to either survive or grow. We cannot visualize a world without plants. Further, since it clearly obvious that the temperature cannot be the same over the entire Earth all of the time, the average temperature must be about one-half way between these two extremes, that is, between freezing and the upper limit for seed germination. In fact, that's where it is at the present time as the average surface temperature of the Earth is about +15C. (59F)(51)

One might expect that this temperature level could be varied somewhat without causing trouble but this is not the case. For example, if the average surface temperature drifted up to +20C it shouldn't be any cause for alarm because 5 degrees does not seem like a very great temperature shift. However numerous scientists have voiced warnings that such an increase would result in widespread chaos and disaster and actually make the Earth less habitable. (19)

We understand that the greenhouse gas factor in the atmosphere is the primary regulatory mechanism that keeps the temperature of the Earth at its present level. Changing the greenhouse gas inventory would therefore change the temperature of the Earth. Change is not welcome. However change is happening and it is happening because the greenhouse gas inventory is changing. The carbon dioxide (i.e. CO_2, the second most influential greenhouse gas) in the atmosphere has been increasing for several hundred years. A small increase in CO_2 can be accommodated by an increase in plant growth because the higher that the level of CO_2 is in the atmosphere, the better plants grow. This is occasionally done in a greenhouse to encourage better plant growth. In this case the products of combustion from a gas furnace will be intentionally released into the greenhouse in an effort to increase plant growth. However if an increase in plant growth (or some other CO_2 regulation mechanism such as an increase in absorption into the ocean) could not handle an increase in CO_2 in the atmosphere then it would increase. Consequently more of the heat radiating from the surface of the Earth would be returned to the surface and the surface temperature would be held at a higher level than it would have been if some of that heat had escaped into space. While greenhouse gases are necessary for our survival, it is also necessary that they neither increase nor decrease. If they increased the world would get hotter. If they decreased the world would get colder. Deviation either way would be a disaster. In particular if CO_2 increased and the Earth got warmer, the warmer air would be able to hold more water vapour. The increase in water vapour would then (because it is also a greenhouse gas and because it is actually the most influential greenhouse gas) cause a further increase in temperature. This further increase would enable the air to hold even more water vapour and a viscous cycle would be underway. The same would be true in the other direction. If the Earth became chilled for any reason, the air would not be able to hold as much water vapour so as the quantity of water vapour decreased in the air, the air would chill some more. It really does not have to chill very much before the freezing point of water is reached. Then, when a skin of ice forms, water cannot evaporate at all so the source of the most important greenhouse gas is cut off. In this case the Earth would continue chilling until any further change in temperature would not result in any further change in the amount of water vapour in the air. Unfortunately this would not happen until the temperature was well below freezing. (87) Our most important greenhouse gas would then be gone and the average surface temperature of the Earth would settle out well below freezing.

If the output from the Sun had originally been about 80% of its present output (as it theoretically would have been 4.5 billion years ago – see the chapter entitled 'The 4.5 Billion-Year-Old-Earth'), the average surface temperature of the Earth would have been well below freezing. Then, if the Sun gradually warmed up to its present level, the surface temperature of the Earth would increase but still be below freezing because it is the combination of the Sun's output and our greenhouse gases that causes the average surface temperature to be at its present level. The current output of the Sun alone would not be able to do this. It further means that the Sun would have to continue warming up until sometime well into the future until it alone was hot enough to raise the surface temperature to above freezing. Then some ice would melt and some water would evaporate. The water vapour thereby produced would cause a further increase in surface temperature which would melt more ice and so on. Assuming that the other greenhouse gas, CO_2, was also present it can readily be seen that the combination of the higher output of the Sun and the greenhouse gases would result in a surface temperature that was now too high. It would be above the comfort zone. The net result would be that the average surface temperature of the Earth would only have been in the comfort zone for a relatively short period of time of possibly a few years at most. This is not good news for advocates of a 4.5-billion year old Earth. In light of the greenhouse gas reality, the idea that the Earth is 4.5-billion years old is not the least bit credible as pointed out in chapter 3.

6.3   Land-Based Animals

Land-based animals are those animals that must spent most of their time on the land or in small bodies of water surrounded by land. In most cases it is quite easy to decide if an animal is land-based or water-based but there are a few cases where the distinction does get blurry. A beaver is a land-based animal. It spends a lot of time in shallow water such as streams and small lakes but goes onto the land to recover most of its food. The same is true for most species of otter. They are just as much at home in the water as they are on land but would not survive if they could not get to shore. The sea otter is a much larger version of the land-based otter and spends virtually its entire life in the ocean hiding in and feeding on kelp. It quite probably could survive quite well if it was placed in the same environment as its land-based relatives but that would not change its ability to survive entirely in the ocean. Most creatures can be placed into one category or another without too much reflection. Squirrels are land-based. Porpoises are water-based. If the sea dried up, porpoises could not survive. If all of the land was washed away squirrels could not survive.

This distinction clarifies one of the tragedies that would accompany the impact of a large asteroid with the Earth. Included among the many miseries which would accompany the impact would be continent-crossing waves. Even a single asteroid would generate numerous waves that would have so much energy that they could cross right over any continent on the Earth. Since these waves would be coming from all directions and returning repeatedly, even the very soil would be washed away taking all of the above-ground sources of shelter as well as all underground sources of shelter away in their wild and uncontrolled movements. There would not be any burrows to hide in because there would not be any soil within which a burrow could be dug. There would not be any trees left because there would not even be any soil left to support trees. Being able to rise up into the sky as a bird would not have helped much either because some of the waves would have been several miles high and there isn't any kind of bird that could have survived trying to hold itself aloft above such waves long enough for things to settle down.

The impossibility of survival for all types of land-based animals is reasonably obvious but the situation for water-based creatures would not really have been very much better. All of the water would have been in motion. The wave formations that criss-crossed the land would also have criss-crossed the oceans. Most of the time the motion would have been very turbulent and most of the time it would have entrained vast tonnages of material that were being relocated. This would have included the particulate matter that would soon be placed in vast layers and harden into sedimentary rock as well as the great tumbling masses of vegetable matter – including entire forests of trees – that would soon become snagged and covered and form the future coal beds. While any creatures trying to survive in this chaotic scene would not have been able to avoid many of these hazards, neither would they have been able to deal with the great amount of heating that the many erupting volcanoes were releasing. In particular, underwater volcanoes caused the ocean to boil in many places. The water would have been extremely turbulent and loaded with material from beneath the crust. Hot molten lava exploded upwards right through the already-heaving surface of the sea and piled it high on both sides of the great fissures that allowed it to escape the interior as another great pressure wave from the far side of the Earth came up - carrying the entire crust of the Earth up with it. The water above these great temporary mounds would have flowed outward away from the highest elevations thereby forming sheet flow which would also have raced around the Earth. Where would any creature have been able to find shelter? At least, if it was water-based it might have been able to ride it out on some passing log-mat but even these formations would have been in turmoil as well. A few water-based creatures would be expected to make it through because it really would have been difficult to kill every last one of them. The land-based animals would not have done so well because they would not have had any source of food nor any form of shelter whatsoever and the turmoil would have continued for months. Such a scenario would have even excluded any stragglers from being able to make it.

An explanation for survival following the impact of a large asteroid has never been offered. Trying to explain how anything survived at all has never been attempted but trying to explain how any wretched land-based animal could have made it might even be described as foolhardy because of the extreme conditions and the extensive time over which they would have persisted. The most obvious and reasonable conclusion would be that land-based animals would have had no hope whatsoever of surviving and water-based animals would have had it only very slightly better.

The derivative conclusion is even more obvious. If it would have been impossible to have survived a single impact how could survival through more than one hundred widely time-spaced impacts be explained? Perhaps this is why the literature is completely void on the matter. There simply is no explanation! With these realities in mind, the arrival of the asteroids must have been a single event where they came in a a shower - spread out over a few months but not over a years and certainly not over millions of years. This conclusion is only reinforced by other corroborating evidence presenting a picture that leaves extremely little room for doubt. The Earth has been hit hundreds of times and many of the hits would have been serious enough to have caused world-wide chaos. Repeated impacts occurring every few million years - where the land-based animals were the prime target every time, is a situation which no amount of evolution or even re-creation can explain. The surviving cohort from one impact would have to be the starting point to get ready for the next impact. By the same reasoning the currently-surviving cohort must have evolved from the survivors of the last significant impact because everything else would have been destroyed. Explanations for such a requirement have never been, and never will be, offered.

6.4   Asteroid Formation

There are currently three ideas being recognized as the possible origin of the asteroids. None of these three ideas are universally accepted as might be expected but there seems to be considerable support for the location of the formation event which is between Mars and Jupiter. The reason is simply that this is the general location where most of the remaining asteroids are still located. Formation has variously been suggested as being caused by an explosion of a small planet or by the impact of some object with a small planet. The other occasionally-suggested explanation is the lack of formation of a planet in the first place with the asteroids being the material that might have formed this missing planet. This third alternative loses credibility when we notice that many of the remaining asteroids both large and (really quite) small are perfectly spherical. If an object is found in space in a perfectly spherical state it must have been cast into space in a liquid state. If a spherical shape is to develop from a group of irregularly-shaped rocks the entire mass of such material must form a sphere almost as large as the Moon. Otherwise the self-gravity of such a mass would not be great enough to bring about a spherical shape. Some of the spherically-shaped asteroids are only a few hundred meters across which means that they do not have enough material to have become spherical by their own gravity unless they consisted of low-viscosity material and obtained a spherical shape before hardening. It has also been noticed that many appear as blobs of material which suggests that they were either molten or semi-molten when they were sent out on their own. Then they cooled down before their self-gravity could make them spherical so they retained their irregular semi-molten appearance. It must also be mentioned that the two moons of Mars are irregularly shaped and have this same appearance. They also have numerous small impact sites which could not have happened if they had been solid when the small impactors arrived because the gravity of these moons is too small to cause anything to approach at more than a very slow speed. While the asteroids do move at considerable speed, most of them are moving in the same general direction which would have been the case after the explosion of a Sun-orbiting planet.

Supporting evidence for the fluidity of an asteroid-originating planet is provided by the asteroid 2005 YU55 which is called a round mini-world at 1,300 feet (400 meters) in diameter. (160) This asteroid passed the Earth in November of 2011 closer than the Moon which enabled amateur astronomers to photograph it. An object this small could not have achieved a spherical shape on its own if it had been solid or composed of small solid pieces. It had to be fluid from the very beginning in order that it's own self-gravity could cause it to become spherical. The shape of this asteroid confirms that the origin of the asteroids had to have been a planet that was at least partially fluid. If such a planet exploded, objects like 'asteroid 2005 YU55' would be the result.

There is another asteroid that tells a very similar story. In this case asteroid Hektor appears as two basically-spherical balls being pushed together. It has sometimes been described as two potatoes stuck together. The pushed-together appearance of the asteroid is very suggestive that both spheres were not quite solid when they came together because the interface between them is quite large. As with the other asteroids they have very little gravity so the impact energy would have been very low. Therefore they could only have achieved their pushed-together shape if neither of them was solid. If they had already cooled down and become solid, coming together would have resulted in an object that appeared like two balls touching with a very small contact area at the interface.

Other examples confirm the semi-fluid assertion. Asteroid Eros has several very large impact marks in comparison to its size. All of them have a soft rounded appearance. Some of the impacting objects appear to have been quite large and have thrust themselves very deep into the asteroid's material. A solid object hitting a larger solid object of this size would not have entered it at all! It would have made some mark and just careened away. (522)

Asteroid Ida provides similar evidence of having had a historically-fluid state. It is oblong with the usual soft impact marks and rounded features. Even more compelling is Ida's moon Dactyl which is described as being egg-shaped with approximant dimensions of 1.0 x 0.87 x 0.75 miles (i.e. almost perfectly spherical) (445) and in some photographs it appears as a sphere. (446) This means that it must have been fluid when it was cast into space. In fact, the heating factor is suggested by the following; ' ... it has been altered by strong heating-evidence so far suggests that Ida was a piece of a larger object that has been severely heated.' (445)

Numerous other objects in the solar system have the appearance of having been liquid when they were formed. 'Recently we have become aware that many objects in the Solar System have a dumbbell shape. If we go to a lava field from an active volcano, we find volcanic 'bombs' which are shaped like dumbbells. These are produced from lava which has been ejected from the volcano.' (351) (Volcanic 'bombs' are blobs of material that fly through the air when a volcano explodes. They are called 'bombs' because they explode when they crash into the Earth.) This type of observation leaves very little doubt that the asteroids were formed from liquid material and the deep soft impact marks reinforces this conclusion.

Ceres is the largest of the asteroids and because it has been orbited recently by a spacecraft named Dawn and a great amount of information has become available. Several of the features of Ceres indicate a fluid history.

There are a significant number of large crater basins which are quite shallow. Many of them have a rebound mound at the center and most of them have a smooth flat appearance. The roundness and the smoothness indicate a molten state at the time of impact.

Vesta, the fourth largest asteroid, also provides evidence of having been fluid. 'During their formation, many planetary bodies in our solar system melted significantly, allowing denser materials to sink to their centers in a process known as differentiation. ... New findings published in the latest issue of the journal Nature suggest that for at least two of our solar system's major asteroids, melting was dramatic. ... The find suggests that both asteroids (that is Vesta and a second unnamed asteroid) experienced widespread melting with more than 50 percent of each object becoming liquid.' (535)

There is another asteroid that also has clear evidence of having been molten when at least two impacts occurred. The one impact is similar to some of the others already mentioned with a depression in the center and a smooth rounded rim. Down near the other end of this asteroid there is a second impact mark which is even more dramatic than the first one. The impact mark is there as expected but instead of a smooth rounded rim there is a frozen-in-space splash formation. Several globules are supported on small diameter stems. Apparently the impact occurred just as the splashed-up material was right at the congealing point. The very limited gravity of this asteroid was not able to pull the splashed-up material back down before it became too stiff. It is well understood that the gravity of a small object like an asteroid of that size would have exerted very little restoring force on the splashed up material. It would have been more like dropping something into a pot of glue because the pull of gravity would have been very week. Therefore the splashed-up material; just stayed up and has remained there to the present time as evidence that the asteroid was once molten. (536)

All of this evidence indicates that a planet that was mostly fluid exploded resulting in the formation of tens of millions of fluid and semi-fluid fragments of material of every-possible size and shape. All of these objects would later be designated as the asteroids. Suddenly there would have been millions of objects, large and small, drifting through the solar system whereas just a moment before there was a planet. This means that a large number of objects suddenly became available to impact the other planets and moons of the solar system. One is reminded of the 1178 impact of an asteroid with the Moon which was followed several months later by an asteroid shower on Earth. (107) That impact would have launched thousands of rocks of every size into space so if some of them were attracted by the gravity of the Earth it wouldn't be a surprise to anyone. Similarly with the sudden appearance of tens of millions of asteroids, asteroid showers on the planets and moons of the inner solar system would be expected during the following months and years.

6.5 Evidence of Asteroid Showers

There are two places in the near solar system that give evidence of an asteroid shower having taken place. If this is what happened at those places, it would be suggestive that it could have happened here on the Earth as well.

6.5.1   The Moon

The Moon shows evidence that an asteroid shower has happened. This evidence includes the maria with their associated mass concentrations as well as both the bulge and chaotic terrain on the far side.

There is no doubt whatsoever that the Moon has been pummelled by asteroids. Even a pair of binoculars or a low-power telescope will enable a person to clearly see many of the craters, each one of which is direct evidence of an asteroid impact. Even without binoculars the maria are visible. Of course they were also visible to the ancients some of whom gave them individual names. Maria is a word from Latin meaning 'seas' which suggests that at the time of the Romans, people were interested and even opinionated that the seemingly flat areas must be oceans. That was not an unrealistic idea because for a short time they were oceans but just not water oceans - magma oceans. The maria are virtually void of impact marks which means they cannot be very old or they would be riddled with impact marks like the rest of the surface of the Moon. Asteroids continue to hit the Moon regularly. 'In February, (2014) astronomers announced that a huge asteroid with a mass of about 400 kilograms (882 pounds) smashed into the Moon last autumn, producing a flash confirmed to be the brightest ever observed from a lunar impact. Another huge impact was recorded six months prior.' (88) Also, in relatively recent history an even larger asteroid impacted the Moon and caused a crater about 13 miles in diameter to form. (107) This happened in 1178 when a group of monks decided to observe the Moon one evening. To their credit they did this without the aid of a telescope but they were well rewarded for their efforts. They reported that there was a bright new Moon, and its horns were tilted toward the east when suddenly the upper horn split in two. From the midpoint of this division a flaming torch sprang up, spewing out, over a considerable distance, fire, hot coals, and sparks. Meanwhile the body of the Moon which was below writhed, as it were, in anxiety ... the Moon throbbed like a wounded snake. ... This phenomenon was repeated a dozen times or more, the flame assuming various twisting shapes at random and then returning to normal. Then after these transformations, the Moon from horn to horn - that is along its entire length - took on a blackish appearance. The Moon was just 1.6 days past the new Moon. (107) The crater that was formed was named in 1970 after Giordano Bruno, a philosopher of the 1500's. There have very likely been other impacts in recent history but if no one was watching at the time when they happened they would not be noticed and would go unrecorded.

The only way to avoid the recency conclusion is for the maria to have been constructed of some type of fluid material that allowed impacts to happen over a period of many years without leaving any trace. In such a case, an impacting object would disappear below the surface and the surface material would reposition itself so that there would be no permanent sign of an impact. However in order for this type of scenario to happen, the material must stay in the fluid state for thousands of years which is not very likely. Material like this does not exist on Earth (water excepted) and it is not very likely that it exists anyplace else either. The obvious conclusion is that the maria were formed recently from molten material from the interior. In fact, it was confirmed during the Apollo visits that the maria consist of volcanic material which would have hardened within a short period of time (111) and there has not been enough time since then for impacts of any significant size or quantity to have made any impressions in them. This immediately implies that all of these features were formed recently and at nearly the same time. The fact that the maria only exist on the near side (125) also implies that the maria-forming objects came in quickly before the Moon could rotate and spread them out over the whole surface. Further, if they were of significantly different ages the maria would show various numbers of impacts. As it stands, none of them show very many impacts at all!

Associated with every one of the maria, but one, are concentrations of mass (i.e. mascons) deep below the surface. (This characteristic of the Moon was discovered by an orbiting satellite that was checking out the Moon prior to sending the first Apollo spacecraft there. (126)) Is it just coincidental that the mass concentrations are associated with the maria? It is more likely that they consist of material that was not originally part of the Moon at all and that they are the remains of impacting asteroids which consisted of higher density material than the Moon and which landed with enough force to break through into a region that was either molten or semi-molten. (Coincidentally, three huge impact craters on Mars including Utopia, Argyre and Isidis also have mascons beneath the surface. (126)) If the interior of the Moon had been completely molten, the impacting objects, being heavier, would have descended down through that molten material all the way to the center of the Moon and would not now be distinguishable or cause the Moon to have a center of mass which was different from its geological center. As it is, they stopped short of the center, are spread out and separated from the center by enough distance to cause the Moon to be gravitationally lop-sided. This, in fact, is one of the lunar features that keeps one side of the Moon facing the Earth all the time. The other factor is the bulge on the far side.

The far side of the Moon is bulged away from the Earth. 'Since the Apollo 15 laser altimeter experiment, scientists have known that a region on the lunar far side is the highest place on the Moon.' (125) Attempts have been made to explain the bulge by tidal effects. If this was the case however, one would expect that there would be a bulge on the near side as well. However there isn't any bulge on the near side. Also the far side is said to consist of chaotic terrain. The surface is all broken up and appears as if it has been totally fractured. This does not present a mystery if we recall that whenever a large object crashes into the surface of a planet or a moon it will produce both an impact mark as well as a shockwave which will travel right through to the other side. With a partially-molten interior as witnessed by the maria themselves, the shockwave would have come up under the solid part of the Moon near the surface on the far side and thrust it upwards. The surface material would have become broken and would have remained broken as it is observed today. Any bulge that was formed by these shockwaves would generally tend to stay in place because of the tidal effect of the Earth. The tidal effect basically means that the pull of Earth's gravity on the Moon is slightly less for the material on the far side than it is for material on the near side. A bulge covered by fractured blocks of material would not have been subject to any restorative force so it would just remain as a bulge. The situation would be basically similar to a person swinging a ball around on a rope. If the ball was the least bit flexible one would expect that the rope side would be stretched towards the person swinging the ball and the far side would be slightly bulged in the other direction. The lunar situation is very similar to what has apparently happened on Mars as well, where the massive Tharsus Bulge rises 6 ½ miles above the average Martian surface and occurs exactly opposite all of the impact activity on the other side. (131) (see discussion of Mars in 6,5,2 following)

The maria and their associated mascons are all on one side of the Moon. (85) This implies that the large maria-forming impacts happened within a relatively short period of time before the Moon could rotate and spread them out. Even at the present time the Moon only rotates once a month and there isn't any evidence to suggest that it ever rotated any faster. In fact since the current rotation is so slow and is caused by the gravity of the Earth acting on the lack of symmetry of the Moon's material, it is quite probable that the Moon never actually rotated prior to being struck. It could quite readily have just orbited the Earth without rotating at all. For any Earth-dwellers at that time there would have been the added benefit of having been able to observe the entire surface of the Moon. The Moon's current libration supports this conclusion. Libration is when a body like the Moon rocks back and forth slightly as it orbits its host planet. (86) This rocking action enables slightly more than 180 degrees of the Moon's surface to be observed. If the Moon had always rotated at one revolution per month, libratory action would not be expected. On the other hand if the Moon had originally rotated considerably faster, becoming gravitationally unsymmetrical wouldn't necessarily cause it to decelerate to one revolution per month.

Certain commentators have explicitly stated that the Moon has been struck by an asteroid shower. The non-uniform cratering of Mars, Venus and Mercury as well as the Moon suggests a catastrophe such as a giant swarm of asteroids moving through the solar system. (108) The same author suggested that there were more impact marks on the far side of the Moon than there are on the near side. (109) However, this ignores the maria effect. If the maria were caused by the near-simultaneous impact of large asteroids and the interior was fluid enough to allow molten material to come to the surface as a result, evidence of other impacts would have been obliterated. The largest scars on the Moon are the impact basins which range in size up to about 2500 km (1600 m) across. The basin floors have been flooded with lava soon after the titanic collisions that formed them. The dark lava flows are what the eye discerns as maria. The wrinkled ridges, domed hills and fissures mark the maria. All of these familiar landmarks indicate volcanic activity. On the Moon there are no mountains like there is on the Earth. Lunar mountains consist of volcanic domes and the central peaks and rims of impact craters. (111) The maria are extensive in area and occupy a very large percentage of the area of the near side. Numerous impacts could have happened on the near side and there would be no remaining evidence because the molten material that welled up creating the maria would have engulfed them. While there are no maria on the far side of the Moon, neither are there any of the mascons below the surface that accompany all but one of the maria on the near side. All of this is suggestive that a swarm of large asteroids hit the Moon before it could present a different surface to spread the impacts out.

If the Moon had been struck by a small swarm of large asteroids, the maria would have been formed and the mascons would have been placed. The shockwaves from the impacts could have formed the bulge on the far side. These two features – the mascons on the near side and the bulge on the far side – would have caused one side of the Moon to continually face the Earth. This means that the Moon would then be rotating (at one revolution every month) and the far side would also have become exposed to incoming asteroids. (This assumes that they were coming from basically one direction in space as would quite reasonably be expected because they probably came in from beyond Mars and would all have been going in the same direction.) Consequently, over the following few months the entire Moon could have been pummelled by asteroids which would have been an asteroid shower indeed!

In support of this contention, it is also a fact of science that the Moon has been impacted a great number of times by smaller objects. There are approximately 200,000 craters with diameters greater than one kilometer on the Moon. (110) One can only wonder how many there would be if the maria had not covered any of them!

6.5.2   Mars

Mars also bears evidence of having been pummelled by a shower of asteroids. The study of both the Moon and Mars is facilitated somewhat by the absence of both water and vegetation. The surface, in both cases, is plainly visible – as long as there isn't a sandstorm in progress as Mars seems to suffer quite often. In both cases we are looking directly at bare rock or some type of soil and the view is unrestricted. By contrast the Earth is almost three-quarters covered by water and the ocean floor is not the least bit visible. In fact it is hard to see even by viewing it through the port of a submersible because only a very small area can be seen at any one location. There is also the question of silt and since the ocean is a reservoir, any silt that enters will simply make its way to the bottom and stay there being relocated only by currents as the water flows along. The currents are invariably quite slow which means that relocation of material is not hasty. On Mars since there is a very thin atmosphere and storms do happen some surface material will relocate but since the atmosphere is so thin the relocation cannot compare to the shifting sands of the Sahara Desert in Africa. Even though the winds do blow across the Sahara, the absence of vegetation enables the surface to be viewed quite well from high elevations. With other parts of Africa we are not so fortunate and the lush vegetation masks the surface making identification of surface features, like asteroid craters, difficult. The Moon has none of these limitations and the only relocation of surface material that can occur is when an object strikes the surface and propels some surface material away. Surface disturbance could also be caused by molten material coming to the surface but one would expect that only a major impact could cause this to happen. In any event the surface of both Mars and the Moon can be seen quite clearly from an elevated viewing location. While Mars has been viewed from afar for many years, it is only quite recently that good close-up pictures have been taken and the surface has now been repeatedly photographed so a great amount of detail has been observed.

Mars has suffered a violent history. The devastation is massive and widespread. There isn't basically any location on Mars that has avoided major trauma. One side of Mars could be called the 'impact' side. The other side could be called 'the results-of-impact side'. The impact side has been hit by thousands of asteroids and over three thousand have left craters that are more than twenty miles across. (76) Also on the impact side there are three very large craters that indicate very massive objects hit that side as well. (77) Further, 'It may be properly asked if, among the 15 largest of the Martian craters, any of them are in the opposite hemisphere. No, they are all in the same hemisphere.' (77) While the opposite side isn't totally void of impact features there are only about three hundred that are over twenty miles in diameter.

As Mars orbits the Sun it rotates on its own axis. In fact it rotates at about the same speed as the Earth with a day on Mars being only a little over one-half hour longer than a day on Earth. In recognition of the concentration of impacts on one side, this suggests that Mars suffered a shower of asteroids that came in so quickly that it didn't have time to rotate to spread the impacts out over the entire surface. The only alternative conclusion is that they came in groups with about a twelve hour lull in between each group. Any suggestion such as this has almost no credibility but acknowledging that they came in one day is very alarming. The collective impulse from all of the impacts would have modified the orbit of Mars. It is curious on this point that Mars has a very elliptical orbit which could partially be explained if the impulses from all of the impacts added up in one direction (which they would if the asteroids came as a shower).

Opposite the impact side there is a scene of utter chaos and destruction. Monster volcanoes have boiled up. A massive canyon has been opened. The surface has been disturbed so violently that it is just a jumbled mass of broken blocks and chaotic terrain over a wide area. Something caused all of this mayhem and it is more than curious that it has happened near the collective antipode of the three huge craters on the other side. Every one of the impacting objects that created those craters would have generated a shockwave that would have travelled through the interior and surged up under the far side. The crust would have been blown open as the surface layers were hurled upwards with unusual force. The crust split open and didn't completely close up again leaving a trench some four thousand kilometres long (2500 m) complete with extension trenches off of both sides. When the surface material settled back down, the massive crack (or fault) did not completely reclose leaving this long, deep trench. (Large impacting objects would have increased the total amount of Mars' material so the original surface would not have been adequate and an open crack would be expected.) In addition, volcanoes boiled up. The largest one is called Olympus Mons and it rises more than thirty-seven kilometres (23 miles) above the surface. There are several others that, by any comparison, are monsters as well and they number more than a dozen. The largest are of these huge volcanoes are; Ascraeus Mons, Pavonis Mons, Aris Mons, Tharsus Tholus, Uranus Patera and Uranus Tholus. Curiously all of them are located directly on or immediately beside a great uplifted region called the Tharsis Bulge. 'The Tharsis Bulge is a large shield or uplift area in the opposite Hemisphere. (i.e. opposite the impact side) It is approximately 6 ½ miles high in the center relative to the general crust of Mars and it is about 3,000 miles in diameter on that small planet; it crosses 90 degrees of latitude.' (131) Exceptional turmoil in the interior of Mars produced all of this chaos which must have had a cause. The only reasonable conclusion is that Mars was hit by a shower of asteroids and it did not have time to rotate to spread the impacts more evenly over the entire surface. 'The three largest impact asteroids, Hellas, Isidis and Argyre, broke open the crust of Mars so wide that ... extrusions of magma resulted. And within a period of about 25 minutes, and in addition to the massive, surging tides of internal magma, Mars received scattershot blasts by about 2800 impact asteroids which were over 15 miles in diameter plus 10,000 more of a smaller diameter. But above all, the impact of Hellas, ... must have been devastating to the innards of Mars. So immense were the pressure waves that ... the Tharsus Bulge resulted ...' (132)

Mars has mascons similar to the Moon. (126) Three mascons have been identified and they are associated with the Isidis, Argyre and Utopia Basins. Hellas, the largest, is not mentioned. In order to be identified, a mass concentration must not be at the center of its host body. It has to be somewhat away from the center or there would not be any way to recognize it as being anything other than just part of the planet.

Isidis and Argyre were monster impacts. The respective diameters of their basins (or craters) are given as 684 miles for Isidis and 481 miles for Argyre. The respective impacting asteroid diameters would not have been larger but they could have been a very significant fraction of these basin diameters. This type of assertion defies the usual wisdom that an asteroid diameter would probably only be about 10% of its respective crater diameter which type of relationship developed from observing the respective dimensions of small craters. However the circumstances for large impacts are dramatically different from small impacts and this relates to the construction of the planet.

The monster Hellas crater on Mars is variously given as being approximately 1000 miles in diameter (138) or 1300 miles in diameter (140) This is even bigger than the Caloris Basin on Mercury which is about 963 miles in diameter. (139) The clue to the nature of the interior of Mars is given by the shape of the Hellas crater. While the average elevation is below the general surface of Mars by about 4 km (2.5 m) (140) the bottom of the crater includes 'diverse landforms, some of which appear volcanic in origin (this means that the basin filled with melt soon after the impact event). Among features believed volcanic in origin are linear ridges similar to the wrinkle ridges found in lunar maria.' (140) The main point of interest is that the bottom does not have a crater shape at all but basically follows the general curvature of Mars' surface. This means that the impacting object punched right through the crust of Mars into an interior region that was molten. The material forming the floor of the crater then resumed the general shape of the planet simply due to gravity. While there might have been broken pieces of both crust and asteroid in the mix, generally it was flexible enough and fluid enough to respond to Mars gravity and minimize its potential energy by resuming the general curvature of the planet prior to solidifying.

At 1300 miles in diameter the Hellas impactor would have had a diameter that was approximately ten times as great as the huge Chicxulub impactor. Its volume, and hence its mass and energy, would have been about 1000 times greater. However, Mars has only about 11% of the mass of the Earth (141) so the effects it would have had on Mars would have been totally overwhelming. The great size and energy in comparison to the size of the planet taken together with the absence of a mascon would suggest that even though the interior of Mars might not have totally consisted of low-viscosity material, the Hellas impactor probably plunged all the way to the center anyway and just stayed there. The other large impactors – in particular Isidis and Argyre, being monsters in their own right - might not have made it to the center or possibly some portion of them broke off and remained identifiable as mascons while the rest sank right down to the center. Since the force of gravity reduces the further one travels into the interior of a planet, it might have taken some time to get all the way down. In any event, Hellas does not have an associated mascon like both Isidis and Argyre.

The concentration of impact evidence on one side and the wide-spread evidence of destruction on the other side confirms that Mars suffered an asteroid shower.

6.5.3   Mercury

Other evidence from the solar system confirms that this same type of thing happens. Mercury has a huge impact site on one side called the Caloris Basin. It is about 800 miles across which is very large by any comparison but significantly large in Mercury's case as the planet is only 3000 miles in diameter. Evidence that the impacting object produced a shockwave which travelled right through the interior is provided by the chaotic terrain on the far side. The two features are related. Sometimes this chaotic terrain is referred to as 'weird terrain'. (78) The surface on the far side consists of massive broken blocks of material arranged in a totally chaotic way. Since the impacting object was so large in comparison to the planet, it is a wonder that it didn't totally blow out the other side. It is even possible that this is what happened. Some of those blocks could have been heaved upwards for several miles before crashing back down again and they have remained in a totally jumbled state right to the present time. Something certainly caused these features and suggesting that the asteroid that hit the far side is responsible is perfectly reasonable.

While Mercury does not have evidence of an asteroid shower having happened the monster impact on one side with wide-spread destruction on the other side is very similar to the situation on both the Moon and Mars where the large impacts hit one side while the other side recorded the destruction.

6.5.4   Ceres

As mentioned above, Ceres is the largest of the asteroids and because it has been orbited by a spacecraft a great amount of information has become available. There are a significant number of large crater basins which are quite shallow. Many of them have a rebound mound at the center and they also have a smooth flat appearance. The roundness and the smoothness indicate a molten state at the time of impact. Since Ceres could not have been molten for more than a short period of time, the consistency of the features of these marks means that they must all have happened within a short space of time. Only an asteroid shower could have done this.

6.5.5   The Earth

If asteroid impacts caused the formation of chaotic terrain on the opposite side to the impact area on both the Moon and Mars, it would be reasonable to expect something similar to have happened on the Earth. Opposite the Vredefort Crater in South Africa, the Laurentian Plateau consists of a more or less circular area of broken chaotic terrain approximately three million square miles in extent, but it doesn't bulge upwards at all. However, the lack of an upward bulge is what would be expected if the crust was relatively thin and there was a fluid interior. In such a case the crust would be thrust upwards by a pressure wave, fracture, and then settle back down and because of the low viscosity of the material below, the entire area would once again resume a generally spherical shape. Gravity would do this and if the crust was seriously broken everything would settle back down until all of the displacement forces were minimized. Since a bulge was retained on both the Moon and Mars, it is suggestive that the interior of these bodies might have a mixture of higher and lower viscosity regions. Displaced material always returns to it's minimum energy level if there is any way for this to happen.

A fluid interior would facilitate such a happening but a semi-fluid interior or a mixture of solid and liquid would result in a slightly different final positioning.

It must be noted at this time that Mars, Mercury and Earth all have a similar characteristic. Opposite Hellas, Argyre and Isisis on Mars the surface has been totally disrupted and is now recognized as chaotic terrain. Opposite the massive Caloris Basin on Mercury there is an area of chaotic terrain or 'weird terrain'. On Earth, almost directly opposite the huge Vredefort impact crater in South Africa we find the Laurentian Plateau. Since everything happens for a reason, associating massive impact sites and their mascons with chaotic terrain on the opposite side clearly satisfies the physics that would accompany impacts by unusually large asteroids with any planet or moon.

An example of a small crater on Earth is the Arizona Crater which is about 4000 ft. in diameter and about 570 ft. deep. (133) Since the material that has been displaced can be estimated, a calculation can be carried out to make a reasonable guess of the size of the impacting object. Of course the density and speed of the asteroid must be assumed but at least there are things that can be measured. A similar approach could be taken for much larger asteroids if the material of the planet was similar all the way down. In the Arizona case, the material is mostly soil with some rock. There is a case in Southern Ontario of an impact into solid rock. In this case the crater (the Brent Crater in Algonquin Park) was formed in solid rock but it has the same basic shape as the Arizona crater; 3 kilometers in diameter and 1 kilometer deep. (The difference in the ratios of the dimensions would relate to the difference in material. The unconsolidated soil of Arizona would be easier to push away than the solid rock of Algonquin.) Actually, in this solid rock case, there is the expected bowl shape but the fragmented rock completely fills it. It has been declared that the crater was formed, the material was ejected and then the sides slumped back down and filled the crater back in again. (3) However, this is not likely simply because the crater is virtually filled right to the brim with broken rock. If the sides had just collapsed there would be a void in the center but the elevation at the center is only slightly below the rim elevation so the mentioned explanation must be set aside. However the idea that a classical bowl shape would form from an impact is valid in the rock case, so a similar calculation can be carried out to make a reasonable estimate of the impacting object's energy and hence its size. However, this type of thinking cannot be extrapolated to large impacts.

One example of a large impact on Earth is Chicxulub at the northern end of the Yucatan Peninsula in Mexico. This crater is about half in the water and half on land. It is virtually a perfect circle with a diameter of 170 km. (106 m) (136) In keeping with the common wisdom, its size is declared to be 106 miles in diameter and 30 miles deep. (136) While the diameter is an observation the depth is speculation. The depth has never been measured and never will be measured because 30 miles is deeper than the solid crust of the Earth extends. This means that if an asteroid was large enough to form a crater that was 30 miles deep, it would have been large enough to punch right through the crust of the Earth and plunge on into the interior. The Chicxulub impact crater is further declared to have been formed by a comet that was about 15 km (9 ½ m) in diameter. (137) This is further speculation and relates directly to the diameter of the crater. The entire situation changes as soon as the impacting object has enough energy to punch right through the crust of the Earth because it would then generate an enormous shockwave that would continue on into the interior and manifest itself on the far side of the world. In fact the disturbance and chaos produced would be manifest all over the world but the far side would experience uplift and fracturing. Since the asteroid's energy would have been distributed so far and wide, the diameter of the crater is not indicative of the diameter of the asteroid in the conventional manner and it would be larger than conventional wisdom would suggest.

A situation like this could be compared to a low power rifle like a 22 being shot at a spherical boulder. The logical expectation would be that a small crater would form in the surface of the boulder and that it would recoil somewhat from the force of the impact. Now if we use the same 22 and shoot at an ostrich egg the bullet will not expectedly make a crater but it will punch through the shell and travel on into the interior. The shell on the far side will register a shock-wave. It might crack open. Either way the entire interior of the egg and hence the entire mass of the egg will experience trauma from the entry of the bullet into the interior.

6.6 Single-Impact Devastation

The destruction that would be caused by the impact of a single large asteroid with the Earth is so overpowering that it is hard to imagine. There would be numerous aspects to the destruction including, of course, great tsunamis that would be powerful enough to cross right over continents and even climb over mountains. (The height-reaching ability of any object is simply dependent on its forward speed.) These monster waves would encircle the Earth in all directions and there would soon be the complications that intersecting wave trains would cause. Numerous tsunamis would result from a single impact and they would result from three basic types of material displacement. First would be direct impact on water. Obviously great waves would be generated by this type of impact but waves would also result from impact on land far from any ocean. Earthquakes would develop from impacts anywhere but an impact on land would cause all of the shorelines of the impacted continent to vibrate, sending a train of tsunamis across every ocean. A third type of water wave would happen at the antipode. At this location on the Earth opposite to the impact site, the shockwave that was produced by the impact would come up under the crust of the Earth. The crust would be elevated. Elevation under water would raise the ocean up but it wouldn't stay up. It would settle back down again. However the water that was immediately above the raised-up area would flow down off of the temporary mound in all directions. Sheet flow would result with great layers of water rushing across the top of stationary water below towards all of the continents. Even more chaos would result when the various water movements intersected. How anything could survive such turmoil is difficult to understand! Chaotic water movement on a world-wide scale is clearly one of the types of destruction that an impacting asteroid would cause.

Another one is the flying-boulder hazard. If an asteroid was large enough and powerful enough to cause a scar in the crust of the Earth that was many miles across, we should expect as a side effect a shower of boulders that would be hurled into the air by the impact. Further it would be reasonable to expect that the material that was blown out of the crater would shower down over an area involving several crater diameters. In the case of the great Sudbury crater in Canada, boulders might be hurled for more than one thousand miles. Such boulders would be called Erratic in later years because they would be completely out of place and remote from their parent material. Erratic Boulders are found in very large quantities all over the world including on the peaks of mountains as well as on islands in both the Atlantic and Pacific Oceans. (80)

Another type of hazard would be the atmospheric pressure wave that would have been generated in the atmosphere as the asteroid hurtled through. It would be devastating to everything that was the least bit moveable - in particular animals and trees. The entire herd of Woolly Mammoths were very likely killed by a pressure wave because they died instantly with food still in their mouths. Also, their blood vessels ruptured under their skin and the blood pooled as they were hurled away. Further, the blood was dark red indicating that it was loaded with oxygen which the animal had not had time to remove. The entire herd died instantly along with rhinoceros, elephants and horses which are found buried with them. (81) Enormous numbers of animals died and were all buried together in a tangled mass which clearly indicates catastrophe. This is particularly evident along the north coast of Russia where the tusks of the Woolly Mammoth are currently being removed from melting permafrost in a 'fresh' state. Fresh in this case indicates that both burial and freezing happened before the ivory could degrade from the 'fresh' or recently-harvested state. Also, the burial material must have been in motion at the same time. Both of these factors are indicative of the violence of the entire affair and this is confirmed by the observation that the tusks are almost never found attached to the rest of the animal. They must have been broken off during the turmoil and carried with the burial material to be found many years later as the ground started to melt. One might suspect that the list of woes from an asteroid impact would now be complete but there are several more.

The next one of immediate interest is the cloud. An impacting asteroid would generate a cloud. This cloud would rise from the impact site – whether it was land or sea and would climb very high into the stratosphere. Then it would spread out. Within days it would envelop the entire world causing total darkness which would persist for months. (79) This type of development is expected from a single impact and could make the impact of a single large asteroid the terminal event for life on the Earth. There is no doubt that globe-encircling waves would be a threat to all forms of animal life and there is no doubt that massive world-wide earthquakes would be much more than unpleasant but hazards to life such as these would have a definite life-span which undoubtedly could last for weeks but would sooner or later terminate. It is expected that the results of world-wide darkness, which lasted for several months or even longer would not terminate when the cloud dissipated but remain indefinitely because the greenhouse effect would have been lost.

The greenhouse gas inventory would have been depleted by a world-enveloping cloud that lasted for months but depleting the greenhouse gas inventory can only be done at the peril of keeping the Earth habitable!

6.7 Positive feedback Loops

6.7.1 Introduction

A positive feedback loop is simply a situation where the results of some action cause that action to continue and increase. This is sometimes referred to as a vicious cycle. If the results of some activity cause that activity to continue and build up independent of the original cause, we have a positive feedback loop or a viscous cycle in operation. If anything upsets the greenhouse effect (on a world-wide basis) we clearly have a serious situation confronting us. Further, if the greenhouse effect was upset there is a possibility that various positive-feedback loops would come into operation and aggravate the situation even more. If this should happen, the end result could be much more exaggerated than the initial upset might suggest. Several possible feedback loops will be discussed but the bottom line is that we are not really interested in any of them because temperature stability is so paramount to our existence here on the Earth.

The greenhouse gas inventory of the Earth has been slowly changing for the last few hundreds of years. While this is well known, as long as there weren't any serious implications involved, there would not be any need for concern. However numerous scientists now recognize that there are implications and that they are very alarming.

The average surface temperature of the Earth is computed after collecting temperature information from all around the world and then processing it in a way that will give a meaningful reading. This has been done now for many years but recently the temperature has been measured to be going up. An increase of a few degrees would not seem, to the casual observer, to be any cause for alarm but numerous scientists are greatly alarmed which means that the situation deserves more serious attention. The average surface temperature of the Earth could increase directly as the result of an increase in CO_2, an important greenhouse gas, but it is the indirect developments which would accelerate any temperature change that are the cause for a higher level of alarm. Taken collectively, the indirect developments would have the effect of magnifying or amplifying the direct development causing the initial change to have several times the effect that the CO_2 increase (i.e. the direct development) would have had on its own. These indirect developments are called positive feedback loops and there are four that are of immediate interest.

6.7.2   Ocean CO2

There is a great amount of CO_2 in the ocean. It has dissolved and is being kept there simply because of temperature. Cold water holds more CO_2 than warm water. (82) If there is an overall increase in the temperature of the world as a whole, the temperature of the water in the ocean will increase. Currently the average temperature of ocean water is very close to freezing and it is only the surface layers near the equator that could really be called 'warm'. (83) An increase in the temperature of the ocean would actually have two effects of immediate interest, the first of which is expansion. Warm water requires more room. In fact a portion of the recent increase in sea level is being attributed to expansion which of course is a cause of great concern for communities that are already close to sea level. The Maldives, a group of islands in the Indian Ocean, come into this category.

The second effect is the reduced ability of warm water to retain its CO_2 inventory. Therefore a warming ocean will release a portion of the CO_2 dissolved in it into the atmosphere. Greater atmospheric CO_2 was the cause of alarm in the first place and putting even more of it into the atmosphere would only exaggerate the situation further. If the original increase in CO2 caused an increase in temperature, then putting in more would only do the same. Then, this further increase would do the same thing again. There would be no end in sight until a further increase in the temperature of the ocean did not result in any further increase in atmospheric warming. Until that level was reached there would be a positive feedback loop in operation without any apparent way to interrupt it.

6.7.3 Pack Ice

Pack Ice is the ice that floats on the water in the high arctic. As long as this ice covers the water surface most of the heat from the Sun will simply reflect off of it and have very little effect on raising the water's temperature. The reflectivity of a surface is called the albedo and ice has a reasonably high albedo reflecting about 80% of incident solar energy. This situation changes dramatically where there is open water. Open water has a much lower albedo of only about 20% which means that 80% of incoming solar energy is absorbed. This is reasonably comparable to the situation when the Sun shines on a snow-covered roof. At first the Sun shines and nothing seems to happen. Then a little patch of bare roof appears. From this first little patch the rest of the snow melts as the patch expands more and more. This happens because the temperature of the roof on which the sun is shining directly is higher than where the roof is still covered by snow. Quite naturally the snow will melt more quickly around the bare patch and continue melting until the entire roof is snow-free.

The same thing is happening in the high Arctic at the present time as the pack ice melts more and more every year with the expectation that all of the old pack ice will be gone within a few more years. This will not cause any increase in ocean level directly but by allowing the ocean to absorb more heat it will augment the general increase in ocean temperature so the overall expansion of the water in the ocean will be augmented. Energy absorption would raise the temperature where the temperature of an ice-covered ocean would not be expected to increase at all. Consequently, if the temperature of the Earth increased, some pack ice would be expected to melt. More uncovered ocean water would allow more solar energy to be absorbed with the expected result of increasing the water temperature causing even more ice to melt. This is another type of positive feedback loop in operation. The end effect of this procedure will not be as dramatic as the general increase in ocean temperature because there is only a limited amount of pack ice. When its gone its gone and any related positive feedback loop will peter out. On a world-wide basis the effect is not very dramatic but where the Arctic Ocean is concerned it is of considerable importance because the pack ice is expected to completely disappear within another few years. This has practical implications including the use of the Arctic Ocean for international shipping as well as easier access to the mineral wealth of the Arctic Ocean that is currently covered by ice. There will also be easier access to the mineral wealth on the high arctic islands which will be of direct benefit to Canada.

6.7.4 Methane

Methane is also a greenhouse gas and it is about twenty-five times as effective as CO2 in retaining atmospheric heat. The total impact however is still quite small but never-the-less is increasing rapidly since the great Russian bogs started to melt in 2006. (84) Vast tonnages of methane are now entering the atmosphere every year and this is not welcome news. The muskeg bogs of North America are also contributing methane to the atmosphere. While the influence of methane is still relatively small on a world-wide scale this is another example of a positive feedback loop in operation. The methane released into the atmosphere increases the greenhouse gas inventory thereby causing the Earth to warm up slightly. The warming trend in turn causes more bogs to melt releasing still more gas. This particular viscous cycle will continue until there isn't any more methane to be released. Then it will stop. While the world-wide effectiveness of methane as a greenhouse gas is admittedly small, it comes at a time when other positive feedback loops are also operating. In other words, it couldn't come at a worse time.

6.7.5   Water Vapour

Water vapour in the air is one of our most important greenhouse gases. CO_2 and water vapour together constitute the bulk of the greenhouse gas inventory. The over-riding problem involved in having water vapour as a greenhouse gas is that the amount of water vapour in the air is determined solely by temperature. This might be intuitively obvious to most people from everyday experience because it is well recognized that as the temperature goes up there is always more humidity in the air. While a parcel of air might be at the 100% relative humidity level, if the temperature of that parcel is increased, the relative humidity will drop. The same amount of moisture might still be present but the warmer temperature would enable the air to simply hold more. As a greenhouse gas, water vapour is a troublesome factor and one that will, as it increases, along with increasing CO_2, cause the average surface temperature of the Earth to eventually spiral up out of control. This particular cycle would only peter out when any further increase in temperature of the air did not result in any further increase in its water vapour content. Unfortunately this level would not be reached until the temperature was well above the comfort level and well above the level at which seeds could germinate. This is, of course, very unwelcome news. The only mitigating factor is that the atmosphere is very large and it takes time to bring about significant changes of this nature.

6.7.6   Summary

Several positive feedback loops have been mentioned and they all have the effect of aggravating an undesirable situation. Ocean CO_2 is held in the ocean by the temperature of the water. As it warms CO_2 will be released into the atmosphere. There is an enormous amount of CO_2 in the ocean and if it all came out the greenhouse gas inventory would be well above an acceptable level. As a contributor to the warming trend, pack ice is not nearly as effective as ocean CO_2. In fact, compared to just a few years ago, there is very little old pack ice remaining and even it will also be gone within another few years.

As a greenhouse factor, Methane will probably be more influential in the long run because there is a very large supply of it remaining. It comes out of the ground as the permafrost melts. While a great amount of it has melted in recent years there is still much more remaining to melt and release its methane load.

The temperature within the permafrost has been measured in several places and the results of these measurements provide another indicator that the Earth is indeed warming up. In fact it is warming up in a way that it has never warmed up before. Otherwise the land formations that are currently held in place because they are frozen would have been washed away long ago. In addition the tusks of the great Woolly Mammoth would not be available as 'fresh'. They would have simply degenerated the same as everything else degenerates if it is left either buried or in the open air for any significant period of time.

It will be a tossup whether water vapour or ocean CO2 will ultimately have the most influence on the warming trend. Of course there is a virtually unlimited supply of water to make water vapour but there is also a very large supply of CO2 in the ocean. Any factor that increases the greenhouse gas inventory is worry-some but those that do not seem to have an upper limit are the most worry-some and both ocean CO2 and water vapour are in this category.

While these positive-feedback loops have been discussed from the viewpoint of over-heating they could also be discussed from the viewpoint of over-cooling. It is most undesirable to drift into either of these directions as the situation, whether it is over-heating or over-cooling can easily be aggravated by these effects. However we note that the greenhouse effect has been maintained right to the present time with the result that the Earth is still habitable. A single asteroid hit, and by association widely time-spaced hits, would have produced asteroid winters which would have destroyed the greenhouse effect every time. The positive-feedback loop effects that would have accompanied its destruction is seriously suggestive that the arrival of the asteroids was a singular affair and not a series of widely time-spaced repeated events as is popularly believed.

6.8 One Mass Extinction

This particular topic has been discussed in section 3,3,4,3 above but rather than refer the reader back to that section it will be repeated here because of its relevance to the present topic.

Even small asteroids are feared much more than their size would suggest. The impacts of objects larger than about 1 kilometer in diameter would be large enough to cause climate change that could kill most inhabitants of the planet but not cause mass extinctions. (525) (Clearly, this commentator didn't equate 'kill most inhabitants' with 'mass extinction' but he should have.) Asteroids which are over one kilometer in size, (and therefore potential causes of mass extinctions) are expected to strike about once every few million years. (314) If the generalization that there is a ten to one ratio between asteroid diameter and crater size, (which is probably reasonable for small asteroids at least) the evidence included in the Earth Impact Database (EID) indicates that the Earth has experienced at least 70 impacts involving objects greater than one kilometer diameter. (2)

With respect to the above declarations, an estimate can be carried out to determine how many animals would survive the repeated impacts that the Earth has endured. For the sake of discussion, if it could be allowed that there would conservatively be only an 80% extinction accompanying all asteroid impacts which produced craters over 250 km. in diameter, there would have been four (reading from the EID (2)) 80% extinctions.

Further, if a 50% extinction accompanied all asteroid impacts which produced craters between 150 and 250 km. in diameter, there would have been one more mass extinction. (This is conservative in consideration of the 80% quoted above.) If all eleven of the craters between 50 and 150 km. diameter indicate 30% extinctions and the 17, between 25 and 50 km. diameter indicate 15% extinctions, the estimate can be carried out. The Manson, Iowa, crater is in this category at 35 km. diameter, and is considered to have been formed by 'a stony meteorite about two kilometers in diameter.' (312) 'An electromagnetic pulse moved away from the point of impact at nearly the speed of light, and instantly ignited anything that would burn within approximately 130 miles of the impact (most of Iowa). The shock wave toppled trees up to 300 miles away…and probably killed animals within about 650 miles.' (311) This description, while consistent with others, does not mention any fatalities resulting from ongoing effects such as the dust cloud. Also, it is highly probable that there are crater sites which have not yet been discovered. Further, the Earth Impact Database indicates that there are at least 40 more with diameters between 5 and 10 km. The following calculation will not include any entry to recognize these last two factors, which would only make things worse.

The oceans of the world account for approximately 70% of its surface area. Therefore it would be reasonable to expect that impacts would also have occurred in the oceans. In particular, a crater was recently discovered under the Indian Ocean. The Burkle Crater is only 18 miles in diameter and is well buried below the bottom of the sea but it makes one wonder how many more there are on the ocean floor. (309) Further, the ocean floor includes numerous areas called Igneous Provinces, many of which are much larger than the largest craters so far discovered. These are places where molten lava from the interior has come up and spread around. (526) Do any of these provinces cover an impact site?

While any of these additional potential impacts would also have resulted in much destruction, the following calculation will ignore them. The portion of surviving species from an 80% extinction is 20%. Similarly from a 50% extinction the surviving species are also 50%. From 30%, 70% remains and from 15%, 85% remains.

Therefore the calculation appears as follows.

20% (Tanitz Basin, Kazakhstan, 350 km.)
20% (Sudbury, Canada, 250 km.)
20% (Vredefort, South Africa, 300 km.)
20% (Czechoslovakia, 320 km.)

These four events would therefore leave 0.2 x 0.2 x 0.2 x 0.2 = 0.0016 or 0.16%. This means that only a fraction of 1% of all animal life on the Earth would be left as a result of the four largest impacts. Between 50 and 150 km. diameter there is only one event. Therefore from the above assumption of 50% survival for this category, only 50% of all animal life would be left after the asteroid in this category hit the Earth.

50% (Chicxulub, Mexico, 170 km.)

It is noted that this particular event is credited with completely wiping out all dinosaur life. Giant waves rolled over shorelines, and fine dust blotted out the Sun casting the world into darkness. Most scientists now accept that the Chiculub impact caused a mass extinction, which included the dinosaurs. (308) The 50% being used for the calculation is therefore very conservative, as mentioned.

There are twelve craters between 50 and 150 km. diameter, to which a 30% extinction was assigned. Therefore, for this part of the calculation we have the following.

70% (Acraman, Australia, 90 km.)
70% (Beaverhead, USA, 60 km.)
70% (Charlevoix, Quebec, 54 km.)
70% (Chesapeake Bay, 90 km.)
70% (Kara, Russia, 65 km.)
70% (Kara-Kul, Tajikstan, 52 km.)
70% (Morokweng, South Africa, 70 km.)
70% (Puchezh-Kalunki, Russia, 80 km.)
70% (Tookoonooka, Australia, 55 km.)
70% (Manitouigan, Quebec, 100 km.)
70% (Popigai, Russia, 100 km.)
70% (Ries, S. Bavaria, 80 km.)

These twelve events only leave 0.7 x 0.7 x 0.7 x 0.7 x 0.7 x 0.7 x 0.7 x 0.7 x 0.7 x 0.7 x 0.7 x 0.7 = 0.0138 which is only 1.38% of all animal life remaining. In a similar manner, the 18 impact craters between 25 and 50 km. would only leave a small portion surviving.

There are also eighteen craters between 25 and 50 km in diameter.

85% (Araguainha, Brazil, 40 km.)
85% (Carswell, Canada, 39 km.)
85% (Clearwater East, Canada, 26 km.)
85% (Clearwater West, Canada, 36 km.)
85% (Kamensk, Russia, 25 km.)
85% (Keurusselka, Finland, 30 km.)
85% (Manson, USA, 35 km.)
85% (Mistastin, Canada, 28 km.)
85% (Mjoinir, Norway, 40 km.)
85% (Montagnais, Canada, 45 km.)
85% (Saint Martin, Canada, 40 km.)
85% (Shoemaker, Australia, 30 km.)
85% (Slate Islands, Canada, 30 km.)

85% (Steen River, Canada, 25 km.)
85% (Strangways, Australia, 25 km.)
85% (Woodleigh, Australia, 40 km.)
85% (Yarrabubba, Australia, 30 km.)
85% (Sahara, Eygpt, 31 km.)

Therefore, these 18 events would only leave 5.28% of all animal life remaining.

When all of the above events are considered together, only 5.28% of 1.38% of 50% of 0.16% would remain. This is approximately 0.00005%. While this does seem pretty pessimistic there is widespread recognition that by far the vast majority of creatures that once lived are no longer living. No one knows how many species of organisms have existed since life began. Whatever the actual total, 99.99 percent of all species that have ever lived are no longer with us. (538)(Of course if the higher number was used, 99.99999% lost or 0.00001% remaining would be relevant.)

In everyday language this is less than one ten-thousandth of the original number of creatures. If this calculation is one thousand times too pessimistic for whatever reason, it would still leave only 0.05%. Since there were at least four impacts even greater than Chicxulub, the above calculation does seem to line up with the following comment. 'Every 300,000 years, a one-to-two-kilometer-wide asteroid crashes, initiating a short-duration global winter. Every 100 million years, a bigger asteroid hits, producing the world-wide calamities that have punctuated evolutionary history. There are 14 known mass extinctions in Earth's geological record.' (27) One is inclined to ask how many times this sort of thing can happen and still have anything left. Of course, a different set of assumptions could be made and a different result would be obtained. It could be argued that some of these impacts occurred prior to any animals being on the Earth. One commentator offsets this idea by declaring that 'evidence ....gathered from around the world related to earlier (i.e. earlier than Chicxulub) mass-extinction episodes, 230, 365, and 445 million years ago.' (11) On the other hand, even more impacts than those listed above were probably involved because it is presumptuous to think that all impact sites are currently known. In particular, the Earth is mostly covered by water. While a few craters have been found underwater, why wouldn't there be even more underwater than there are on land? In addition, more craters will certainly be discovered on land. In particular, the Chiczulub Crater and the Chesapeake Bay Crater are very recent discoveries. Unknown sites, like these two, might be so completely buried that they will not be found until some totally unrelated activity leads to their discovery, even though they may be as large as Chicxulub. No matter how many craters were actually involved in mass extinctions, the fact will still remain that following every major impact there would be very few, if any, survivors. Then, when the next impact came along, only a small percentage of that number would be left and so on. The so-far-unmentioned problem is that the surviving cohort would not include any large animals but would be dominated by vermin and other small creatures which, simply because of their size, were better able to avoid destruction.

Since the Earth has experienced several impacts, which are in the very large category, life must have been extinguished or virtually extinguished, several times. This is suggestive that life did not really result from evolution because there would not have been enough time for a diverse range of species to have evolved prior to the first major impact. Then, starting with the surviving cohort from the first impact, another diversity of species had to be in place prior to the second impact and so on. Mass extinctions every 50 or 100 million years would have made it very difficult for evolution to proceed properly, partly because, after every extinction event, a different mixture of survivors would have been the starting point for the next period. On the other hand, if life on Earth was originally created, the setback of every extinction would have necessitated virtually total creation all over again, repeatedly. Whatever explanation is accepted must account for the present diversity of life on the Earth. In fact it must account for the known diversity which was even greater in the immediate past because of the numerous species that are known to have recently become extinct.

There are several very large impact sites on the Earth. With each of these impacts, the "Window of Life" would have closed, if not from the impact events directly, at least from the subsequent post-impact dust cloud. Could it have closed repeatedly? The probability that a diversity of life could have spontaneously developed prior to the first impact, is mathematically virtually zero, so the possibility that it could redevelop after subsequent impacts is just another zero. Any declaration that life repeatedly redeveloped into a multitude of different species from a few vermin-like and other stragglers, is preposterous. Life is just too complex. Could a great diversity of life forms have been repeatedly created? While this certainly is not very likely, neither is there any evidence to support such a hypothesis. Therefore it is concluded that; THERE HAS ONLY BEEN ONE MASS EXTINCTION OF LIFE ON EARTH.

6.9   Greenhouse Effect Maintenance

The greenhouse effect is increasingly being recognized as a necessary 'window of life'. The Earth absolutely must have the greenhouse effect in operation or it simply would not be habitable. Temperature control is paramount to our survival and if the temperature at the surface of the Earth drifted either way from its present level of about 15C (59F), the Earth would become increasingly less hospitable. It follows immediately that any idea or theory that is put forward trying to explain what has happened on the Earth absolutely must recognize the necessity of keeping the greenhouse effect in operation. This includes the idea that the Earth has suffered widely time-spaced impacts by asteroids. Whenever ideas like this are offered they must recognize the greenhouse effect or they will not have any validity.

As discussed in chapter 4, these criteria also apply to theories of the Great Ice Age. In particular, The Astronomical Theory of the Ice Age totally ignored the greenhouse effect and must be set aside because of this. All aspects of our survival must be dealt with whenever an explanation involving the operation of the Earth is offered, but very commonly, ideas are offered in an isolated sense where we are to accept them without being concerned about other inter-connected realities of our existence. Unfortunately for the proponents of numerous theories of the Earth, this has not been done so their ideas cannot be seriously considered.

Every major asteroid that hit the Earth would have produced an Impact Winter which would not have been survivable. Once the Earth chilled enough to cause hoar-frost or snow to cover the surface, most of the incoming energy from the Sun would have been reflected away as soon as the Sun was able to poke through the cloud cover. Consequently, the Sun's energy would not have been useful for heating the Earth. It would have been reflected away for two reasons. First, the visible light part of the solar energy would remain as visible light. This is simply because the surface, being ice-covered, would not have been able to heat up. It is partly the conversion of a portion of the visible light energy to heat energy that would have heated the surface and given the greenhouse gases something to reflect back to the Earth. The greenhouse gases are transparent to visible light energy whether it is coming toward the Earth or going away, so unless some of it is converted to heat energy and held near the Earth (by the greenhouse gas effect) it will not provide any heating benefit at all. Secondly, it would be reflected away simply because it is white. This would really be a lose-lose situation. Consequently, every impacting asteroid which produced a world-wide cloud would have chilled the Earth into an uninhabitable state.

On the other hand, while each and every impacting asteroid would have caused a devastating chill factor thereby upsetting the greenhouse effect, a shower of asteroids would have caused both a chill factor and a heat factor thereby preserving the greenhouse effect. A single impact would have produced a globe-enveloping cloud which would have darkened the entire world. Additional impacts would have added to the mass of the cloud but would have had only a marginal effect on increasing the darkness. On the other hand, while a single impact would have stirred up volcanic activity, a shower of impacts would have stirred up proportionately-more volcanic activity. This is the basic reason that the Great Ice Age happened. The repeated and numerous impacts caused the release of an immense amount of heat from the interior of the Earth thereby providing the heat that was necessary to evaporate the ocean enough to provide the moisture for the great ice fields. A single impact would not have released enough heat to have enabled this to have happened. Hence an ice age would not have resulted from a single asteroid hit but it would have as the result of an asteroid shower. The shock-waves from an asteroid shower would have split the crust of the Earth open in numerous places and molten material from the interior would have welled up. The ocean was effectively placed on a hot plate but such a development was necessary or the several thousand feet of ocean that had to evaporate for the ice fields to form would never have happened. An impact winter would have happened instead. Further by releasing these immense quantities of heat, the Earth retained the greenhouse effect. The heat and the cold offset each other so an overall balance was retained. The darkness chilled the land but the under-water volcanic activity warmed the ocean.

During the years following the asteroid shower, the dust cloud dissipated and then the moisture cloud thinned out as the ocean cooled and could no longer produce enough cloud to keep the Sun from shining through. Through it all a balance of heat and chill was retained so that the greenhouse effect was not irreversibly lost as it would have been with a single hit. This was paramount to our existence and has enabled the Earth to have ongoing habitability or else animal life would have been terminated and the Earth would have locked up cold and become uninhabitable indefinitely.

The greenhouse effect must have been retained throughout the entire Ice Age event. If this had not happened the Earth would have joined the other planets of the solar system as being uninhabitable.

6.10   Conclusion

Every time a large asteroid hit the Earth, it would have been driven into an uninhabitable state. Recovery from such states would not have been possible. Therefore the arrival of asteroids at widely time-space intervals would have kept the Earth in a perpetually uninhabitable state. Since this has not happened, the idea that the asteroids arrived, one at a time, over long periods of time is not valid. Alternately, since the Earth is habitable, the greenhouse effect must have been maintained and since this would have been possible with an asteroid shower but not with isolated impacts, the conclusion must be reached that THE EARTH HAS SUFFERED AN ASTEROID SHOWER.

7.0   The Asteroid Theory of the Ice Age

7.1   Comparing an Ice Age to a Winter

An ice age has some of the characteristics of a seasonal winter so it can be compared to other types of winter of which three will be mentioned; nuclear winter, volcanic winter and impact winter. Almost any type of winter would be tolerable as long as spring came again within a reasonable period of time. Unfortunately with all three of these types of winter, spring would never come. In contrast with the Great Ice Age, spring did come again so there must be something very distinctive about the type of winter that the Great Ice Age brought.

7.1.1   Nuclear Winter

A few years ago during the "Cold War" there was serious concern that the whole world could become entangled in a war which included atomic bombs. The devastation from any war which involved atomic bombs would be widespread and could leave the world in such a state that there really wouldn't be any winners. The immediate results of this type of war include the usual damage that might be expected in any major war such as broken bridges, cratered highways and demolished buildings. The nuclear factor adds radioactivity which is unseen and which continues its devastation during the succeeding years. Among the survivors there would be people with terrible burns, and loss of body functions including sight and mobility. Radiation sickness would be ongoing and mutants would show up among the children which were born during the next few years. Radioactivity near every bomb site would not die out for many years. Many of these effects would not be visible but they would be quite real and cause untold mayhem for a long time. There would also be a winter which would be quite visible and it too would last for a considerable time, - time in this case being measured in years and possibly indefinitely.

A nuclear winter would be the period of cold following an all-out war during which numerous nuclear bombs were dropped. Like any other winter, it would be cold because the cloud cover produced by the nuclear explosions would block sunlight from reaching the ground. Most of the sunlight would reflect from the cloud tops. Satellite measurements have revealed that clouds do exert a major influence on the surface temperature. An increase in cloudiness causes cooler surface temperatures. The most new research reveals that there is a lot of surface cooling from increased cloudiness. (442) With extensive cloud cover, the ground would not be warmed by the Sun and it would freeze. As with any winter north of the temperate zones, when the weather gets cold, the ground will freeze. Gardens cannot be planted. Roads cannot be built. Rivers and canals crust over with ice so shipping comes to a halt. If it got cold enough, cars would not start unless they had been plugged in to an electrical source of power. All of these things are expected during any winter in a northern country. By taking precautions and planning ahead, winter conditions can be managed, even if they last for several months. The main difference with a nuclear winter is that it would happen quickly and spring would be seriously delayed and might not come at all. It would remain winter right through the expected springtime. There would be no letup in the summer and fall would be the same. It would be winter all year and it could continue for years.

A "nuclear winter" is comparable to the effect of very large prehistoric volcanic eruptions, and to the asteroid extinction of the dinosaurs. A nuclear winter would involve mid-latitude temperature drops in continental interiors to below freezing within a few of days - even in summer. (425)

When any bomb explodes, heat is produced. As the chemicals which make up the bomb react, heat is produced and produced very rapidly. The heat raises the temperature of everything around, so if there is anything nearby that can ignite, it will catch fire. The smoke from these fires will ascend up into the air and form a great black trail high into the sky. If the bomb is strong enough, it will blow a bowl-shaped crater opening in the ground. The material which is blown out of the way to make this crater will be splashed around the area as well as being pulverized into dust and blown into the air. An atomic bomb is very powerful and when it explodes, an enormous cloud results. The pressure wave from the explosion will form a crater in the ground but instead of making a relatively shallow depression a few feet deep, an enormous opening hundreds of feet across might be formed. As with a smaller bomb, some of the material from the crater would be splashed around the area. However a lot of it would be pulverized into dust and would rise into the air to augment the smoke. One atomic bomb can generate significant atmospheric loading from these two effects. The fear related to a war is that there could be a large number of bombs exploded with associated amounts of material becoming airborne and contributing to the cloud cover. Great plumes of smoke would rise into the air for thousands of feet and then start drifting downwind. Wherever they drift, they would cast a shadow on the ground. Anyone standing in one of these shadows would be shielded from the Sun and immediately feel a chill. The smoke from different bombs would coalesce to produce a widespread cloud. All of the area under this cloud would be in shadow and the Sun's rays would be blocked from reaching it. A nuclear war could result in thousands of tons of smoke and dust being placed in the atmosphere and it would drift on for miles and miles. The entire sky could become overcast. The chilling effect would be very widespread and weather would be affected. It would simply cool down and stay much cooler than usual. If summer was expected or in progress it would suddenly be fall instead. While the fear factors anticipated from a nuclear war certainly include the devastation which would show up around the site of each explosion, and the radiation problems which would last for years, the additional concern is the chilling effect of the widespread clouds which would develop from the smoke and dust rising from ground zero. If a large number of nuclear bombs were involved, an extended winter would result. The effect of this winter would be very widespread. Plants would not grow and everything would be continually frozen, so people could not survive and the devastation from the war would reach much farther than the site of the conflict and could, in fact, affect the entire world. With such a no-winners-at-all scenario the major powers seem to have backed away from seriously considering nuclear bombs in future conflicts and have made efforts to dismantle some of them.

7.1.2 Volcanic Winter

A volcano could produce a winter. Volcanoes come in various forms but most of them produce a great deal of smoke, dust and ash. The smoke might have similarities to the smoke from something burning but in many cases it will be water vapour mixed with dust. The solid matter, which we refer to as dust and ash, is mostly small disintegrated particles of rock. However, it doesn't really matter what the exact nature of the particles are as much as their effect, which is to cast a shadow on the ground. Some volcanoes produce an enormous amount of material with the result that the cloud they produce is very extensive. The effluent from volcanoes has occasionally been estimated and these estimates are sometimes given in cubic miles. It is hard to imagine how much dust and other sun-blocking material there is in a cubic mile of mountain.

A volcano named Tambora on the island of Sumbawa in Indonesia erupted in April, 1815. Two other major eruptions preceded this one. Soufriere on Saint Vincent in 1812 and Mayon in the Philippines in 1814 were also considered to be especially intense volcanic eruptions. Tambora was the most productive. It was estimated that Tambora ejected an incredible 35 cubic miles of rock, dust and debris into the air. Collectively, the material from these three eruptions created a thick layer of volcanic dust in the atmosphere that girdled the globe and lasted for years. (426) The Sun could not shine on the ground and the Earth cooled down. The cooling effect was serious in far away New England. A history of Madison County (New York State) included the following report. 'There was frost every month. The crops were cut off, and the meager harvest of grain was nowhere near sufficient for the needs of the people. The whole of the newly settled interior of New York (i.e. New York State) was also suffering from the same cause. The inhabitants saw famine approaching. ... Every resource of sustenance was carefully husbanded; even forest berries and roots were preserved. The spring of 1817 developed the worst phases of want. In various sections of the county families were brought to the very verge of starvation.' (427)

New England is far away from the source of the problem, which actually occurred about half a world away, but it never-the-less was not immune. The effects of these eruptions were felt worldwide. 'Throughout the northern hemisphere, the weather was abnormally cold. In England it was almost as cold as the United States, and 1816 was a famine year there and in Germany and France. Actually the year 1816 was just one of a series of cold years from 1812 to 1817. Everywhere, temperatures were lower than usual, and in the United States the depression of summer temperatures was the lowest on record.' (427) Cold weather is often accompanied by snowstorms and there was no exception in this case either. 'No account of major blizzards or extraordinary periods of cold is complete without at least passing reference to the year 1816, the famous "Year Without a

Summer". In that year the problem was not so much a cold winter but an abnormally cold summer. It was preceded by a late and cold spring and followed by an early and cold fall, so the frigid summer of 1816 brought widespread hardship and misery. From June 6 to 9 frost occurred every night from Canada to Virginia. Ice was an inch thick on standing water in Vermont. Everywhere, people shivered, broke out their winter clothing, and watched helplessly as their gardens and crops blackened in the cold. Newly-shorn sheep died, and millions of birds perished. A light snow in much of New England and Western and Northern New York State on June 6 was followed by a moderate to heavy snow in New England on June 7 and 8. In Vermont there were drifts 18 to 20 inches deep.' (427)

All of these miseries resulted from the eruption of just three volcanoes. Reports of some other eruptions follow a similar pattern. On June 6, 1912, Mount Katmai in Alaska exploded. This eruption is considered by some to have been the most violent of the twentieth century. It is only fortunate that it occurred in a relatively-remote area. Otherwise the destruction would have been costlier. This volcano produced very serious darkness and in this case it lasted for several days. '... Mount Katmai ejected tremendous quantities of ash into the atmosphere.' A ship was in the harbour at Kodiak which is 100 miles from the mountain. 'At about 4 P.M. on the day of the eruption, Perry (the captain of the U.S. Coast Guard cutter Manning) noticed a peculiar black cloud approaching from the northwest. Soon course gray ash began falling, accompanied by thunder and lightning, and the sky turned dark. Five inches of ash had accumulated on the deck by the next morning, and by noon the ash fall resumed. Now everyone was becoming concerned. At 2 P.M. pitch darkness set in. The radio receiver gave forth only static. Through a sleepless night, the crew anxiously awaited dawn. It failed to appear; all sunlight was obliterated. The cloud of falling ash was so thick it was impossible to see a lantern at arm's length. ... The decks had to be constantly shoveled and sprayed with water... On the afternoon of June 8, two days after the arrival of the cloud of night, the ash fall decreased, the sky turned reddish, and objects became dimly visible.' (428) At Kodiak, the total darkness lasted 60 hours. During the next few weeks, a reddish haze from the fine dust in the atmosphere was observed throughout the world. (423)

The explosion of Tambora was the most powerful on record up until Krakatoa in 1883 which is considered to be the most powerful of all time. This is, of course, a very serious claim. The effects of that explosion were seen, heard and felt literally around the world. In May of that year, activity started with fountains of liquid lava and a giant column of steam. Apparently people came to the island just to get a closer view. On August 26, a series of sharp explosions occurred and they continued all night. Then during the morning of the next day, the mountain blew up. The sound traveled all the way to Rodriguez (an island east of Madagascar in the Indian Ocean) in four hours and sounded like the distant roar of heavy guns. Subsequently, the sound waves traveled all the way around the world three times. Monstrous waves were produced and swept many villages away. Of more immediate interest was the darkness which followed. 'The eruption, it is estimated, blew one cubic mile of material to a height of 17 miles, and the dust was carried completely around the Earth several times by the high-altitude currents. Soon areas within 15 degrees of the equator, then gradually the entire Earth, were witnessing brilliant glows in the skies after sunset and before sunrise because of the dust particles in the air.' (429) There was an accompanying decrease in solar radiation reaching the Earth. 'Measurements of the amount of solar radiation reaching the Earth's surface show a marked decrease (to less than 88% of the average) in 1884 and

1885, corresponding with Krakatoa's eruption. This supports the idea that the low global temperatures of the late 1880s and early 1890s were due to the dust from Krakatoa.' (429) The drop in temperature was not as dramatic as the drop associated with Tambora and the difference is attributed to the lower volume of material ejected into the atmosphere.

Darkness was also involved. 'The enormous load of ash sent hurtling into the air brought days of darkness. At a distance of 103 miles, the darkness lasted 22 hours, and at a distance of 50 miles, for 57 hours.' (430)

Other major eruptions involving large quantities of material could also be mentioned. 'When Bandai erupted in Japan in 1888, it cast up almost three billion tons of material and blew off one of its peaks. (431) Pelee, a volcano on Martinique in the West Indies, blew up at 7:50 on the morning of May 8, 1902. (432) The eruption in the Aegean Sea about 1400 BC has left the Islands of Santorin as the remains of the rim of the cauldron. Fourteen cubic miles of island disappeared in that event and it is thought to have caused the darkness reported in the Egyptian Papyrus Ipuwer which says that 'the land is without light.' Further, in 'the Papyrus Anastasi IV the years of misery are described, and it is said: "The Sun, it hath come to pass that it riseth not"' (431) This might also be the darkness which accompanied the Children of Israel after they left Egypt and travelled through the desert. (424)

Mount St. Helens is a more recent example of the darkness which will result from volcanic dust. The smoke and ash rose to a height of sixty thousand feet in just a few minutes and the ash turned day into night and got into everything. (443)

The dust in the atmosphere not only causes darkness, which is unwelcome, it also results – because of the darkness – in cooling, which is even more unwelcome. Any surface that is isolated from the Sun will be immediately cooler than any nearby surface exposed to the Sun. The Sun simply must shine directly on the Earth in order for it to be warmed into a temperature range which is suitable for normal life activities to be carried on. Extensive cloud cover will result in extensive cooling. Persistent cloud cover will result in persistent cooling. Since there is only a narrow temperature range wherein normal life activities can be carried on, it is immediately apparent that the Earth and its residents cannot tolerate any significant upset in energy balance. The source of the Sun-blocking material is of secondary consequence to its effect. That effect will always be to cause darkness and cooling and neither of these developments is tolerable for life on Earth.

When the number of active volcanoes on the Earth is considered, it is a wonder that there isn't more trouble with temperature control. The number of occasionally active volcanoes is between four and five hundred and there are several hundred more that are considered extinct, some of which apparently have not been extinct for very long. Stirring up any significant percentage of these volcanoes at the same time would altogether likely be catastrophic for life on the Earth because the cloud cover produced would chill the Earth into an intolerable temperature range.

7.1.3   Impact Winter

Any object which can produce a high pressure wave near the surface of the Earth, would stir up a great cloud of dust and debris. As higher and higher energies are involved, more debris would be stirred up. The exploding object would not necessarily have to touch the ground to produce the dust but if the ground was contacted, the debris production would be worse. Any explosion at or near the ground would result in a massive cloud being formed. It is not surprising then, that atmospheric dust production, followed by darkness and cooling is expected to accompany all asteroid impacts. In fact this expectation is so high that every time major asteroid impacts are discussed, nuclear winters are declared to follow.

The Yucatan Peninsula has been stuck by a very large asteroid (i.e. the Chicxulub impact crater) involving the energy of 100 million megatons and with the injection of trillions of tons of dust into the atmosphere. This would have resulted in complete world-wide darkness and cooling which would have lasted for months. (433)

Further comment on the darkness and its accompanying cold is offered by another author. The surviving creatures below (the vast layer of dust, smoke and debris from the impact) probably thought that night had come early, but it was night without a moon, without stars. The dinosaurs could not see their own claws in front of their faces. Morning would not come for several months. A few animals had avoided the initial destruction, and at first they seemed to manage surprisingly well. Most of the plant eaters could still find food, although the settling dust added a gritty texture to it all. Some of the carnivores were accustomed to hunting in the dark, although they had never experienced blackness such as this. But the ultimate source of all food, the Sun, had been effectively blocked out. Without sunlight, there was no photosynthesis, no creation of sugar and starch from carbon dioxide and water. Unseen by the animals, the plants were turning from green to yellow, and then to brown. Without sunlight, the temperature of the Earth began to drop. On much of the land the temperature soon fell below freezing. Only those fortunate few animals that had already begun to hibernate didn't notice. Virtually every animal and plant died … it was a miracle that any of the higher life forms made it through … The reptilian dinosaurs, both on land and in the sea, had vanished forever. (434) In this discussion the author is simply recognizing that whatever creatures were not killed by the impact, would die soon thereafter because of the darkness and cold. If there is nothing to eat, it only takes a few days to die.

It is understood from comments such as this that the period of extreme darkness and cold was long enough to ensure that no animal would survive. In spite of this understanding some commentators like to hedge their bets and suggest that less than 100% would be killed. 'Hurtling at velocities up to twice that of the asteroids … a comet … could hammer the Earth with such force that most of the life forms on the planet would be annihilated. This is exactly what seems to have happened … when the dinosaurs were extinguished along with 70% of all other animal and plant species.' (435) Similarly another commentator offers; The greatest mass extinction occurred at the end of the Permian period when up to 96 percent of all species died. (436) However, justification is never offered to explain why a few would survive nor which types would have had the best chance of survival.

The Chesapeake Bay impact site is a relatively recent discovery. As the name indicates, it is located at Chesapeake Bay but its large size causes it to extend, not only across the Bay but up onto the mainland as well as out into the ocean. It is a very large impact site at 85 km (53 m) in diameter, which indicates that it was formed by a large object. Various suggestions have been made concerning the impactor size and they usually fall in the 3 to 5 mile diameter range. A lot of material would have been stirred up and projected into the air from the impact. In particular, a large amount of water would have been vaporized. Vaporized and fragmented rock and sediment would be entrained with the steam explosion. The Chesapeake Bay impact released 100 times the energy of all existing nuclear weapons! Such an impact is estimated to have the kinetic energy equivalent to approximately 10 trillion tons of TNT, while the total potential energy yield from the world's entire nuclear arsenal is only 100 billion tons of TNT. (i.e. The Chesapeake Bay impact was 100 times stronger.) Though the impact occurred in one region of the world, its environmental ramifications would have been worldwide and would have included a drop in temperature the same as a nuclear winter. (437)

In this example, one of the results of an asteroid impact is compared to a nuclear winter which is referring to the results of exploding numerous atomic bombs in the atmosphere as discussed earlier. Whatever effect the asteroid would have is being compared to exploding all of the nuclear bombs in the entire world and then ordering up another 99 times as many and exploding those also. If all of the atomic bombs in the world were exploded in one day, a nuclear winter would certainly be expected. Actually it seems reasonable to expect it if only a fraction of the total arsenal was exploded. If 1000 atomic bombs were exploded, world-wide cloud cover would be a reasonable expectation. As the number of exploding bombs increases, there would expectedly be even more cloud cover. Supposedly the more cloud cover there was, the more difficult it would be for it to dissipate and let the Sun shine once again. The above comparison, however, is being made to 100 times the nuclear arsenal to emphasize that this impact was very serious and would have resulted in the annihilation of life on the Earth. More than 100% of animal life cannot be annihilated but if it could have been, in the case of the Chesapeake Bay impact, it certainly would have been.

Any animals not destroyed by the impact and its immediate results would be destroyed a short time later when their food supply ran out. How long would that take? If the food supply ran out in one month, most animal life would perish within the next month. Occasionally people go on hunger strikes or extended long-term fasts. In many of these cases, a person can continue to live for more than a month after their last meal. If this were also true for animals, we might expect a few to still be alive after a month. We would not likely expect any to remain alive after two months. With this line of reasoning, within two months of an impact, all animal life would have perished from starvation. Any further suggestion that some would still be alive only recognizes that mass extinction is really expected to follow an impact such as the Chesapeake Bay impact.

The notion of mass extinctions and nuclear winters seems to be no stranger to the popular press and its contributors. After discussing the Tunguska event, one commentator offers the following: 'Every 5,000 years, a 200-meter-wide rock strikes with the force of a 100 megaton bomb, which is capable of destroying a large city. Every 300,000 years, a one to two kilometer-wide asteroid crashes, initiating a short-duration global winter. Every 100 million years, a bigger asteroid hits, producing the world-wide calamities that have punctuated evolutionary history. There are 14 known mass extinctions in Earth's geological record. So far, only one – the Chicxulub impact – has been linked to an asteroid or comet.' (19)

Even much smaller asteroids are thought to have produced widespread cooling. The Ries Crater of Southern Bavaria, is about 24 km. in diameter and is thought to have been formed by an asteroid about one mile in diameter. However, the effects are expected to have been world-wide. The blast wave would have destroyed all life to a radius of hundreds of miles, while the earthquake shock must have been felt around the world. Of the overwhelming amount of material blasted from the crater, a significant portion of it would have remained suspended in the atmosphere and would have darkened the sky all around the globe. (438)

An asteroid 1.5 km. in diameter would be expected to produce a crater in the 15 to 20 km. diameter range. There is a certain expectation that even an asteroid this small would have devastating results worldwide. '… wondering what newly-located 1997XF11 might do. He knew it was a near-Earth asteroid by its motion. Some early, simple calculations suggested that in the year 2028, the 1.5-kilometer-wide rock would approach precariously close to Earth. If it hit, millions would die from the impact and subsequent famine produced by a decade-long, planet-wide winter.' (439)

The winter that would follow a major asteroid impact would actually be a combination of an asteroid winter and a volcanic winter. Following the impact, hundreds of volcanoes would erupt all over the Earth. The dust and ash from every volcano would augment the dust produced directly by the impact. The chill resulting from the asteroid dust cloud would be reinforced by the effluent from all of the volcanoes. As discussed previously, there were three types of volcano; igneous provinces, cone volcanoes and underwater ridges. Since all of these types of volcano are found world-wide, the dust developing from an asteroid impact would have been quickly augmented by volcanic clouds all around the world. Within a few hours after an impact, the entire world would have been in total darkness which would have been the beginning of a post-impact winter.

7.1.4   Repeated Asteroid Winters

The Earth has experienced numerous asteroid impacts similar to or even greater than the Chesapeake Bay event as evidenced by the following. Tektites are small tear-drop shaped lumps of glass and are found over much of Central Europe. Several other tektite fields have been recognized around the world, each apparently corresponding to the melted debris from a very large impact. (448) The tektite evidence has also been used by another group to identify major impacts and their accompanying mass extinctions. 'In the Geological Survey Office in Ottawa, … Wayne Goodfellow studies evidence he has gathered from around the world related to earlier mass extinction (i.e. earlier than the Chicxulub event in Mexico) episodes, 230, 365, and 445 million years ago. He finds … small glass-like spheres believed to be produced only at the high temperatures created by an impact. Statistically, a comet of this size should clobber the Earth every 100 million years.' (449) In other words, five major impacts are being recognized by these comments and more are expected to arrive on a regular basis.

From these particular sources, opinion varies from five major impacts, to "a dozen or so", to "14 known mass extinctions". (19) A review of crater data generally supports these comments. The Earth Impact Database (EID) which is maintained by the University of New Brunswick in Eastern Canada lists six craters which have diameters equal to or greater than Chesapeake Bay crater discussed above. These include the Acraman in Australia at 90 km., the Chicxulub in Mexico at 170 km., the Manicouigan in Canada at 100 km., the Popigai in Russia at 100 km., the Sudbury in Canada at 250 km., and the Vredefort in South Africa at 300 km. This database does not include the Ishim Crater in Kazakhstan at more than 350 km. but does include six more with diameters exceeding 50 kilometers. Supposedly, a post-impact winter would have accompanied all of the large ones as well as, perhaps to a lesser degree, the others as well. This line of reasoning implies that between seven and thirteen mass extinctions have happened on the Earth due to the combined effects of direct asteroid impacts followed by cold post-impact winters lasting months or possibly years.

7.2 The Transition

7.2.1 Introduction

A post-impact winter is expected to follow a major asteroid impact. However, a post-impact winter would not have been survivable. Life did survive as witnessed by the diversity of life on the Earth at the present time, even though many local eco-systems have been lost and many species have been lost. Therefore it may be concluded that a post-impact winter began but it did not continue long enough to totally terminate both plant and animal life. The Great Ice Age happened instead. While a post-impact winter and an ice age do have several characteristics in common, the developing post-impact winter was truncated and transformed into the Great Ice Age because of the unusually warm ocean and also by the timely demise of the massive dust cloud.

7.2.2 Dust Cloud Development

As discussed above, a post-impact winter is expected to immediately follow the impact of a large asteroid with the Earth. The direct cause of this winter would be atmospheric and surface cooling. This cooling, in turn, would be caused by the very large, dense, dust cloud that would form as one of the immediate effects of an impact. When an asteroid smashes into the Earth, some of the material of the asteroid and some Earth material would be pulverized. Surrounding material would be stirred up. The net result expected is the formation of a large spreading cloud, which will rise many thousands of feet into the air. 'The explosion produced by such a large impact would itself be damaging, but even more important would be the consequences of lofting 100 trillion tons of dust into the atmosphere. Scientists have calculated that this quantity of dust would have produced a pall over the whole globe with a duration of at least several months. During this period of global darkness, neither light nor heat was available to sustain life so there was a great dying. (450). From this understanding, the extent of the asteroid-produced cloud was the whole world. There would be no getting out from under such a cloud.

Secondary effects of the impact include volcanic activity which would also produce an abundance of cloud. The clouds produced by a single volcano can be even darker than a thunderstorm and can be so dark that a person would not even see his own hand when it was right in front of his face. An experience like this would be similar to being in a mine when the lights are turned off. When this happens, it simply is not possible to see anything at all. Since the post-impact volcanic activity included the production of igneous provinces, cone volcanoes and underwater mountains – all of which would have produced cloud, it is easy to understand that, following a major impact, the entire world would have been darkened. It would not only have been virtually pitch black, the area covered would have involved the entire world. A more discouraging scene is hard to imagine.

7.2.3   Moisture Cloud Development

There is abundant evidence that the Earth was once universally warm. This evidence includes the trees of Axel Heiberg Island, the coal at Spitzbergen Island and the coal and plant remains found on Antarctica. When referring to the ancient fauna of Axel Heiberg Island, which is an island close to the North Pole, it has been referred to as a tropical paradise. 'An ancient tropical paradise, complete with turtles and crocodile-like lizards … a hot steamy world … anywhere a champsosaur is found could not have been very cold.' (451) The authors further state that the deep ocean was warm at 18 Celsius. Others agree with this conclusion. Immediately prior to the Chicxulub impact, the Earth was warm. This was the Cretaceous Age of Chalk which is believed to have terminated when the Chicxulub impactor crashed into the Earth and killed all the dinosaurs. Other commentators agree with this understanding and offer that the average annual temperature was 25C (77 F) and that the general fall in temperature since the Chalk was deposited has been confirmed from pollen and other plant remains as well as by the oxygen-isotope method. (452)

While the ocean is recognized as having been warm, the effect of the post-impact volcanic activity would have warmed it even more. Hot lava oozed out of the Earth and spread around forming Igneous Provinces both above water and below water. Some of these provinces were similar in size to Europe. Also, the floor of the Atlantic Ocean was ruptured, hot lava welled up and formed the long ridges on either side of the mid-Atlantic rift (i.e. the great fissure or crack in the ocean floor). Numerous other rifts accompanied by large ridges on either side also formed. Further, the ocean is home to hundreds of cone volcanoes, many of which are located close to the major faults (i.e. cracks) in the Earth's crust. Much of this activity would have caused the ocean water to locally boil with the overall result being an ocean with an average temperature that was much warmer than it was prior to the impact. If it had been 18 to 25 Celsius prior to impact, it could only have become warmer than this as a result of the volcanic activity and would have stayed warmer as long as the volcanic activity kept happening.

Warm water evaporates. The warmer it is, the more it evaporates. Therefore one further result of the impact of a large asteroid was to cause the entire ocean surface around the world to evaporate just as if the whole ocean had been placed on a stove with the burner turned on. The burner was in fact turned on when the hot molten lava from beneath the crust was forced up forming the underwater igneous provinces, the mountain ridges and the cone volcanoes all of which effectively heated the entire ocean. As it was heated, it simmered and evaporated at rates directly dependent on its higher temperature. (420)

The prodigious evaporation caused the air over the ocean to become extremely humid. Almost immediately above the ocean surface moisture clouds would have formed in great banks covering thousands of square miles and extending for thousands of feet into the air. As these extensive cloud banks drifted away they were continually replenished by a seemingly endless supply of moisture from the over-heated water. Any parcel of air that wasn't quite humid enough to form cloud right away would have done so as soon as the moisture-saturated air drifted over the cooler sun-deficient land.

An example of how the rapidly evaporating ocean might have appeared is provided by the Niagara River. The Niagara River flows from Lake Erie to Lake Ontario and partway along its journey it plummets over a precipice forming both Niagara Falls and the American Falls. The violence of the drop causes a great cloud of mist to continually form and rise several hundred feet into the air. Much of the surface of the ocean would have appeared very similar following the asteroid impact and would have continued this way as long as it was over-heated and evaporating at a high rate.

Another example is the formation of the Island of Surtsey. This island formed in the Atlantic Ocean about seventy kilometers south of Iceland in 1963. As an example, it is much more representative of the situation following the impact because in this case both moisture and dust rose from the ocean surface as a dense breath-stifling cloud. The Icelandic National Tourist Office took numerous photographs of the island as it formed. When the ocean started to evaporate following the impact of the asteroids, much of the ocean surface would have appeared like it did during the formation of Surtsey. The fact that Surtsey formed right on the mid-Atlantic Rift makes it even more representative of the post-impact situation.

When moisture evaporates from a body of water, it will rise unseen into the air and simply drift away. However, as soon as its temperature drops to a critical level, the previously-invisible vapour becomes visible as a cloud. There are several ways in which this phenomenon can be observed. On a warm summer day, as the Sun shines on the ground, it will be heated and moisture will be released into the air. In particular, if the Sun shines on a freshly-plowed field of dark earth, the moisture in the disturbed and uneven surface will be released into the air and rise upwards. As it rises it will cool because air expands as it rises and expansion results in cooling. At a few hundred feet up, a critical temperature will be reached and a cloud will form. This type of cloud is commonly called a cumulus cloud because of its pillowy appearance. Such clouds indicate an updraft and both birds and sailplane pilots make use of such information to locate an updraft and gain altitude. Another example can be observed during the winter when mist can be seen rising from a lake or pond. In this case, even though very little moisture is being released from the water, the air is cold enough to cause a mist to form. A third way the moisture-forming activity may be observed is in the mountains. As air rushes up the side of a mountain it will chill. If the critical temperature is reached before it gets to the top, a cloud will form. Often such clouds form at a very distinct elevation even allowing a person to have his head in the cloud and his feet under it. From a distance, it may appear that the cloud is stationary on the mountain top whereas in reality the air might be moving at considerable speed.

Following every asteroid impact, a great moisture cloud formed when the ocean heated up and caused the surface layer of water to evaporate into the air. The air was also warm over the ocean and so clouds might not have formed right away. However, the land had been chilled by the great dust cloud. Therefore as soon as the moisture-laden air drifted over the land, clouds would have formed in abundance. The water temperature was much higher than room temperature and massive quantities of moisture were released. (It was already at room temperature before the impacts so the volcanic activity just raised its temperature more.) Extensive cloud production and torrential rains resulted. This would have been very similar to the situation in the tropics at the present time where warm moist air rises and tropical cloudbursts result. As the Ice Age developed, the cloudbursts would have been virtually continuous as soon as the cooler land was beneath them. (419) But, a little further north where the temperature was below freezing, snow would have fallen instead of rain and it would have fallen in great abundance.

7.2.4 Two layers of Cloud

Immediately following the impact, the dust cloud formed. It rose to high altitudes and then spread out to cover the entire world. When the fireball reached the top of the atmosphere it spread out over the entire world. (453) Similarly, volcanic clouds will also rise to extreme altitudes as discussed previously with respect to Tambora and Krakatoa.

The moisture clouds on the other hand did not rise to extreme altitudes but formed and drifted along much lower down. Never-the-less these moisture clouds were also very effective at blocking the light from the Sun from reaching the surface of the Earth. Occasionally when a thunderstorm passes overhead, the sky becomes dark. The clouds appear almost black. But clouds do not really have various shades from bright white to dark gray as they appear. Instead, the shade of a cloud is determined primarily by its thickness. The thicker it is, the darker it is because the greater thickness prevents the light from getting through. The darkest period of a thunderstorm occurs when the thickest part of the cloud is directly overhead. The reason that it becomes so dark is simply because these clouds have great vertical extent and sunlight cannot penetrate all the way through. For example, 'The moisture-laden air from the south began to mushroom upward until, within an hour, the tops of the thunderheads had reached an altitude of 60,000 feet in the evening sky, darkening the land before sunset.' (454)

Due to the impact it was already dark because of the dust cloud higher up so the addition of a moisture cloud only added to the totality of the darkness. Normally a moisture cloud appears with a degree of darkness dependent on how thick it is and how much light is shining on it. In this case there wasn't any light shining on these moisture clouds at all because they were in the shadow of the dust cloud higher up. Therefore they were dark all the way through with the result that no light from the Sun reached the Earth at all. It would be hard to imagine the utter despair that would have accompanied such a scene.

7.2.5   The Warm-Cold Anomaly

The consequences of the asteroid impact involved both heating and cooling. The volcanic activity heated the oceans until they were, on the average, considerably warmer than room temperature. The direct contact of molten magma with the water would actually have caused it to boil in many areas. The ocean would have become like the pools of water that form near hot springs. Some of these pools are warm enough to allow bathing but some are too warm to permit bathing as their temperature exceeds what a human body can stand. The volcanic activity would have raised the average temperature of the ocean into a range which was well above room temperature, resulting in prodigious evaporation.

On the other hand, since the double layer of cloud blocked the heat from the Sun as well as the light, there would have been cooling. In particular it would have been cool over inland areas because they were further from the influence of the warm ocean water. Since the heat from the Sun was cut off, the only heat that would have reached the inland areas would have been from the relatively-warm moisture-laden clouds that drifted in from the over-heated ocean.

The situation could therefore have been considered anomalous. While it was dark over both the water and the land, it was warm over the ocean and chilly over the land. The entire Earth had suffered the unthinkable trauma of an asteroid impact and was now in a state that was untenable for animal life. Further consequences of the impact would depend on the longevity of both of the two different types of cloud. As long as the dust cloud persisted, heat from the Sun would have been blocked. If the resulting chilling trend continued until the ocean cooled down, the Earth would have gone further and further into the deep freeze. This would have been the post-impact winter that so many commentators expect from a major impact. Even though the impact occurred in one part of the world, the environmental ramifications would have been worldwide and would include a drop in temperature similar to a nuclear winter. (455) Such comments appropriately acknowledge the chill brought on by the cloud but do not recognize the heating affect that the asteroid impact had on the ocean. While the chilling conditions were suggestive of a developing post-impact winter, the unusually-warm ocean would not have permitted this to happen but would have produced the second NECESSARY condition for an ice age(i.e. heat) so an ice age would have happened instead. Further, if the dust cloud could have been brought down within a few months, at least in some areas it would have been survivable.

The great dust cloud produced an immediate and significant chill over the land. This effect of cloud cover can be experienced even on a warm summer day. If direct sunlight is suddenly cut off by a thick bank of cloud, air temperatures will immediately dip and a sweater might be needed. Or simply standing in the shade of a cloud or a tree can be welcome relief on a hot summer day. Since the post-impact-winter-causing cloud would have cut sunlight off completely, temperatures would have plummeted and freezing would soon have followed. With these developments, a post-impact winter would have been underway within hours of a major asteroid impact. However, in order for it to have been sustained, the cloud cover must have remained intact not allowing any significant amount of sunlight to get through. If either layer of cloud started to break up, any discontinuities in the other layer would have allowed sunlight through, the Earth would have been warmed and perpetual winter conditions could not have been maintained.

A world-wide cold-causing dust cloud would have been a necessary condition for a post-impact winter. However in order for this winter to have developed properly, it would also have been necessary that the ocean not have been warm. An unusually warm ocean would have partially offset the chilling condition produced by the absence of sunlight and relatively warm air drifting in from the ocean would have tempered the chill caused by the sunlight-blocking clouds. The dust cloud would have been expected to persist for several years because the dust-clouds produced from recent volcanoes have persisted for that long. 'Measurements of the amount of solar radiation reaching the Earth's surface show a marked decrease (to less than 88 percent of the average) in 1884 and 1885, corresponding to Krakatoa's eruption. This supports the idea that the low global temperatures of the late 1880's and early 1890's were due to the dust from Krakatoa.' (456) However if the ocean warmth also persisted as long as the dust cloud, the offsetting factors of chill from the dust cloud and heat from the warm water would have prevented a post-impact winter from becoming established.

Significantly, the chill factor which would initiate a post-impact winter is one of the factors which would be NECESSARY for the initiation of an ice age. It is therefore to be expected that these two phenomena are occasionally confused. 'We do not know how long this dust pall remained, or what effect it might have had on the world's ecology. If it were thick enough, it could have had long-term influence on the climate even triggering an ice age.' (457) While ice ages are not really triggered, this type of comment captures the expectation that both a post-impact winter and an ice age involve cold.

7.3   Ice Age Requirements

7.3.1   Water and Heat Requirements

It would be appropriate to examine the requirements for an ice age and in particular to identify how much water had to evaporate along with the amount of heat needed to bring about so much evaporation.

The water reservoir for the Earth is the ocean. All of our water comes from the ocean and all of the water that formed the great ice fields of the ice age also came from the ocean. In fact various estimates of the amount of water involved in the ice age have been advanced and they always refer to the amount of ocean water that would be involved. The ice in the great remaining glaciers of the world would raise the level of the ocean if they melted. For example, Antarctica has enough to raise the level of the ocean by around 200 feet. The great Greenland glacier would add about another 20 feet. These are enormous quantities of water. Of course during the Ice Age even more water was involved. Various commentators have offered estimates which usually fall in the 300 to 500 foot range. It follows that sea level would have been that much lower at ice-age maximum. This means that much of what we look at now as ocean would have been bare ground. Some commentators even refer to it as 'dry land' or 'land bridges'. (441)

With these estimates in mind it seems that, at ice age maximum, there would have been a total of 500 to 700 feet of ocean tied up in the great glaciers. In order to get to the ice fields all of that water had to leave the ocean. As demanding as this would have been, in reality the situation would have been even worse. Every gallon of water that evaporated would not have formed ice. Some of it would have precipitated right back down on the ocean. Some of it would have fallen as rain on the land and run directly back into the ocean and some of it would have fallen as snow that didn't accumulate enough to be identified later as a glacial ice. Therefore, only a fraction of the water that evaporated from the ocean would have actually formed the ice of the great ice fields. It would be safe to assume that if 600 feet of ocean evaporated and made its way to the ice fields that four or five times that amount must have evaporated altogether. This does not mean that all of it had to be away from the ocean at the same time but it certainly would have had to evaporate or the Great Ice Age would never have happened. Therefore the total amount of water that would have risen above the surface of the ocean would have been in the range of 2,000 to 3,000 feet (or even more).

In order to evaporate such an enormous amount of water a great deal of heat was required. The Astronomical Theory of the Ice Age does not recognize this necessity at all but certain commentators do. '... cold alone would not have been sufficient to produce the continental ice covers.' (419) '... the oceans must have steamed.' (420) The 600 feet of ocean that was required to form the ice fields would only have been a fraction of the total amount that evaporated. Where it wasn't quite cold enough for snow to fall rain would have fallen. Just as the snowfall would have been overwhelming in magnitude so also the rain would have fallen as it does during a hurricane. Other investigators have recognized this effect as well. 'And, in fact, numerous scientists who conducted their field study in various areas outside the former ice cover came to the conclusion that these areas had experienced periods of torrential rains that were simultaneous with the glacial periods in higher latitudes.' (419) How much ocean would have evaporated when we include these rains? Further, the surface of the Earth is more than 70% water. Some of the evaporant would have simply fallen right back down on the ocean and would never even have reached land! When these rains, the great extent of ocean surface and the estimated amount of glacial ice are recognized it is clear that the above suggestion of 2,000 to 3,000 feet of ocean is not over-stating the case. Amounts in excess of this could readily have been involved and every ounce would have required heat to change from liquid form to vapor form enabling removal from the great ocean basin.

The oceans are deep at about 2 ½ miles (4 km) (421) so would never have been empty but they would certainly have appeared different than they do now.

7.3.2   Chill Requirements

In addition to heat, there must have been cold simply because we need snow to make ice and both ice and snow only exist when the temperature is below freezing. Therefore it was necessary that both heat and cold be present or the Great Ice Age would never have happened. It had to be cold where the snow was falling or it would not have stayed. And since the Ice-Age ice was spread all across both the northern part of North America as well as Europe and Asia, all of these areas had to remain below freezing for as long as the ice was accumulating. This is not the case at the present time as most of these areas rise above freezing during the summer months. If this had happened during the Great Ice Age there wouldn't have been an ice age but only a period of snowy winters. Snowy winters are not uncommon but ice ages are very uncommon and require that the temperature remain below freezing right up to ice-accumulation maximum.

7.3.3   Temporal Requirements

An enormous amount of heat was required and a sustained period of cold was required. The heat had to persist until all of the required water had evaporated and the cold had to persist until ice age maximum was reached. This could have taken several hundred years but certainly not the thousands of years that have been declared to have been involved with the Astronomical Theory where extended periods of time of tens of thousands of years are declared to have been involved. What would have kept the ocean warm enough to evaporate at an unusually high rate for tens of thousands of years and what would have kept the snow from melting right through the summers for thousands of years? The response might be that it was cold so snow could fall all of that time but if it was cold, what caused the ocean to keep warm and keep evaporating at such an unusual rate for such a long time?

7.3.4   Greenhouse Effect Maintenance

It might seem contradictory that both heat and cold are required at the same time but in order not to irrevocably upset the greenhouse gas effect, the Earth cannot be allowed to suffer either prolonged cold or prolonged heat. From either of these conditions there would be no recovery. In the case of prolonged cold, ice and hoar-frost would form on vast areas of both land and sea. This would be intolerable because of the increased reflectivity of the surface. Incoming light would be reflected and if it was reflected it would not warm the surface. It is the warming of the surface and the radiation of that heat energy back into the atmosphere that gives the greenhouse gas portion of the atmosphere something to reflect back to the Earth's surface. If there wasn't any heat to reflect, the greenhouse gases would be of no value so they might just as well not even be present. It would be like taking the cover off of the greenhouse. Effectively there would not be any greenhouse, or by comparison any greenhouse effect for the Earth! Without practically any greenhouse effect, the surface of the Earth would descend into a deep cold which would only extend the layer of ice on the surface even more and reflect even more of the Sun's energy away. Even though the Earth would still be in the habitable zone of the Sun, it would not be habitable and there would be no escape from such a condition.

A situation like this can be compared to the present situation on Mars. Mars has a day-night cycle very similar to the one on Earth as the planet continues to rotate through a day-night which is slightly longer than Earth's at 24 hours and 37 ½ minutes. (422) Therefore, the time during which any particular area would be exposed to the Sun is about the same as it is on the Earth. Mars is very cold however, with a range in temperature from about -112F (-80C) at the poles to freezing at the equator. (112) However, the surface is quite dark so when the Sun comes up, it heats up. This would not happen if the surface was ice-covered. A white surface of ice and snow would reflect most of the incoming heat and light so it would not heat up very much at all. The advantage that the Earth has is that it is closer to the Sun but this would still not enable a habitable temperature to be reached if large areas were covered by ice.

In order for the Earth to remain habitable, the greenhouse effect cannot be violated. This means that during whatever has happened on the Earth in the past the greenhouse effect must have been retained. Otherwise, nobody would be living here at the present time. Any theory of the Earth that is put forward absolutely must recognize the greenhouse effect necessity or it will not have validity.

7.3.5 Summary

The NECESSARY and SUFFICIENT conditions for an ice age are heat (to evaporate the ocean to provide moisture for snow-making) and cold (to chill the moist air enabling snow to fall). These two conditions are also simultaneously required to retain the greenhouse effect which is one of the necessities enabling the Earth to be habitable. Heat and cold are both NECESSARY because an ice age will not develop unless both conditions are present and they are SUFFICIENT because when both of them are present an ice age will develop.

With both the heating and cooling conditions satisfied an ice age would proceed but since the greenhouse effect would not be lost it would be possible for life to become re-established once everything settled down again.

If an asteroid shower impacted the Earth, both moisture clouds and dust clouds would form and they would augment each other to darken and chill the land. Underwater volcanic activity caused by shockwaves from the impacts would have heated the ocean, producing the moisture clouds which would also bring widespread darkness and cooling as well as copious precipitation. An abundance of dust clouds from the impacts directly as well as from the volcanic activity would augment the cooling effect. Excessive rain would result in flooding and erosion while cooling would shift the environment into the freezing zone enabling snow to fall instead of rain. With excessive snowfall, all activity would grind to a halt. Excessive rain would do a lot of damage but it would drain away whereas snow would not drain away but would keep right on building up wherever it landed. When everything was buried in snow, life would not have been able to carry on.

Fortunately, ice ages are not triggered. The environment is not all wound up tight, in an extremely unstable state just waiting for some minor development to occur so an ice age can get underway. Things happen in nature because the NECEEARY and appropriate physical conditions are in place so they can happen. Then, they not only can happen, they will happen.

The importance of heat to effect the great Ice Age has been recognized by several commentators. 'If we take into account the area occupied by ice during the glacial epoch, much larger than the area of the present polar ice … The usual estimate of its thickness is between six and twelve thousand feet … the water must have come from the oceans … which must have been at least three hundred feet lower … an enormous quantity of heat was necessary …" Some eminent men have thought, and some still think, that the reduction of temperature, during the glacial epoch, was due to a temporary diminution of solar radiation; others have thought that, in its motion through space, our system might have traversed regions of low temperature, and that during its passage through these regions, the ancient glaciers were produced ... many of them seem to have overlooked the fact that the enormous extension of glaciers in bygone ages demonstrates, just as rigidly, the operation of heat as well as the action of cold. Cold alone will not produce glaciers."' (467)

An ice age doesn't happen because the Earth simply turns cold. It is a fact of nature that if the temperature drops too far, a crust of ice will form on water. This crust of ice will prevent any further escape of moisture into the air and so snowfall will drop off. This happens very predictably in locations which are downwind of large lakes. Snowfall can be abundant up until the lakes freeze over but then it will drop off substantially. An example of this type of phenomena has been observed at Buffalo, New York. Being downwind of Lake Erie, Buffalo often experiences heavy snowfalls which have even amounted to more than three feet in one storm. In one case early in the season 40.5 inches of snow fell on Buffalo during a four-day storm in early December 1976. (458)

The second effect from extreme cold relates to the water vapour capacity of air. As air gets colder, it has less capacity to retain moisture. This is actually the reason that snow falls. It falls because the air chills and cannot retain the moisture. Very cold air retains very little moisture so if it should be chilled a little further to extract whatever remaining moisture it has, very little snow would result. Therefore if the Earth turned cold and the oceans froze or became covered with drifting ice, the ocean water would be unable to evaporate and the cold air would be unable to transport any significant amounts of moisture. In such a situation not only would the moisture source be turned off, but the moisture transport mechanism would also be turned off. Snowfalls under these circumstances would be minimal. An ice age would therefore not result.

VAPOUR CAPACITY OF AIR

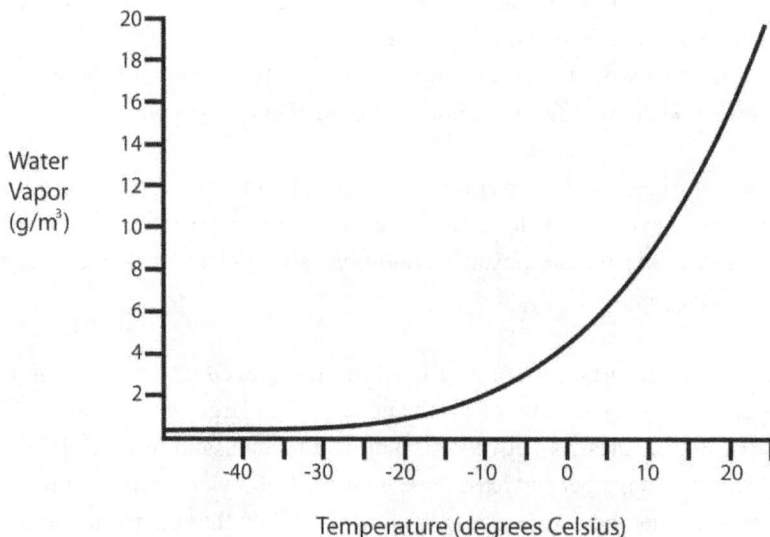

Water Vapor (g/m³) vs Temperature (degrees Celsius)

234

Mathematical models of the Earth's orbit (used in the Milankovitch or Astronomical Theory) indicate that, over long periods of time, the Earth will drift a little closer to the Sun as well as a little farther away. The variation in solar energy received by the Earth during these excursions would be in the range of one or two percent. (459) A drop of two percent would certainly chill the Earth and an increase of two percent would heat it but when heat regulation and transfer mechanisms (including atmospheric circulation and ocean currents which operate throughout the world) are factored in, the actual temperature variation would be very small. (459) Such circumstances would not be sufficient to cause and sustain an ice age. 'For past orbital changes, both the seasonal and the mean annual model fail to produce glacial advances of the magnitude that are thought to have occurred.' (460) It was therefore concluded that the Milankovich Theory did not offer an appropriate explanation of the Ice Age. 'It is very unlikely that the Milankovich cycles can start or end an ice age... the Milankovich cycles are too weak ... to explain the start and end of ice ages ...' (461)

It must also be noted that the small changes anticipated by this theory would occur over extended periods of time involving thousands of years. (462) The evidence however, contradicts this assertion and indicates that the Ice Age ended abruptly and that melt-water formed massive flows with rivers being hundreds of feet deeper than they are at present. '... a larger stream ... must have flowed when the ice cover melted ... prodigious floods of almost unimaginable magnitude accompanied the melting of the ice cover.' (464) Clearly long periods of time were not required for the ice to melt and as mentioned with the Lost Squadron report they were not necessary for the ice to accumulate either.

As mentioned earlier, the further inhibition inherent in the 'cold Earth' approach relates to the greenhouse gases. Water vapor is a major greenhouse gas so if the entire world became chilled, the warming affect of water vapor would be lost. This means that the Earth would become even colder and only stop getting colder when any further reduction in temperature did not result in any further reduction in water vapor in the atmosphere. The end result would be that the entire world would freeze solid and simply stay that way indefinitely.

A further aggravation with the whole approach relates to the reflectivity of snow. If the land should become chilled enough for even one inch of snow to cover it, the heat from the Sun would be reflected away. This would further establish the chill and an ice age would not result.

An ice age would be a period of time when snowfall continued for most or all of the year. There would be no intervening summer periods, which would be warm enough to melt all of the snow, which accumulated the previous winter. As long as a winter's accumulation of snow could melt again the next summer, an ice age would not be happening but only snowy winters. In order to have an ice age, more snow must fall than the amount which melts on an annual basis. This would result in a build-up of snow and an ongoing build-up, year after year, would be an ice age.

The two NECESSARY conditions, heating and cooling, occurring simultaneously, provide SUFFICIENTCY, for a world-wide ice age. The heating criteria involves keeping the ocean water warm enough to evaporate the prodigious quantities of water required to build up the vast glaciers. The water must be initially warm enough and kept warm enough to evaporate moisture at the enormous rates required to enable hundreds of feet of ice to accumulate. Ironically the water had to have been heated at a rate which would maintain the moisture-cloud cover sufficiently continuous to prevent the Sun from getting through and melting the ice. A certain minimum abundance of clouds would have been required to prevent the Sun from melting the snow. In other words, a certain minimum cloud cover was required to keep the ice-fields cold which means that evaporation had to proceed at a certain minimum rate to provide the cloud-continuity which was necessary to keep the land chilly so the ice could accumulate. The heat provided the evaporate for ice buildup at the same time that it generated cloud cover to keep the land cold. As soon as the evaporation rate fell below the critical rate, the dependant moisture clouds would have thinned out and allowed the Sun to melt the snow as fast as it fell. Alternately it could be stated that the evaporation rate (i.e. the heating rate) had to be above a certain critical minimum to keep the ice-fields cold enough to cause snow to fall instead of rain. If rain fell instead of snow, the ice would have stopped accumulating. Then with a further slight decrease in ocean temperature (together with the resultant decrease in moisture production and cloud cover), melting would have increasingly exceeded accumulation and the Ice Age would have been over. It might seem contradictory that both heating (evaporation) and cooling (Sun-blocking) were necessary but without these two factors being present to the necessary degree, the Ice Age would never have happened.

Further, the absolute necessity of both factors being simultaneously present is understood with respect to the greenhouse gases. Chilling was necessary or snow would not have fallen. However, simply cooling the Earth would have been a disaster because the entire Earth would have locked up frozen solid. If the Earth were chilled for any reason, water vapour in the atmosphere would be lost and a chilling spiral would develop where the chill further reduced the greenhouse gas effect resulting in more chilling. The oceans would eventually have frozen right to the bottom. Life on Earth could not have existed. Therefore, on a world-wide basis, heating had to balance cooling so the greenhouse effect could be maintained.

It might initially seem contradictory to require both heating and cooling to effect an ice age but there is significant recognition that it has been far from simple and obvious how the Great Ice Age occurred. Numerous theories have been advanced but none of them have withstood serious scrutiny. In some cases, investigators have basically thrown up their hands in despair of finding an explanation. The cause of the Ice Ages is one of the greatest riddles of geological history, despite the work of numerous astronomers, biologists, geologists, meteorologists and physicists. (468) Others are in agreement. Numerous methods of accounting for ice ages have been proposed but nothing is generally accepted and the contributors are hopelessly in contradiction with one another. (469) More than 60 different hypotheses of the origin of the Ice Age have been proposed and more ideas are published every year but serious contradictions continue throughout the literature. (469) The conditions for an ice age do not readily recur, and it has proven very difficult to postulate the actual environmental circumstances, which resulted in the Ice Age. There certainly is no ice age developing now, even though some ice fields are growing and some are diminishing. 'Actually the screening of the Earth by clouds of dust of volcanic origin was one of the theories concerning the origin of ice in the glacial epochs; however, like heat alone, cold alone would not have sufficed to produce the continental ice covers.' (470)

During the Great Ice Age, volcanic activity was ongoing, even as the snow was falling and the ice was accumulating. (471) While underwater volcanic activity continued to heat the ocean, the dust cloud from the volcanism continued to chill the land. As long as this volcanic activity continued, both of these effects would be taking place. It was the combination of a warm ocean and cool land that enabled the Great Ice Age to happen.

In addition it was absolutely necessary to have a warm ocean at the same time that the land was chilled in order not to irreversibly lose the overall greenhouse effect. Otherwise the Earth would never again have been suitable for animal life.

The heating factor and the chilling factor were both NECESSARY to have an ice age and together they were SUFFICIENT to have an ice age. Therefore an ice age happened. Further they were both NECESSARY and collectively SUFFICIENT to retain the greenhouse effect or there would have been no recovery from the Great Ice Age.

7.4 Heating Factors

7.4.1 Heat Quantity

In order to build up the massive quantities of snow and ice of the Ice Age, a very large quantity of water had to evaporate. While much of the land had very recently been flooded with puddles and lakes being everywhere, a much greater quantity of water than this was needed. Also, there is always a certain amount of moisture in the air and recent volcanic activity would have added more, but such sources could only have supplied a very small portion of the total moisture required. Only the ocean could have supplied the prodigious quantities of water that were needed. In fact, the amount of water which is still bound up in the glaciers of the world is always compared to a certain number of feet of ocean.

In both Greenland and Antarctica, the two major remaining ice-fields of the world, the ice is up to approximately two miles thick. If Antarctica alone melted, the level of the oceans of the world would increase by about 200 feet. (472) This means that in order to form only the Antarctica ice-field, a quantity of ocean equivalent to a layer of ocean water, 200 feet deep around the entire world, had to evaporate. When the water requirements of the Greenland glacier are added, another 28 feet of ocean are required. This great quantity of water evaporated, became moisture in the air and then fell as snow over Antarctica and Greenland until these great glaciers were formed.

During the Ice Age, other ice-fields besides Antarctica and Greenland were formed and they would have required an additional equivalent layer of ocean water of similar or possibly even greater magnitude. It is generally thought that sea level was 300 feet lower during the Ice Age. (473) Other commentators suggest that even more ocean water became tied up in the ice. 'An early and obvious theory was that … sea level was about 125 meters (412 feet) below present sea level.' (474) Based on these comments, 228 + 300 = 528 (or possibly 228 + 412 = 640) feet of ocean must have evaporated to form the glaciers. However, this would just be the ocean water that actually found its way directly into the massive ice sheets. Most of the water that evaporated would not have become glacial ice. Only a portion would have gone directly into the ice-fields. Since the Earth is more than half covered by water, it would be reasonable to expect that more than half of the water that evaporated simply fell as rain or snow directly back on the ocean. It would also be reasonable to expect that only a portion of the precipitation over the land actually became part of the great glaciers. Some of it would have been rain and some of it would have been snow which did not become deep enough to press itself down into glacial ice. If only half of the precipitation over the land became glacial ice, this line of reasoning means that less than one-quarter of the water that actually evaporated became ice. (The surface of the Earth is about 70% water.) The total amount of ocean water that evaporated to form the great glaciers of the Ice Age would therefore have been at least four times the amount that became tied up in glacial ice. Therefore, the total amount of ocean water that evaporated was at least 528 x 4 = 2112 feet. Obviously, this estimate is clearly conservative. Any theory of an ice age must explain the source of the enormous amount of heat that was required to evaporate such an incredible amount of ocean water.

It is understood that a large amount of ocean evaporated but that all of this water was never missing from the ocean at any one time. Much of it evaporated and immediately returned to the ocean, either directly as rain and snow or indirectly from rivers (as it does at the present time).

7.4.2 Heat Source

If the Earth was universally warm prior to a major asteroid impact, as has been declared, (475) the oceans all over the world would have been warm. (476) Warm oceans mean that a great deal of heat has been stored and if the atmosphere of the Earth chilled into the freezing zone, it would require a considerable length of time for this heat to dissipate. The air temperature in the northern regions could fall below the freezing point and remain there for years before the temperature of a warm ocean would drop to freezing. Until the water reached the freezing point, a warm ocean would be an enormous reservoir of heat, which would cause ongoing evaporation. Moisture would have been added to the air continuously to provide an ongoing source of rain and snow.

While some of the heat required for evaporation was available directly from the water because the water was warm, it would not have been enough to provide more than a small amount of snow. The lion's share of heat was available from freshly-exposed material from the interior. The crust of the Earth had been faulted (or cracked) by the shock waves from the impact. These faults allowed volcanic activity to proceed in three different ways. First, molten material came directly up through the faults and formed ridges such as those down the middle of the Atlantic Ocean. This activity would have been accompanied by much turbulence and the sea would have boiled on contact with the red-hot molten rock welling up from below. Secondly, we note that the major faults of the Earth are always accompanied by cone volcanoes. In fact the great majority of the volcanoes of the world occur in the vicinity of a fault. Supposedly the fault-causing shock and stress weakened the crust allowing hot material from within the Earth to ooze upwards bringing a lot of heat with it. Thirdly, the oceans are home to many 'Large Igneous Provinces'. These are lava flows that have oozed up from below the crust and spread around both on the ocean floor and on land and have occasionally piled up into mountains. (e.g. the

Deccan Traps of India) (477) Wherever these enormous quantities of hot material became exposed to ocean water, it too became hot and evaporated at enormous rates.

7.4.3 Ongoing Heating

In fact, these volcanic sources of heat kept right on supplying heat even as the snow was falling and the great glaciers of the Ice Age were accumulating. In particular the ice cores recovered from both Greenland and Antarctica show layers. (465) Since there are about 110,000 of these layers and they are quite thin (i.e. they only represent about ten inches of snow each), they could not possibly represent years as has been declared. They could however, indicate volcanic activity and that it was ongoing throughout most of the time that the ice was accumulating.

Other evidence supports this conclusion. The ice age volcanoes left huge deposits of ash. In the Western United States alone, more than 68 ash falls, coinciding with the Ice Age have been recognized. The size of the ash beds indicates that some of these eruptions were gigantic. An exceptionally large ice age eruption was recently discovered in New Zealand. This eruption would have spread a thick layer of ash over at least ten million square kilometers of the South Pacific Ocean. Based on correlations with modern volcanic eruptions, the dust and aerosol loading was similar to the worst nuclear winter scenarios during which almost all sunlight is blocked out around the entire world. (478) The source eruptions could have been on land or underwater. In any event, if there were numerous eruptions on land, one would expect that even more would also have occurred underwater and been a source of heat for ongoing evaporation.

Initially warm water, (i.e. at 18 to 25 degrees Celsius) without additional heating, would not have been sufficient because of the incredible amount of heat required. It requires 540 times as much thermal energy to evaporate one unit of water, as it does to change its temperature one degree F. Cooling the entire ocean 25 degrees C would not have been sufficient to evaporate the 2000 feet plus of ocean water required for the Ice Age.

Further, the ocean had to evaporate at a certain critical minimum rate or the ice accumulation function would simply have terminated. This would have necessitated a certain associated minimum average temperature which would altogether likely have been above room temperature. Also the maintenance of the minimum-necessary cloud cover would have required a very high evaporation rate and room temperature water would not have been warm enough to effect so much evaporation.

Not only was an enormous quantity of heat required but it was required at a rate which would actually cause the ice to build up. Things had to happen quickly. If the snow had not been produced at a prodigious rate, melting would have matched accretion, as it soon did anyway. During the time of the Great Ice Age, the Earth was cooling down toward a new climate equilibrium. The last of the dust clouds were dissipating and the Sun was shining through more and more. The snow clouds themselves kept the Earth cool enough a little longer but they could only shield the Earth from the Sun if they were thick enough and continuous enough. Reduced evaporation rates would have caused the clouds to thin out and indeed that is how ice accumulation tapered off at Ice Age maximum. The cooling ocean simply could no longer supply cloud-making moisture at a fast-enough rate. Ice accumulation therefore maxed out and the Great Ice Age peaked and went into its recession phase. It was the prodigious rate at which evaporation proceeded and continued to proceed due to ongoing volcanic activity coupled with the chill brought about by the sun-shielding clouds that enabled the Great Ice Age to happen in the first place. Without plenty of heat being added in a timely manner to an ocean that was already warm, significant quantities of glacial ice would not have formed before the ocean cooled down too much.

The Sun could never have been that source of heat because heat was required deep in the ocean. Further, solar heat would not have come quickly enough but when it did come, it was counterproductive and it melted the snow. On the other hand, underwater volcanic activity would certainly have followed a major asteroid impact but it had to continue long enough to provide the necessary heat in a timely manner. While the volcanoes could have oozed and gurgled for years they would have eventually settled down because the initiating shock-waves would have dissipated. When the volcanic activity subsided enough, evaporation dropped to the critical level and ice accretion ceased.

It was heat from the interior of the Earth that provided the enormous amount of thermal energy required to evaporate the water from the ocean that enabled the Great Ice Age to happen. The asteroids fractured the crust and the shock waves from the impact caused molten material from the interior to explode, surge and ooze up onto the top of the crust. This over-heated the ocean which evaporated at a prodigious rate providing the moisture which enabled the massive glaciers of the Great Ice Age to accumulate. Thereby one of the two NECESSARY conditions to have an ice age, were fulfilled.

7.5   The Chill Factors

As the Great Ice Age progressed, several factors contributed to the chilling of the land.

7.5.1   Vapour Canopy Collapse

When the asteroids rushed through the atmosphere, they caused the destruction of the Vapour Canopy. The Vapour Canopy enclosed the oxygen-nitrogen layer of atmosphere completely. Locally, it was a second layer of atmosphere floating on top of the oxygen-nitrogen layer. It had long-term stability because it was both lighter and warmer than the air underneath. However, when the asteroids came rushing through, stability was lost. The initial stability-destroying factor was the forced air movement directly caused by the asteroid. In front of the asteroid, air was forced aside. Behind the asteroid, air rushed back in. A short-term circulation pattern developed with air from the lower atmosphere moving vertically and then horizontally to fill the void left behind the speeding asteroid. This circulation pattern was localized and would have extended outward for several diameters of the asteroid itself. For example, if the asteroid diameter was 5 miles, the circular pattern of air movement may have extended outward for 20 or 30 miles. As this movement of air took place, the lower, oxygen-nitrogen layer was forced up into the H2O layer. Unfortunately, air cannot hold very much water vapour, which effect would now be referred to as relative humidity. When water is in vapour form, it can remain that way indefinitely as long as the temperature is high enough. However, when water vapour is mixed with air, there is a very definite limit to how much water can remain. The rest will precipitate out right away. The mass of water in the Vapour Canopy layer would have far exceeded the allowable upper limit for the mix. Therefore it would have rained. The rain would have been a downpour. Violent volcanic activity would have had a similar effect. Wherever a volcano blasted into the upper atmosphere, mixing of air and water vapour would have resulted and that portion of the atmosphere would then have been supersaturated with water vapour. Nature would not allow this and rain would result. With the arrival of the first asteroid the Vapour Canopy started to collapse. Collapse would have been augmented by the pressure waves propagating outward from both the asteroid pathway as well as ground zero and whatever portion of canopy wasn't destroyed by one factor would have been destroyed by the other. As soon as any portion of the canopy disappeared, its insulating value was lost and local cooling resulted which would not have been reversible. Due to the world-wide development of volcanoes following the impact of the asteroids, the entire vapour layer of the pre-existing atmosphere was destroyed resulting in the loss of its climate-moderating and heat-insulating characteristics. Extremes of climate for the whole world followed. Temperature reduction near the poles would have been most dramatic while the temperature near the equator would not have dropped at all and might, in fact, have risen.

7.5.2   The Dust Cloud

The impacts produced a massive sunlight-blocking cloud, which was augmented by volcanic clouds. While volcanic clouds come with various characteristics, depending on the nature of the particular volcano, the impact activated numerous volcanoes so all types would have been involved. Even one major impact would have stirred up dozens of volcanoes. The darkness produced by the impact clouds would thereby have been reinforced. While it cannot get any darker than dark, the additional activity from the volcanoes would guarantee that the whole world was included in the darkness. While total darkness would be most unpleasant, it could be survived until the interruption of photosynthesis killed the plants. However, darkness also means lack of sunlight which means cold and cold cannot be survived.

The aggregate cloud cover produced by the impacts and the resulting volcanoes would reflect some sunlight and block the rest. It is the constant inflow of sunlight toward the Earth that keeps the Earth warm. The drastic cooling would have soon caused all the water to freeze, cutting off drinking supplies. Also, the plants would have frozen cutting off food supplies. All creatures need both food and water on a continual basis. Even short-term interruptions would jeopardize survival, but longer term interruptions would be fatal. How cold it would have gotten due to the great dust cloud is speculative but dropping below freezing would be expected to happen within hours. Without sunlight, the temperature of much of the Earth would drop and fall below freezing within hours. (479)

7.5.3 The Moisture Cloud

As the volcanic activity warmed the ocean, it commenced to evaporate. The evaporant formed great banks of moisture clouds (i.e. storm clouds) which blocked the Sun as effectively as the dust cloud higher up. Any openings in one layer would have been covered by the other layer. The resultant darkness and cooling would have been continuous and world-wide.

7.5.4 CO2 Chilling

As discussed previously, the ancient Earth was universally warm. There is evidence that the Earth's polar regions were a lot warmer at one time than they are today. This is confirmed by the Antarctic expedition of Shackelton during 1907-09, when he found seven seams of coal between three and seven feet thick. He also found preserved remains of warm-water coral. (480) There were even forests on Axel Heiberg Island, Spitzbergen Island and the New Siberian Islands. (192) If lush forest thrived in these places, one can only imagine what it was like elsewhere. The key to all of this plant growth was warmth and illumination. Since everything happens for a reason something caused it to be both warm and illuminated. The Vapour Canopy was a key ingredient in this arrangement but the extensive forest growth indicates that the level of CO2 was probably much higher as well. (53) CO2 also has a blanketing effect (i.e. the greenhouse effect) but would not have favoured areas near the poles particularly. However, as the high level of CO2 was reduced, its accompanying greenhouse effect would have been lost. The CO2 wasn't lost from the Earth but it was lost from the atmosphere by being absorbed into the ocean and absorbed into freshly-exposed rock. At room temperature, CO2 will dissolve in water at the rate of '0.9 volume of the gas dissolving in 1 volume of water at 20 C.' (481) However, as the temperature drops, water will dissolve more. 'At 0 C, 1 volume of water dissolves 1.79 volumes of the gas.' (i.e. CO2) (482) In other words, water near the freezing point can absorb twice as much CO2 as it can at room temperature.

After volcanic heating subsided, the ocean cooled due to its own evaporation. The earlier collapse of the vapour canopy caused the atmosphere to cool which in turn further helped to cool the ocean. The ocean started to take up CO2. All during the build-up phase of the Great Ice Age the atmospheric level of CO2 dropped. (53) As the level of CO2 was reduced in the atmosphere, its associated greenhouse effect was reduced. Thermally, this would have been similar to the loss of the Vapour Canopy but it would have taken longer. Consequently the atmosphere cooled a little more and the ocean cooled a little more, resulting in more CO2 absorption into the ocean which resulted in further chilling of the atmosphere. This type of arrangement is called a positive feedback loop wherein the results of the activity make the activity happen even more. (Unfortunately the opposite is happening now where atmospheric heating is warming the ocean and forcing some of the CO2 in the water to come out (or possibly not allowing it to be absorbed in the first place which would have had the same effect). This positive feedback will accelerate the warming of the Earth.) CO2 was continually absorbed over the ensuing years as the Earth further settled into a lower temperature climatic condition.

7.5.5   Snow Cover

As discussed in a previous section, the post-impact winter which commenced within hours of impact would have quickly resulted in lowering temperatures over northern land areas to below freezing. Precipitation would soon have turned from rain to snow. The collapse of the vapour canopy would have resulted in a massive amount of rain. As the impact-produced cloud expanded and spread, larger and larger areas of the world chilled. The falling rain might have gone through a period of freezing rain before becoming entirely snow or it might have switched back and forth as the temperature drifted down past freezing. A post-impact winter developed and expanded as the cloud cover thickened and spread.

Even if it wasn't snowing over 100% of the land, wherever the post-impact winter developed, it would probably have turned the land white anyway due to hoar frost. When the temperature and humidity are just right, hoar frost will develop. Crystals of ice form right out of apparently clear air and a pattern of ice crystals could readily have formed on everything just as they do now during certain times in the winter. The result from a heat loss perspective would have been the same as if snow had fallen and covered everything. The ground would have been very reflective either way and the loss of the Sun's heat due to it being reflected would have augmented the chilling effect.

The thick, black, dust cloud was the initial cause of cooling because it prevented sunlight from getting through. Subsequently, the moisture-filled storm clouds had the same result. When the clouds eventually thinned out and some solar energy was able to penetrate, the snow cover would have reflected almost all of it away. As the snow and ice started to form, particles from the dust cloud would have been settling out, darkening the snow surface. (probably creating a layered effect as noted in the Greenland ice cores) The darker the surface was, the more heat it would have been able to absorb. However, as the dust and particulate matter in the cloud kept settling out, the snow cover kept building up. The Earth was continually being covered by bright fresh snow. As more and more of the dust settled out, the snow became less and less darkened by the dust. Eventually it would have been quite bright and would have assisted the cooling trend because it would have reflected most of any solar energy that penetrated the clouds. Reflection would have been most effective at the higher latitudes where the albedo may have resulted in more than three-quarters of incident solar energy being reflected. The albedo (i.e. the reflective factor) of fresh snow in an area free of trees and other vegetation, which can create an uneven surface, can exceed 0.7 which would leave less than 30% for any other purpose. (483) As the snow cover increased, vegetation would have been completely buried anyway and the air would have been increasingly free of dust so the reflectivity of the snow would have been near maximum. In this way snow tends to preserve itself. Solar radiation is mostly reflected from a snow surface so the air above an extensive snow cover is colder. Also atmospheric pressure decreases rapidly with altitude in the colder air and creates an upper cold trough. (484)

7.5.6   Orbital Change

The direct cause of the Great Ice Age was the impact of numerous large asteroids with the Earth. All of these objects would have approached the Earth from behind and every impact would have nudged it forward in its orbit. The forward momentum of every asteroid would have been added to the forward momentum of the Earth pushing it forward and into a slightly higher orbit. This in turn means that the length of the year would have been slightly increased. (This is discussed in the chapter entitled; The 360-Day Year.) It also means that the Earth would have received a little less energy so it would have settled into a slightly cooler equilibrium which, like the loss of the Vapour Canopy, would have been irreversible.

7.5.7   Ongoing Chilling

These developments; Vapour Canopy collapse, CO2 absorption into the ocean, orbital change and snow field development would ensure ongoing chilling after the initial chill producers, the clouds, had dissipated. The Earth would thereby have been prevented from returning to its pre-impact level of warmth.

The chill factors not only caused chilling, they also introduced different types of chilling. The dust cloud was a transient chill producer. The great dust cloud caused the Earth to chill for a few months. Then, as it dissipated, that particular chill factor was gone. The storm clouds were also a transient type of change but they would have lasted for several years instead of just a few months. The collapse of the Vapour Canopy brought a unilateral step-function change. This type of change happens once and remains in place. The circumstances producing the change do not go back and forth. Once this type of change is introduced, it remains that way for all future time. CO2 absorption, on the other hand, was an exponential type of change. This is actually the type of change that usually occurs in nature. In particular, when a change follows from an initiating event, the dependent change might follow an exponential relationship. With an exponential change, the same percentage of change takes place for every similar time period of interest. For example, if there was a 10% change the first month, there would also be a 10% change the second month. The difference is that at the beginning of the second month, there is less of the initial value to work with than there was at the very beginning. With such a constant percentage relationship, the parameter undergoing change, changes less and less all the time. For example, 10% of 10 is 1, which only leaves 9 to start the next period. Then 10% of 9 is 0.9, so the change the second month is less. Similarly, during the third month the change would be less again. CO2 absorption into the ocean would be expected to approximately follow an exponential type of change. Rather than being abrupt like the collapse of the Vapor Canopy, CO2 absorption settled down to its final state quite gradually many years later. It would remain that way until some distant future time when the oceans started to warm up again. The type of change brought by snow cover would be different again. In this case it would build up toward a maximum as the snow covered more ground and then it would taper off years later when the snow melted. Orbital change was not as abrupt as the loss of the Vapor Canopy as it occurred over a period of a few months but it was just as irreversible.

Both types of cloud caused cooling for as long as they existed but when they dissipated their cooling effect disappeared with them. The cooling effect caused by Vapour Canopy collapse would occur within weeks and be irreversible. Absorption of CO_2 into the ocean would continue for many years and also be irreversible – at least in the short term. (i.e. hundreds to thousands of years) The chill factor introduced by snow cover would taper off as the snow melted. The chill factor brought by an orbital change would have been relatively sudden and irreversible.

Collectively these chill factors fulfilled the second NECESSARY condition to have an ice age. This, together with the heating factor, provided SUFFICIENCY for an ice age to happen. Therefore the Great Ice Age happened.

7.6   Snow Field Development

7.6.1   Atmospheric Circulation

With the collapse of the vapour canopy, temperature discrepancies developed between the equator and the poles with the result that global air circulation patterns developed where previously there had not been any. Currently, global air circulation is understood as consisting of three major circulation cells both above and below the equator. These are the Polar Cells, (extending from the polar regions respectively to about the 45th parallels of latitude) the Ferrel Cells and the Hadley Cells (which divide the remaining space between the 45th parallels and the equator). The Hadley Cells operate near the equator. Starting at the equator, Hadley Cell circulation rises upwards until it approaches the tropopause which is usually about seven miles up. However, during the months immediately following impact, these air currents rose much higher because ozone had not yet formed in the upper atmosphere producing a tropopause. It had been prevented from forming because the water vapour in the canopy had prevented ultraviolet energy from reaching the oxygen in the atmosphere and forming it. 'The amount of ultra-violet light that reaches the surface of the Earth depends on … water vapour. On heavily clouded days, very little ultra-violet light gets through to the ground.' (485) The warm moist tropical air therefore rose quite high, released its load of moisture and drifted north. (south in the southern hemisphere.) The falling rain partially cleansed the atmosphere. The air moved north and when it descended, it brought more dust down to where it too could be rinsed out of the atmosphere. In this manner the Hadley Cell circulation was most effective in removing dust from the atmosphere over the tropics, allowing the Sun to shine once again and heat the Earth. Post-impact winter (as well as the Ice Age) was not very long in the tropics.

While Hadley cell circulation is certainly still in existence, it cannot remove dust from the upper atmosphere anymore because its movement is restricted to the lower atmosphere below the tropopause. This restriction results from the presence of ozone. After the Vapour Canopy collapsed, ozone started to form and accumulate and the atmosphere then became layered into the troposphere near the Earth and the stratosphere higher up with the demarcation being the tropopause. If this arrangement had been in place immediately following the asteroid impacts, vertical movement of air would not have been high enough to effect cleaning and the dust would have remained airborne much longer (as it does now following a major volcanic eruption which has enough power to raise dust until it is much higher than the tropopause).

Ozone development did not happen until after the Vapour Canopy was destroyed. Ozone develops when UV radiation from the Sun strikes oxygen. However UV radiation is restricted by moisture. Prior to the first impact, the Vapour Canopy was in place and the oxygen in the atmosphere was effectively shielded from it. Similarly when the massive moisture clouds were allowed to circulate to great heights, their moisture also shielded the oxygen and significant development of ozone was delayed.

Atmospheric Circulation

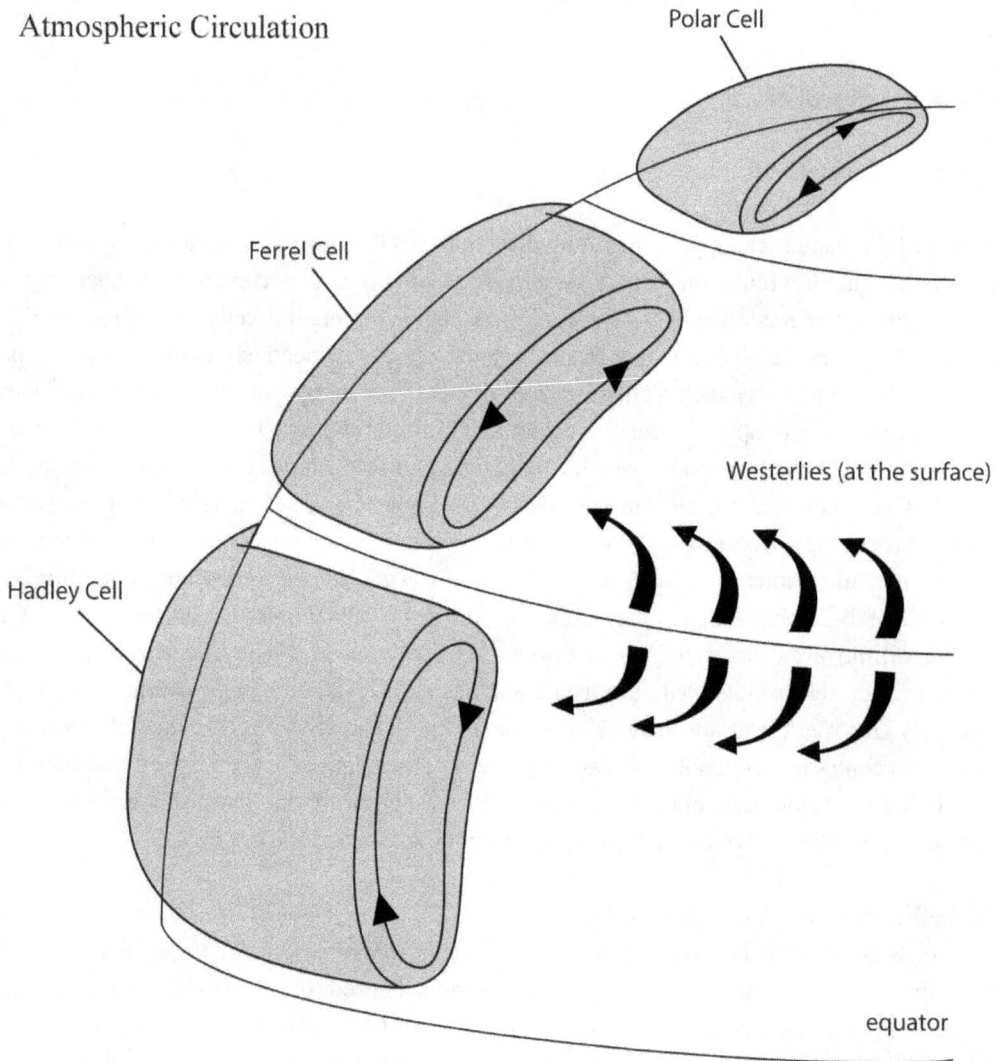

Polar Cell

Ferrel Cell

Westerlies (at the surface)

Hadley Cell

equator

246

Instead of Hadley Cell circulation, the mid-latitudes have Ferrel Cell circulation, which proceeds in the opposite direction. In the Hadley Cells, before air moves away from the equator, it will have risen until it is high above the Earth. The entrained moisture then condenses and rain falls right away. With Ferrel Cell circulation, northward movement proceeds initially along the surface of the Earth. The moisture entrained in the moving air will therefore move along the surface and remain suspended in the clouds until an area cold enough to condense the moisture is reached. With this arrangement, moisture from warm ocean surfaces is carried from lower latitudes to higher latitudes. This meant that during the Ice Age, moisture-laden air was continually carried northward along the surface of the Earth until it cooled and dropped its moisture load. As soon as it reached areas below freezing, snow fell. The moisture-carrying clouds would also have helped keep the ground cold but some of the moisture would have fallen as rain before turning to snow farther north. It took years to cool the oceans and as long as they remained warm and the land remained cold most of the time, the snow would have kept falling. It would have been a continuous conveyor belt of snow.

In particular, it is Ferrel Cell circulation that brings air from the Gulf of Mexico across the USA right into Canada. During the Great Ice Age, this pattern of air movement would have continually brought snow to the northern USA as well as the eastern half of Canada. The snow that dropped over the northern USA would have piled up and formed long tongues of ice, extending well down into the northern states. (This would be expected if the air currents repeatedly followed the same pathways as they commonly do at the present time.)

As storms from the Gulf of Mexico move north across the USA, they drift to the east due to the Coriolis Effect (the effect introduced by the rotating Earth). In this manner, moisture from the Gulf of Mexico would have provided the snow for the north-eastern USA and eastern Canada. The great ice sheet covered more land than this however and extended further west than the moisture from the Gulf of Mexico could have supplied on its own. (486) However, it wasn't on its own. Sea level prior to the Ice Age was probably about 200 feet higher than it is at present. (487) The ocean therefore extended well up into central North America on either side of the present Mississippi River. Along with the Gulf of Mexico, this vast surface of warm water was available to feed the snow-producing engine and it remained this way until enough water became tied up in the snowfields to drop ocean level and thereby reduce the surface area available for evaporation. (In fact, from the Gulf of Mexico all the way to Hudson Bay and the Arctic Ocean, with minor exception, the elevation of the land is much less than 1000 feet above sea level even now. (488))

Moisture for glacier build-up also came from the endless puddles and lakes that were left over from the inundations resulting from the impact. One effect of a major impact is to produce massive waves which would have, in many cases, rolled right over the land. Most of this water would have returned to the ocean but any location which was the least bit depressed, would have been filled and stayed filled until the water either evaporated or left for some other reason. At the beginning of the Ice Age the surface of the land included numerous lakes, ponds, and puddles and all of them were available to contribute to glacier accretion. Later when the glaciers were melting, some of these bodies of water reappeared.

One difference between present Ferrel Cell circulation and the circulation during the Ice Age relates to extent. At that time the Arctic Ocean was warm. Therefore the air above the Arctic Ocean had no tendency to drop to help form the Polar Cell so there probably wasn't any Polar Cell of any consequence. It is therefore possible that Ferrel Cell circulation would have continued much further north and right out across the Arctic Ocean. The situation might have stayed that way for years until the Arctic Ocean chilled and crusted over with ice. When this happened, the northern Polar Cell would have been able to develop. At the South Pole, the Polar Cell developed much sooner because there is land at the South Pole which chilled as soon as the great dust cloud appeared overhead. This enabled descending air movement and the development of the South Polar Cell much sooner than the North Polar Cell.

After cooling and dropping its moisture load, the northward-moving air probably continued out across the warm Arctic Ocean. There another load of moisture would have been picked up. Then, the flowing air mass would have risen and folded back over the land on its way to completing its Ferrel Cell loop. As on the way north, it cooled once again while passing back over the land on its way back south, so a second moisture load was dropped and snow fell across northern Canada and Greenland. In this manner, the Arctic Ocean contributed to ice-field accretion.

A similar return flow enabled the Arctic Ocean to also contribute to the ice-fields of northern Europe and Asia. In the European case, the primary source of moisture was the North Atlantic, over which the air currents veer to the east the same as they do over North America. Eastern Siberia was spared from massive ice accumulation because it is too far from the north Atlantic as well as from the warm Indian Ocean to the south.

Above an ocean surface there is no restriction to air movement so the air is free to rise away from the surface and drift over the land. Any warm moisture-laden air, which passed over colder land, would chill. As soon as it reached its saturation temperature, precipitation would fall and would continue to fall for as far as the air from over the ocean was enabled to circulate over the land. The development of the Ferrel Cell air circulation patterns was like having a great conveyor belt bring water, in the form of humidity, from the ocean and deposit it, in the form of snow, over the land.

7.6.2   Critical Evaporation Rate

The rate of water evaporation from a water surface is directly related to the temperature of the water. This phenomenon is always discussed with respect to vapour pressure. When water is at room temperature, its vapour pressure is four times as high as it is at freezing. (489) This means that warm water readily evaporates with water leaving the liquid surface and being forced into the air right above the surface. As water gets warmer, it evaporates faster. When the boiling temperature is reached, water will not get any warmer and will evaporate just as fast as its source of heat can supply the energy.

In order to form the vast ice-fields of the Ice Age, ocean water evaporated. However, for any of the evaporated moisture to contribute to glacial accretion, the ocean water had to evaporate at a rate that was within a certain critical range. There was both an upper limit and a lower limit to this critical evaporation rate. If evaporation was too fast, the air would have become super-saturated with water. Whenever the relative humidity goes above 100%, air is said to be supersaturated. The atmosphere never stays super-saturated. Whenever moisture is forced into the air above the saturation (or 100% relative humidity) level, rain will fall immediately and it will keep falling until the air is back below 100% relative humidity. This is just basic physics.

As earlier discussed, when the asteroids struck the Earth, numerous volcanoes erupted. Everything that came out of them was hot whether it was gas or ashes or molten magma. All of these hot materials caused the ocean to heat up and in many cases the ocean would have locally boiled. All of these activities would have sent tremendous amounts of moisture into the air but very little of it would have been useful for ice-field build-up. As soon as the hot evaporant rose above the ocean surface, it would have rained. There would have been a torrential downpour in the immediate vicinity of many eruptions. (This happens currently when the effluent from a volcano includes a lot of water.) None of the water involved in these downpours would have made it to the ice-fields and therefore none of it would have been useful for ice-field construction. Hence, there was an upper limit to the evaporation rate that would have been useful for ice-field build-up and that limit was just below the rate that would cause the air to become saturated. As long as the air did not become saturated and its relative humidity did not reach 100%, the evaporated moisture might have contributed to glacial accumulation. Active under-water volcanoes would certainly have warmed the ocean. In many cases, the ocean would have been above the boiling point, just as it is now at underwater vents where the temperature is commonly around 400F. However the only portion of heat that might have contributed to forming snow was the heat that didn't force too much moisture into the air too fast.

There was also a lower critical limit to the required evaporation rate and it is even more important than the upper limit. While initial Earth cooling was caused by dust clouds, the snow-fields had to be kept cool long after the dust clouds were gone. Even at maximum evaporation rates, it would still have taken months to build up enough snow to cause significant glacial ice to form. Therefore the storm clouds had to take over the function of keeping the Sun from shining on the snow until enough had accumulated to form the glaciers. Certain commentators declare that the ice became as thick as the present Greenland glacier. Others think it was actually much thinner. (490) The actual thickness does not concern us. The important factor is that the conditions had to remain near optimum for snow accumulation. As soon as the storm clouds thinned out, two circumstances counter to snow accumulation would have developed. Thinner clouds would have carried less snow and thinner clouds would have allowed the Sun to shine on the snow. As the clouds thinned out, snowfall dropped below the lower critical rate (which was simply the rate at which the snow no longer accumulated). Thereafter, during every warm season the same amount of snow melted as had accumulated the previous winter. Just before this happened would have been the time of glacial maximum. Soon thereafter, more snow melted every year than fell every year and the Great Ice Age was in its recessional phase. (Note; This would not have occurred simultaneously over the entire Earth.)

In order to sustain adequate cloud cover, the ocean had to evaporate at or above the critical level. As soon as the evaporation rate fell lower than this, the storm clouds would not have been adequate to sustain the snowfall fast enough and glacial maximum would have been reached. While the ocean started off warm and was warmed further by volcanic activity, it would have been impossible for the critical minimum evaporation rate to have been maintained if the average ocean surface temperature dropped below room temperature although some commentators suggest a lower level. (68) While this temperature would be considered warm by today's standards it would probably have been barely warm enough for ongoing snow-field accumulation. (The reason for this suspicion is that evaporation is not directly related to temperature. As previously mentioned, vapour pressure for water is four times as high at warm-room temperature as it is at freezing. Half way between freezing and warm-room temperature, vapour pressure is less than one-half of what it is at room temperature. (489) In other words, water evaporates a great deal less below room temperature than it does above it.)

In general, the warmer the ocean was kept, the faster it would have evaporated. This means that the higher the volcanic activity was in the ocean, the faster the glacial ice would have accumulated. One consequence of high volcanism would have been rapid glaciations because moisture would have evaporated into the air at a higher rate. Consequently, ice sheets would have grown rapidly. (496)

In order to form the great snow-fields of the Ice Age, the ocean had to maintain an evaporation rate that was above the critical rate and it had to maintain this rate until the glaciers built up. Anything less and the snow would not have accumulated any farther. Of course as water evaporates the reservoir of water cools and the massive evaporation that was happening would have been cooling the ocean at the same time. Once the source of heat was cut off, it was only a matter of time until the evaporation rate dropped below the critical level. After this, some snow would have continued to fall but even more melted and the great glaciers of the Ice Age started to reduce in size until some of them completely disappeared.

As a corollary, it is clear that the massive glaciers of the Ice Age could not have been built up over a vastly extended period of time (e.g. 100,000 yrs.) because the critical evaporation rate could not have been sustained for such a long time.

7.6.3   Modern Examples

A.   Lake-Effect Snowfalls

There is nothing surprising about the basic way that snow is produced. During winter, snowfalls are a common occurrence and they are always caused by the same circumstances. An open body of water evaporates moisture into the air above it. Air currents carry this moisture-loaded air to the land. If the temperature of the land is cold enough, snow will fall. The greatest snowfalls happen when the evaporating body of water is several degrees above freezing because the warmer it is, the higher its vapour pressure is with more water being forced into the air and being available to fall as snow. As a result of these factors, certain coastal areas commonly receive snowfalls of several feet from a single storm and these higher amounts occur during the fall before the water temperature approaches freezing. In particular, while not currently widespread, snowfalls of more than three feet can occur in one night. (495) 'The Buffalo area is periodically buried under what meteorologists call lake-effect snowstorms. These are intense local snowfalls to the lee of the Great Lakes that result when … air picks up large amounts of moisture as it passes over the lakes. Single-storm snowfalls of three feet or more are common.' (494) While three feet of snow is a significant amount, it is equivalent to only about three inches of rain and it is not the least bit uncommon for a summer storm to bring three inches of rain in one day.

B.   Volcanic Examples

As with the very recent mini-volcanic-winter, which developed after the eruption of Tambora in 1816, the chilling factors of the Great Ice Age did not distinguish between winter and summer. 'There was frost every month … A light snow in much of New England and Western New York State on June 6 was followed by moderate to heavy snow in New England on June 7 and 8. In Vermont there were drifts 18 to 20 inches deep.' (427) Similarly, as the Ice Age developed, it would have kept on snowing through June, July and August.

The deplorable situation in New England developed even though the ocean temperature was about the same as it is now. As the great Ice Age commenced however, the entire ocean was much warmer. The initial availability of moisture was therefore almost limitless and snowfall would not have tapered off until the temperature of the ocean had dropped significantly. In the meantime the blizzards raged on and the snow accumulated until it was thousands of feet deep.

C.   The Lost Squadron

The case of the Lost Squadron was discussed above in, 4,2,4 Faulty Time Assumption, and illustrates very well how quickly snow can accumulate. In that case 250 feet of blue ice and snow accumulated in just 50 years and in fact continues to accumulate right to the present time.

The present day conditions in Greenland are certainly not ideal for the build-up of ice. The ocean around Greenland is certainly not warm and it isn't even open all year. Still, the ice, which covered these WWII planes, was 250 feet thick and it only took fifty years for it to accumulate. Under similar conditions the great glaciers that accumulated during the Ice Age, would only have required one thousand years to reach their maximum thickness. How long would it have taken if the conditions for ice accumulation had been ideal?

7.6.4   The Ice-Cores

Certain ice cores recovered from Greenland have been named NorthGRIP, Grip2 and Grip. Also some have been recovered from Antarctica and have been named Dome Fuji, Vostok and DomeC. 110,000 layers have been counted in the Greenland Grip2 ice core down to the 2800 m level (9200 ft.). (54) Between the 2300 m level and the 2800 m level 25,000 layers were thought to be identified. Each layer is declared to represent one year. (492) The average ice thickness for each of the 25,000 layers would therefore only be (2800-2300)/25,000 = 0.02 m or about two cm. (0.8 in.) Since it requires about ten inches of snow to make one inch of ice, the average snowfall each year for these 25,000 years would only have been about eight inches! The upper 2300 meters included 85,000 layers which would represent 85,000 years. The average ice thickness for the upper region would therefore have been 2300/85,000 = 0.027 m or about 2.7 cm (a little more than one inch). Using similar reasoning the average snowfall every year for 85,000 years was only about ten times one inch or about ten inches. The snowfall which covered the Lost Squadron, produced 250 feet of ice in just 50 years. Since almost five feet of blue ice will result from about fifty feet of snow, the average snowfall every year during those 50 years was about fifty feet. With this in mind, the conclusion that each of the 110,000 layers represents one year of snow and that snowfalls continued year after year for 110,000 years, producing eight or ten inch snowfalls every year, does not hold any credibility whatsoever. Why would snowfall, during the present time, when circumstances are not even optimum for producing snow, result in fifty feet of snow in one year compared to ten inches? The only possible conclusion is that the layers do not represent time at all but only the conditions under which the snow was deposited.

As the land chilled, it started to rain. Very soon it was cold enough to snow. As with all ongoing blizzards, snowfall will not be constant day after day and week after week. There will be intense periods of snowfall interspersed with quieter periods. Neither is there reason to expect that volcanic activity had been constant or that the wind had always remained in one direction. The net result of all of these variations would be layers of snow. The layers would roughly correspond to the variations in activity. Also several periods of snowfall might not even be distinguishable if there wasn't some marker included. On the other hand, markers could show up any time – even several times in one month.

The evidence which is shown in the diagram, 'Greenland Grip2 Ice Core, Oxygen Isotope Ratio Variation' is more suggestive that the build-up of snow during the Ice Age, continued up to the 1500 m level. It would be reasonable to expect that volcanic activity would be generally quieting down and that soot and dust in the air would be disappearing by the time the ice had built up to this level. At this thickness of ice, (1500 meters (i.e. 5,000 feet)) about 60,000 feet of snow would have fallen. This is a vast amount of snow under any set of circumstances, but under the favorable snow-producing conditions of the Ice Age, it was not too much to expect. Actually, a situation like that can be compared to the recent (fall 2008) snowfalls in south-western Ontario, where more than two feet of snow fell each day for two days in a row. If this type of snowfall continued and none of it melted, 60,000 feet of snow would fall in about (60,000 ft./730 ft. per year) 82 years.

Oxygen isotope ratios were also measured throughout the depth of the entire ice core. As shown in the diagram, these ratios have remained fairly constant from about the 1500 m level (5000 ft.) upward. Below this level, the ratio was within a wider range of a slightly different constant. This is suggestive that subsequent to the accumulation phase of the Ice Age, (i.e. the 1500 m level) circumstances changed. Some circumstance introduced a step-function change so that the isotope readings from the 1500 m level upward became much closer to the way they are at the present time. The 1500 m level could have been the end of the overall build-up and accumulation phase of the Great Ice Age on a world-wide basis. Thereafter ice and snow – particularly

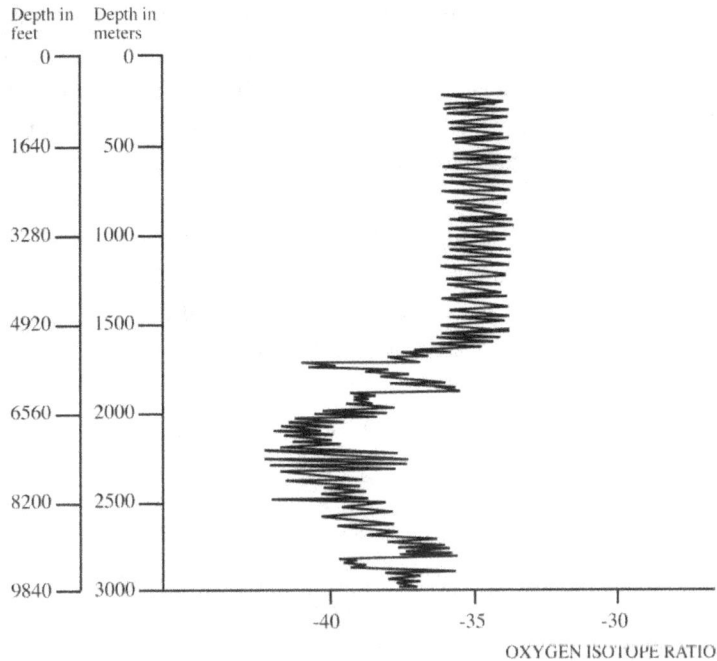

Greenland Grip2 Ice Core Oxygen Isotope Ratio
Variation With Depth

The variation of the oxygen isotope ratio with depth in the ice appears as two distinct categories. The portion above 1500m appears to have been deposited under one particular set of environmental circumstances, while the portion deeper than 1500m appears to have been deposited under a different set of environmental circumstances. This is suggestive that the bottom portion was deposited during the Ice Age, and the top portion since the Ice Age.

at the mid-latitudes – commenced to melt at accelerating rates. At the same time the great Greenland glacier continued to increase in depth even as it continues to do at the present time (491) even though it is melting all around the edges.

7.6.5   Snowfield Attrition

7.6.5.1   Canyon and Coolie Formation

The massive quantities of snow would have kept accumulating up until the ocean cooled down so much that the evaporation rate dropped off and cloud cover became intermittent. Less than continuous cloud cover would have allowed the Sun to shine through and at the mid-latitudes (at first) it warmed up for the first time since the globe-enveloping cloud developed. Thereafter snow melted as much as it fell on an annual basis so the overall accumulation phase of the Great Ice Age was over. Soon melting exceeded accumulation and the great mountains of snow melted at high rates forming rivers that raced to the ocean with unimaginable quantities of water. Across the western plains of North America the flowing water not only included the melting snow but also included unconsolidated surface material leaving behind valleys adjacent to the rivers. Many of these valleys are called 'coolies' and the water flow rates must have been high because the material that has been removed to form these structures is nowhere to be found. There is absolutely no trace of it downstream.

In Arizona the Grand Canyon was formed at this time and it shares the same characteristic as the coolies in that the material that has been removed to form the canyon is nowhere to be found either. It is not downstream which indicates that the flow rates were so high that it was simply carried away to the ocean without leaving any trace behind. In eastern North America the Hudson River is understood to have flowed 200 or 300 feet deeper than it does now. (497) These immense flow rates would have been fed by melting snowwater pouring in from all sides so being in the area at that time would have been extremely dangerous.

7.6.5.2   Submarine Canyons

Submarine Canyons are under-sea canyons that extend from most of the major rivers of the world. These features are not to be confused with Submarine Trenches which is a name given to under-water valleys that run parallel to certain coastlines around the world. Submarine Trenches are always associated with Plate Tectonics Theory which theory is dubious and which requires that the Submarine Trenches be Subduction Zones where ocean floor is being subducted or pushed down under continental material. Unfortunately for the theory if this was happening, the trenches would not be empty and observable. Instead they would be full of all of the ocean floor material that was carried across the ocean floor from the fissures in the ocean floor where it was originally produced. It is also unfortunate that the 'production' regions do not always occur parallel to the trenches but just as often are at right angles to them.

Submarine Canyons are quite well defined physically as they always proceed from the mouth of an existing river - except in a few cases but even in these cases it is obvious that a river had once flowed and emptied into the ocean where the 'orphaned' canyons begin. The canyons themselves always run from a shoreline out across the continental shelf to the ocean deep beyond the continental shelf. This makes it quite obvious that they were formed by flow from these various rivers respectively and that the flow that occurred was quite massive. Clearly and obviously the water that flowed down the rivers continued out into the ocean and flowed across the continental shelves right to the abyss. Another feature of these formations is that at their terminus there are always alluvial fans (material that has spread out) which further indicates that water flow down the canyons spread out at the ocean bottom when there was nothing to restrict it to a narrow well-defined valley. Explanations for these features of the canyons, has not been provided by the interested commentators. This is due to a lack of identification of one characteristic of the Great Ice Age that has been over-looked by most scientists because they subscribe to the Astronomical Theory of the Ice Age whereas they should recognize by now that that theory is incorrect. On the other hand the Asteroid Theory of the Ice Age recognizes that the ocean was warm and at the time that the canyons formed it had not yet completely cooled down. (498) This was after Ice Age maximum had been reached and the snow was melting at prodigious rates. Of course melting snow would be at or very near the freezing point whereas the ocean was still well above the freezing point so the melt-water simply flowed to the bottom of the ocean because it was heavier than the relatively-warm ocean. Being cooler than the water which was already in the ocean was the key factor to the formation of the Submarine Canyons but the extremely-high flow rates also helped because the water did not have time for its temperature to equalize with the temperature of the water in the ocean before it reached the bottom. If the flow rates had been miniscule the melt-water would have acquired the same temperature as the ocean but since the flows were great, there wasn't time for this to happen. As a result the Submarine Canyons formed and provide confirmation that melting rates after ice-age maximum had been reached were extremely high which in turn tells us that once the land started to recover from being covered by snow and ice, it recovered rapidly.

7.7   Conclusion

7.7.1   Ice-Free Areas

More snow accumulated inland than near the ocean. Since the ocean water was still warm, many areas near the ocean remained above freezing. The islands of the high Arctic, as well as parts of Greenland, remained free of significant ice accumulation. (499) 'And what is no less surprising, the northern part of Greenland, according to the concerted opinion of glaciologists, was never glaciated, "Probably, then as now, an exception was the northernmost part of Greenland; for it seems a rule that the most northern lands are not, and never were glaciated," writes the polar explorer Vilhjalmur Stefansson. "The islands of the Arctic Archipelago," writes another scientist, "were never glaciated." (500) In fact, any area which benefited enough from the moderating effect of the warm ocean did not experience the Ice Age. (If coastal areas were ice-free it may also have been partly due the fact that the ocean was initially deeper by both the amount that would soon become tied up in the developing glaciers as well as the shrinkage that would occur as the ocean cooled. These two factors would have caused sea level to be more than 200 feet higher than the present level.) This would have included certain coastal regions in the far north as well as mid-latitudes further south. The Ice Age was not necessarily a period of extreme cold. The warm ocean would have been a large heat source for the atmosphere. Winter temperatures, even over the ice sheets, would not have been extremely coldr.' (501) Everywhere that the ice fields formed had to be below freezing, but not necessarily extremely cold. Snow forms and falls more abundantly when the temperature is just below freezing than when it is much colder because very cold air cannot hold very much moisture or even pick it up in the first place.

While the extent of the Ice Age glaciers is commonly understood to have been restricted, excluding certain arctic islands for example, this does not mean that such areas were not snow covered. Glacial ice will not form until about 80 feet of snow has accumulated. Glacial ice may have been restricted to the areas commonly mentioned when these matters are discussed, but additional areas could also have been covered by up to 80 feet of snow or more. If this had been the case, there wouldn't be any ice-indicating sign left so the matter is inconclusive. However it would be surprising if the only snow that fell was deep enough to form glacial ice while nearby areas were left snow free. It is more likely that deep snow cover was very extensive and that it was simply deeper where glacial ice actually formed.

7.7.2 Climatic Transition Period

The ocean water was initially warm and readily evaporated. The temperatures of the oceans were warm and warm ocean temperatures are associated with increased rainfall over continental areas. (502) Thus the great snow-producing engine was continually fed, enabling the blizzards to continue. However, as time passed, the temperature of the water in the oceans dropped and evaporation reduced. The colder water absorbed more CO_2. The greenhouse effect of atmospheric CO_2 was thereby being reduced. All other effects of elevated atmospheric CO_2 were also gradually lost with the same time factor as the oceanic absorption of the CO_2.

It is suspected that as the ocean water cooled, atmospheric CO_2 was absorbed and the absorption generally followed an exponential relationship with time. Absorption would have been facilitated by movement in the cooling ocean. As the water cooled, it would start turning over. This would continually bring previously-unexposed water to the surface, exposing it directly to the atmospheric CO_2. The climate gradually became colder as this was happening until by the time most of the atmospheric CO_2 was absorbed, the temperature of the ocean had dropped to near its present level and the climate of the world settled out similar to the way it is now. CO_2 absorption tapered off as ocean temperature approached its present level (which is only a few degrees above freezing (503)) because the water had reached its CO_2 saturation level. It could not hold any more and the temperature could not drop any further (or the water would have frozen). The time period from initial asteroid impact until the pre-impact CO_2 was absorbed down to a much lower level and ocean temperature settled out near freezing, was the climatic transition period interfacing the pre-impact climate with the present climate.

Fortunately, there is one set of ancient data available to us which appear to have this familiar exponential relationship and which therefore appear as a possible indicator that some significant factor of nature was being adjusted and that other factors of nature were tracking the adjustment – in this case that the climate was undergoing some dramatic transformation. These data are shown here graphically as 'Genesis Human Age Data'. These data do come close to an exponential relationship. It is certainly not likely that the recorders of this ancient data had any knowledge of exponentials. It seems that they were only keeping records. However, if these data were indirectly tracking the absorption of CO_2 during and following the Ice Age, the implication would be clear that the Ice Age occurred during the time that humans were on the Earth and during a time when human life spans were dramatically shortening.

Genesis Human Age Data

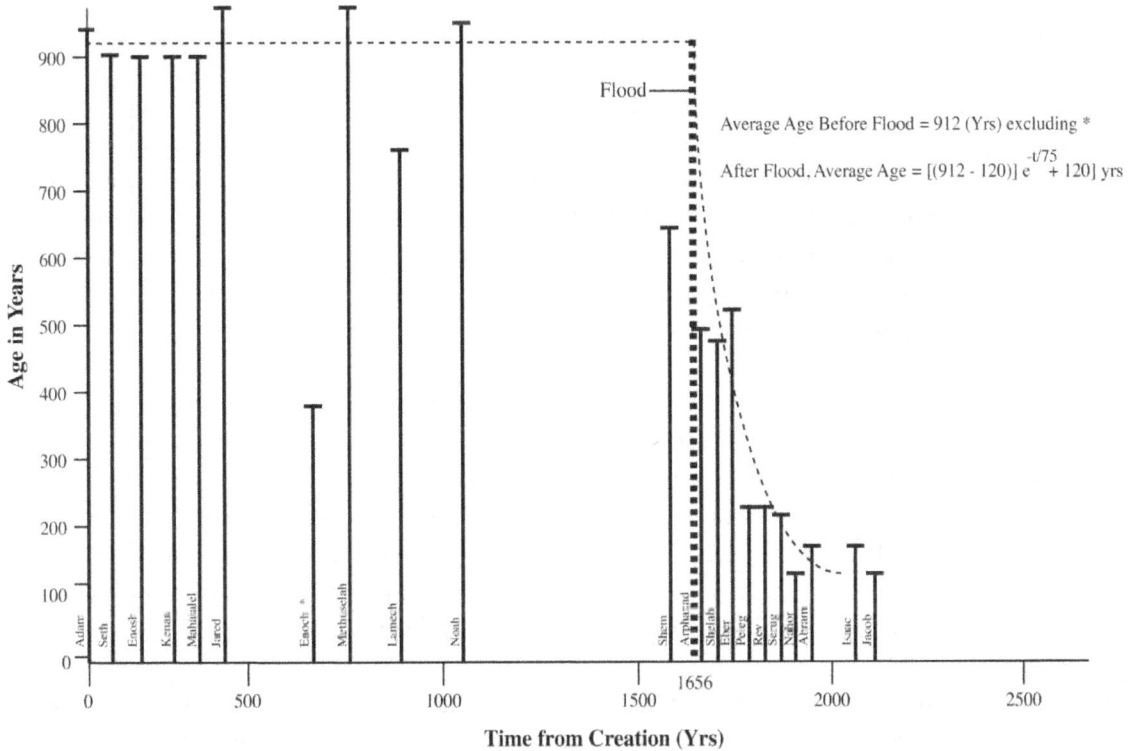

Age in Years

900 —
800 —
700 —
600 —
500 —
400 —
300 —
200 —
100 —
0 —

Flood —

Average Age Before Flood = 912 (Yrs) excluding *

After Flood, Average Age = $[(912 - 120)]\, e^{-t/75} + 120]$ yrs

Adam, Seth, Enosh, Kenan, Mahalalel, Jared, Enoch *, Methuselah, Lamech, Noah, Shem, Arphaxad, Shelah, Eber, Peleg, Reu, Serug, Nahor, Abram, Isaac, Jacob

1656

Time from Creation (Yrs)

0   500   1000   1500   2000   2500

The understanding that the Earth cooled down significantly following a major impact is supported by evidence of cooling from plant remains and the oxygen-isotope method. Immediately prior to the Chicxulub impact, the Earth was warm. This was the Cretaceous Age of Chalk which is believed to have terminated when the Chicxulub impactor crashed into the Earth. This understanding is captured by the following comment which is paraphrased from Sir Arthur Holmes. Annual temperatures were quite warm at 25C. This was the age of Chalk and since the Chalk was deposited the temperature dropped (estimating from pollen and other plant remains as well as by the oxygen-isotope method) and the cooling affected the bottom waters of the Pacific until they were reduced nearly to the freezing point. Today the oceans are cold but at the onset of the Ice Age they were warm and then they cooled off. (504) It is clear from these comments that the Earth underwent a major climatic shift from being universally warm to the climate that exists at present. The cooling being described is much more drastic than the temporary cooling which is commonly thought to follow major impacts. A post-impact winter following a major impact would terminate within a few years and probably start tapering off much sooner. There could certainly be ongoing effects but to suggest that the entire world including the ocean to its very bottom, cooled right down to the freezing point is recognizing a major worldwide unilateral climatic shift.

7.7.3   Ice Age Duration

The Great Ice Age was of short duration. It is physically impossible for the Great Ice Age to have lasted for more than a few hundred years because of the greenhouse effect. It is the greenhouse gases along with the incoming energy from the Sun that keeps the surface temperature around the world in the comfort zone. This arrangement cannot be violated. A worldwide reduction in the major greenhouse gas, water vapor, would result in the entire world going into the deep freeze and staying there indefinitely. In fact it would stay there until the Sun had increased its energy output enough to supply the warming effect that the water vapor had previously provided. According to nuclear physics theories this would require at least another billion years. Who can wait that long? Having greenhouse gases in the atmosphere is not optional. It is absolutely necessary or the temperature on the surface of the Earth could not be within the range necessary for life to exist.

During the Great Ice Age the temperature over the land was reduced which enabled snow to fall and ice to pile up. This reduction in temperature was caused by the two layers of cloud cover. One layer was a dust layer caused by the impacts and the volcanic activity. The second layer was moisture caused by the rapidly evaporating ocean. So the Earth chilled over the land but it did not chill over the ocean. Hence the average temperature around the world did not plummet. If it had plummeted all hope would have been lost and the water vapour levels in the atmosphere would have dropped and the associated greenhouse gas effect would have been lost with them. An irreversible deep freeze for the entire world would have resulted.

Instead the oceans remained warm and evaporated at prodigious rates providing snow for the ice fields. However within a few years underwater volcanic activity subsided. Thereafter the ocean cooled due to evaporation. As it cooled, cloud cover became intermittent and the Sun shone through and melted the ice. The temperature over the land increased during this time. It was the combination of dropping ocean temperature and rising land temperature that kept the greenhouse effect, on the average, within the range which regulated surface temperature and kept it within the comfort zone. The greenhouse effect was preserved and without that having happened the habitability of the Earth would have been irreversibly reduced and life as we know it would have been curtailed.

All of this means that the Earth has never suffered an ice age which lasted for thousands of years. If this had ever happened, the Earth would have locked up cold and would still be frozen solid. The energy from the Sun is simply not sufficient to keep it warm enough on its own without the moderating effect of the greenhouse factors.

The duration of the ice age was therefore directly tied to the temperature of the ocean. As the ocean temperature dropped the land warmed up and the greenhouse effect remained intact. How long did it take for the ocean to cool? As evidenced by the Submarine Canyons which are explained by a warm ocean, the ocean was still much warmer than glacial melt water at the time the glaciers were melting. (498) However, evaporation requires a great amount of heat and as the ocean evaporated, it chilled. It would have cooled substantially by the time the glaciers were melting but it was still warm enough to enable the cold melt-water from the glaciers to form the Submarine Canyons.

Volcanic activity would have started to subside after the asteroids stopped falling and the great shockwaves stopped bouncing through the interior of the Earth, disrupting the crust. Thereafter the chilling effect of evaporating several thousand feet of water would have caused the temperature of the ocean to drop. Since CO_2 is held in the ocean by temperature, a drop in temperature would have enabled the ocean to take up atmospheric CO_2 as it cooled. All through the entire event the dropping ocean temperature was balanced out by rising land temperature so greenhouse gas stability was retained and the average temperature of the Earth stayed within the habitable range.

At the very most the Great Ice Age could not have lasted for more than a few hundred years. All during that time heat and cold on a global basis was held in balance so that the greenhouse effect was not lost because if it had been lost, the Earth would have become uninhabitable and would have remained that way to the present time.

With these realities in mind it is clear that the Astronomical Theory of the Ice Age is a fairytale but that the Asteroid Theory of the Ice Age is valid.

7.7.4   Summary of Main Events

The sequence of circumstances, which characterized the Great Ice Age, may therefore be summarized.

1. The Earth was universally warm, and the oceans were universally warm. Temperatures were between 12 and 25 degrees C. (54 and 77 degrees f) (539)
2. The first of a series of large asteroids crashed into the Earth.
3. A great deal of dust was projected into the atmosphere, which blocked solar energy from reaching the ground resulting in a serious chilling effect.
4. The protective atmospheric Vapor Canopy was destroyed with the result that heat was lost to space.
5. Hundreds of volcanoes erupted contributing to the dust cloud as well as heating the entire ocean until it was much warmer than room temperature.
6. The warm ocean evaporated water at a prodigious rate. Moisture clouds formed, helped to block out sunlight and carried thousands of tons of moisture landward.
7. Over the chilled land, snow formed in the moisture clouds and fell at very high rates.
8. As both the Hadley and Ferrel atmospheric circulation cells initially developed, they rose to very high levels in the atmosphere.
9. At these levels they entrained the dust, and brought it down to much lower levels where rain and snow removed it from the air.

10. The anticipated post-impact winter was thereby short-lived and ended much sooner than it might have otherwise.

11. The warm ocean surface continued to force prodigious amounts of moisture into the air.

12. Underwater volcanic activity, which was repeatedly reactivated for several months by ongoing asteroid impacts, kept adding heat to the water which caused the high evaporation rate to be maintained.

13. Snow continued to fall until huge mountains of it built up and under their own weight they pressed down and formed the great glaciers of the Ice Age

14. Widespread blizzards repeatedly developed and raged on week after week and month after month.

15. The snow continued to accumulate until the oceans cooled substantially and could no longer evaporate water at a fast enough rate to generate continuous cloud cover, thereby block the sunlight, keep the land chilly to enable ice-field accretion.

16. The loss of the vapour canopy, and the extensive snow cover caused the temperature of the polar regions to become and remain cold.

17. Ocean temperature dropped in response to a colder atmosphere, as well as the massive evaporation which had been going on for months.

18. As the ocean water became colder, it absorbed more CO_2 which further chilled the atmosphere, enhancing the further chilling of the ocean. (This could be called a negative feedback loop because the reducing result further diminishes the cause.)

19. Cooler oceans resulted in less evaporation, less cloud cover and less snow production.

20. Solar energy increasingly reached the surface of the Earth in the mid and upper latitudes.

21. However, since the moderating effect of the water vapour layer was lost, the greenhouse effect of CO_2 was reduced and large areas were snow covered, atmospheric temperatures did not and could not return to their pre-impact levels.

22. Soon, the quantity of snow that melted during the summer matched and then exceeded the amount that had fallen the previous winter. Summers were quite short at first but quickly became longer causing the snow at the mid-latitudes to melt at high rates.

23. By this time, ice field accumulation had maximized and the great ice fields, which had formed in North America and Eurasia stopped accreting and started to diminish. While the northern continental Ice Age was over when the bulk of the continental glaciers melted, the Greenland and Antarctic Ice Ages are still not over and will not be over until these areas are also free of year-round ice.

24. Due to the large number of impacting asteroids, the Earth was nudged into a slightly higher orbit which changed the year from 360 days to 365 ¼ days. (see par. 13.20) This further reinforced the generally-cooler climate of the Earth.

8.0   Extraterrestrial Life

8.1   Introduction

The idea that there is life elsewhere in the universe is very popular and does have some rationale to back it up. The most common argument recognizes that there are a very great number of stars in the milky-way galaxy and with such a large number there must surely be places where life does exist. This idea is not new and has been around for years. It has simply expanded as our knowledge of the universe has expanded. It has only been a relatively few years since the whole idea of there being other galaxies was identified. Prior to that, the Milky Way Galaxy was the extent of our collective understanding with the Solar System being the only assembly of planets so attention was naturally restricted to it but that did nothing to hamper the conviction that life must surely exist elsewhere. Mars was a prime suspect and there was great excitement when Giovanni Schiaparelli identified canals on Mars in the late eighteen hundreds. (Scienceblogs.com) This idea was reinforced when Percival Lovel – a well respected astronomer – also thought that he saw the canals of Mars and so he drew a diagram of them. Those were the days when astronomers actually looked through the telescope. This practice has long been abandoned and now cameras are located at the focus points and pictures are taken instead. Large telescopes are very precise instruments and it was always a cause for concern when an astronomer actually climbed onto the instrument and took his/her position. This was commonly at the primary focus at the upper end and the uncertainty of the weight of the astronomer and his/her instruments always necessitated a rebalancing of the instrument. Neither was it a pleasant job. All night long they would be sitting in a very cramped position with their backs exposed to the chilly night air so it was actually a relief when the practice was abandoned. Further the variation in the size and weight of the astronomer was occasionally compounded by the uncertainty of his personal equipment. There was a case involving the giant 200 inch telescope on Mount Palomar in California when an astronomer brought with him an instrument that was too large to fit into the small space at the upper end. At the insistence of the astronomer the instrument was lashed to the outside of the telescope frame and it projected beyond the usual outer limit. In this incident the instrument projected from the telescope for almost one meter (about 3 feet). Things went all right until the telescope was swung around in the darkness of the observatory and the lashed-on instrument crashed into the spiral staircase on the inside of the dome. The instrument came loose and fell to the floor below. It missed the mirror which at that time was the largest one in the world and was practically irreplaceable. The technician was soundly lectured and the astronomer was no longer welcome.

Even as it is commonly the case at the present time that the motivator for a project involving the solar system or beyond is the hope of finding evidence of life out there someplace so it was the motivator for numerous astronomers at that time (1960's and 1970's) that some evidence of life would be found. While hope springs eternal there is, from a scientific viewpoint, no hope whatsoever that life will ever be found apart from the Earth.

The basic conviction of numerous scientists at the present time is that the Earth is about 4.5 billion years old. This idea fits in quite well with the accompanying conviction that the universe is some 12 billion years old. This later idea is based in part on the observations that certain objects in the cosmos are so very far away that it must have taken 12 billion years for the light to get from them to the solar system. There is certainly rationale for such ideas but they do depend on several other factors being supportive of the basic idea. One of these factors is the speed of light. The speed of light has been measured repeatedly and forms the basis for many of the notions in astronomy because it is assumed to be a constant. In fact it is commonly referred to as the constant of the speed of light. Constants are important. We do need references for our personal stability so the more constants that can be identified the more comfortable we feel.

Sea level is a constant. Even though actual sea level at any particular location will drift up and down due to tides, winds and currents there is a particular elevation of the surface of the sea that is recognized as 'sea level'. The idea that it is actually changing at the present time due to global warming is upsetting enough for those who will be directly and seriously affected but it is upsetting for everyone (which includes most of us) who needs such things to be perfectly reliable and just stay the same continually. The 'speed of light' is certainly in this category and the thought that is might have been different in the past or might be different in the future is quite unnerving. We need all of our constants to stay constant. If the speed of light is actually changing, numerous other ideas will be immediately seen to be in jeopardy and this would also be upsetting. The 'speed of light' idea is intimately connected to our concept of time. Unfortunately, time is not very well understood. The last person to do serious exploration in this area was Albert Einstein about one hundred years ago. He understood that time would not always be the same for all locations and all circumstances. It is not really a constant of the universe even though the general thinking up to that point was that it was a constant. (Stephen Hawking also wrestled with the time question and has presented a discussion in his book 'A Brief History of Time'.) If this lack of constancy is actually the case, much of what we think about the universe comes unglued. Did it really take 12 billion years for light to travel from remote regions of the universe to our solar system or did it just take two weeks? What would that do to our concept of the 'Speed of light'? To say the least it would be upsetting. Similarly if our Earth isn't really 4.5 billion years old then how old is it? Many derivative ideas are closely linked to this basic notion and would have to be totally reworked if the Earth was much younger. Unfortunately this turns out to be the case (see chapter 3) which immediately means that a lot of rethinking will be required to once again have a reliable framework for our ideas. Prior to Einstein a clock was a clock and time moved at the same speed no matter where you were or under what circumstances you might be operating. After Einstein this basic way of thinking was abandoned. Something similar will be required when it becomes clear that the Earth isn't really 4.5 billion years old at all.

The question of extraterrestrial life must be approached from a probability perspective. This means that a final, complete and irrefutable conclusion can never be drawn from a scientific viewpoint. This might be disconcerting to some but when a probability is objectively calculated and it turns out to be so infinitesimal that a small leap of faith from there to 'there isn't any life elsewhere in the universe' becomes quite reasonable. Someone can always say 'but'. A probability conclusion might be satisfying to an abstract mathematician or a research scientist but to most of us, including the extremely small step required to draw a firm conclusion will be much more satisfying. In a case like this, absolute certainty could never really be reached unless every star in the universe was investigated. Even then we might get fooled because we could never be certain that something didn't happen just after our backs were turned. Technically therefore, the matter can never be settled. However in order for most of us to live our lives from day to day we must work with more definitive conclusions. The present discussion will work from a probability approach and leave the final step to a firm conclusion up to the reader.

8.2   Life As We Know It

The argument has commonly been raised that there must be life in the universe – especially if we allow for forms of life besides our own. While this has a certain ring of logic about it, there isn't any science to support the idea. 'Life as we know it' is life based on carbon. The carbon atom has a set of characteristics which enables the complex formations and transactions that occur throughout the biosphere to exist and carry on with their many and complicated transactions. No other atom exists that can do these incredibly complicated things. Life - even at the simplest level - is complex beyond our currently-available ability to describe. In fact the simplest assembly capable of reproduction involves hundreds of different kinds of proteins, each of which is a most incredibly-complex structure in itself. There is no such thing as a 'simple' form of life and carbon is necessary for all of this to happen.

For a time, silicon, the closest atom to carbon, was promoted as a possible basic atom on which complex life-forms could develop. This idea was abandoned when it was realized that silicon could not really carry out the needed transactions to get the job done. Only carbon could do this!

Neither can appeal be made to some unknown atom or to some unknown set of atoms. The Periodic Table includes the only types of atoms that exist and it is understood that the entire universe only has these atoms available. The signatures for many of them have already been detected in distant stars and the idea that the particular set of atoms that is currently known is the only one that exists has never been challenged. Therefore carbon is it. No other life-form based on anything else is possible. The only escape from this restriction is to enter the realm of science fiction. In this realm anything is possible and the only limitation is one's imagination. Our current pursuit is much more serious however, so the realm of science fiction will not be included.

8.3   Finding a Suitable Sun

There are several factors bearing on whether or not we have identified a suitable Sun and so the discussion will begin from that perspective. It is common to recognize from the outset that there are many stars in the Milky Way Galaxy. For some time it has been believed that there are about ten to the eleventh power or more plainly stated, about 100 billion stars out there. Of course there are many other galaxies which would initially appear to swing the argument towards convictions like 'it is absolutely certain that there is life in the universe'. We will sidestep that factor for the time being and concentrate on the factors that have an immediate bearing on the discussion. Extending the argument to include other galaxies can always be done at a later time.

8.3.1 The Galactic Habitable Zone

Just as there is a habitable zone for a planet with respect to its immediate star so there is a habitable zone for a star with respect to its galaxy. (A habitable zone is a location where human or animal life could possibly exist.) Unfortunately for the believers in extra-terrestrial life this takes a major fraction of the possibilities out of the picture. The galactic habitable zone is understood to occur at the co-rotation radius. The Sun orbits the center of the galaxy at the co-rotation radius where a star's orbit speed matches that of the spiral arms. Otherwise the Sun would cross the arms too often and be exposed to the hazards that occur within the arms (i.e. such as supernovae). Also it won't go too near the inner galaxy where supernovae are more common. (146) This restriction results in there only being a very narrow band of stars that are possible candidates. Part of the rationale for such a restriction is that further in towards the center of the galaxy the concentration of stars increases quite dramatically and the proximity of any particular star to any other star is increased. This means that radiation throughout the entire area will be increased. A situation like this would never work for living creatures. The radiation count on the Earth at the present time is just about all that animal life can deal with even when we are in a location where other stars are really quite far away. Also we are living on a planet that has a magnetic field that provides critical protection. Otherwise our reasonably good location wouldn't work either.

A certain amount of excitement accompanies the discovery of a planet orbiting a star in the galactic habitable zone. Recently a planet was thought to have been discovered orbiting Gliese 581. The planet Gliese 581g is both in the galactic habitable zone (being close to the Earth this followed logically) and in its stars own thermally-habitable zone. (151) Unfortunately in this case the 'discovery' turned out to be an error in interpreting the data so it had to be concluded that the planet did not actually exist!

Numerous novas and supernovas occur throughout the inner regions of the galaxy and we could not venture there and live for any period of time even if we could get there in the first place. Fortunately our location is far from the center of the galaxy and very close to the co-rotation radius so we can remove that particular hazard from our list. However this restriction dramatically reduces the number of the possibly-suitable-for-life stars to a small fraction of the total that exist and it might be optimistic to suggest that there might be only one hundred million possibilities left. After starting with several hundred billion this does seem like a major letdown but this is the reality of the situation. So instead of starting our probability discussion with several hundred billion we can properly start with one hundred million instead. The procedure will therefore be to multiply one hundred million by the various other fractions to arrive at a probability conclusion.

8.3.2 The Brightness Factor

The brightness factor is one factor which can be identified quite readily. Stars must have a certain minimum brightness (i.e. surface temperature) to be acceptable as a host sun for a life-bearing planet. The star must be hot. It must have a certain minimum temperature to ensure that any planet which might have even a remote possibility of supporting life can remain at a safe distance. We do need heat from our star but we must receive this heat from a safe distance. The distance must be so far away that the star cannot pull up any significant tide. If a planet was too close to its star, the gravity of the star would pull up a tide on the planet that would be too disruptive for life to exist on the surface. Large solar-induced tides are destructive enough on their own but they would also cause the planet to stop rotating. This alone would be the kiss of death. If we can imagine a planet rotating and spreading out the heat that it is getting from its star it might be almost within the realm of feasible to consider it as a possible home for some form of life. However if the planet stopped rotating one side would then continuously face the star and overheating would result. On the other side over-cooling would result. 'Theoretical models predict that volatile compounds such as water and carbon dioxide, if present, might evaporate in the scorching heat of the sunward side, migrate to the cooler night side and condense to form ice caps. Over time, the entire atmosphere might freeze into ice caps on the night side of the planet.' (333) A narrow strip near the terminator might have an acceptable temperature but with all of the water frozen solid on the cold side the situation would not be worth pursuing any further.

If any planet should form within the thermally-habitable zone of a red dwarf star it would, within a very short amount of time, become tidally-locked to the star. Then the temperature inequity around the surface of such a planet would render it inappropriate as a place for life. This is the main reason that astronomers reject any star that is below a certain temperature. In order to be in the thermally-habitable zone, a planet associated with such a star would simply have to be in too close where the tidal effect would be devastating for both surface tranquility and long-term heat distribution. In fact, by the time such a planet was discovered, the de-rotation factor would already be in place and the planet would be locked up, with one side continually facing the star. All red dwarf stars are in this category and there really isn't any point in considering red dwarfs as possible host stars for any life-supporting planet.

The Sun is included in a very small group of stars (approx. 5%) which are bright. This means that the heat produced enables an Earth-like planet to be a considerable distance away and still receive enough energy to enable life to exist. At these greater distances, the tidal effect of the Sun is reduced and, in fact, barely noticeable. On Earth for example, the tidal effect of the Sun on the ocean can only be measured in inches.

Unfortunately, most stars are low-mass stars and are dimmer than our Sun. This means that a life-enabling planet must be much closer in order to receive sufficient heat. However closer means that tides will be higher. 'Science Daily reported that "Tides can render the so-called habitable zone around low-mass stars uninhabitable." Astronomers at the Astrophysical Institute Potsdam studied the effects of tides on planets around low-mass stars (the most numerous stars in the galaxy) and found that the increased volcanism would make them uninhabitable due to the tidal effects. The lead researcher, Rene Heller, stated that remarked that if you want to find another Earth first you need to find another Sun. (542) In other words, in order to avoid the tidal problem of having a planet near a dim sun, an Earth-like planet must orbit a bright sun. This reduces the number of stars as possible energy suppliers for Earth-like planets to less than 5% of all existing stars.

In spite of such scientific realities the inappropriateness of dim stars as hosts for life-enabling planets is occasionally deliberately over-looked. A recent report (winter 2017) recognized an extremely dim red dwarf star as a possible host for life. In this example it was determined that the host star had seven planets in orbit around it and that at least three of them were in the thermally-habitable zone. In this particular case the point about being too close to the star was deliberately over-looked in favour of the observational advantage. 'Trappist-1 is an ultra-cool dwarf star. Small dim stars like this are great candidates for detecting near Earth-sized planets because when one passes in front of the star , temporarily blocking it – a "transit" – the starlight reaching Earth dips more dramatically than it would for a very big bright star like the Sun.' (543) In other words the observational advantage out-weighs scientific truth. The tidal-reality problem would be particularly severe in a case like this because the planets of interest are understood to be orbiting very close to the star. 'Trappist-1 may be small and dim but because the planets all orbit very close to it – their orbital periods range from 1.5 to about 20 days - ... at least six of them sit in the habitable zone.' (543) They might sit in the thermally-habitable zone but they certainly do not sit in the structurally-habitable zone! In this case they are so close to the star that they are almost touching it, meaning that the tidal problem would be over-whelming!

It is well understood that Red Dwarf stars are very numerous and hence should provide an abundance of planetary discoveries. '"Dwarf stars are the most common type of star in the galaxy". said Jayawardhanna. "So if they produce terrestrial planets in abundance, that means yet more opportunities for locales for many many more potentially-habitable worlds in the galaxy."' (543) Searching for extra-solar planets is slow, tedious, boring work. However if there is a significant likelihood of finding something the boredom would be relieved. However, from a scientific perspective this is not valid and relevant scientific realities should not be over-looked in favour of observational advantages.

By eliminating stars that are too dim (i.e. red dwarfs) only about one star in twenty remain as possible candidates. (120) As a probability factor this is 0.05 so we multiply the possible number of stars in the galactic habitable zone by 0.05 and this yields 100,000,000 x 0.05 or 5,000,000.

Some commentators have taken a different approach and have offered suggestions on the number of sun-like stars that exist and decided that there's probably about a billion. (332) No restriction of location was placed on this number but if it is referring to the number in the galaxy, which is thought to have about 300 billion stars, it would only represent about 0.33%. Then this amount must be multiplied by the percentage in the galactic habitable zone, which is probably not more than 0.5% of the number in the galaxy. This procedure also reduces the possibilities to about 5 million. While the actual number can never be known, it is clear that the number of possibilities, by any line of reasoning, will not be a very large number.

8.3.3   The Singularity Factor

A host star must be singular. Multiples of any kind including doubles, triples, and quadruples are to be avoided. The reasoning for this conclusion is that a multiple-star assembly would prohibit any planet in the vicinity from having a circular orbit. A circular orbit is a minimum necessity for temperature stability and it is imperative that if any planet is to be a home for living creatures it must have very tight temperature stability. Even variations of a few degrees would not be acceptable. In fact, the average temperature of the Earth must be exactly where it is at the present time or devastation will result. The current average surface temperature is +15C and the concern among many scientists is that it might rise to +17C. Great devastation would follow a rise to +20C. These really are very small temperature deviations which clarifies that non-circular orbits for a life-supporting planet are simply not an option. Consequently star assemblies of two or more stars are not possible candidates for a life-supporting planet. In spite of this, as with the brightness factor discussed above, multiple-star assemblies are occasionally identified as possible hosts for a life-supporting planet. Recently it was announced that astronomers had discovered a planet with a mass similar to Earth's orbiting the habitable zone of Proxima Centura, the star closest to ours. (543) Proxima Centura is actually a Red Dwarf - part of an assembly of three stars. However since the star being orbited is a Red Dwarf there is no possibility of habitability.

The fraction of stars that this leaves is not exactly known but certain commentators have weighed in on the matter and suggest that most stars are multiples. 'A few years ago Levy and I decided to determine more carefully the frequency of doubles and other multiple stars among Sun-like stars. Levy and I observed all 123 Sun-like stars visible to the unaided eye in the northern hemisphere. Some of the systems observed were triple or even quadruple. Among the 123 primary stars 57 were found to have one companion, 11 to have two companions and three to have three companions. Thus more than half (57%) of the stars have at least one companion. The preceding inventory is only part of the picture, because we are unable to detect all the companions that are likely to be present.' (94) These comments were made more than thirty years ago and since then the percentage of multiple stars observed has increased. 'In fact, … single stars are rare; in spite of the appearance of the night sky to the unaided eye, most stars are actually binary … Our Sun, as a single star, is part of a minority population.' (95) Other commentators offer similar or higher estimates. 'About 75 percent of all stars are members of a binary system.' (96) In addition to the binaries there are triples and quadruples. When these are also included, the proportion of multiple stars is upwards of 80 percent.

It is an absolute necessity that a sun be singular before there is any possibility of an associated Earth-like planet having a circular orbit. (Massive spheres of gas such as Jupiter do not qualify in this case as Earth-like). Most stars are binaries. Instead of just one glowing mass of hot gas, there are two. The two stars revolve around each other and from a great distance it appears that there is only one object. It requires the optical advantage of a telescope with its associated instrumentation to determine that a star, which appears like a single object, is really two objects circling around each other. In some cases, there are three objects circling around each other. It is reasonably understandable that two objects could circle each other for a period of time, but it is not really understandable how three objects could do this for any period of time without their orbits becoming unstable. However, whether this happens or not is quite secondary to the fact that two or more massive objects, (like our Sun) which circle each other, have, associated with them, a gravity field, which includes massive fluctuations. These fluctuations guarantee that no nearby object, like a life-enabling planet, could ever have a circular orbit. In fact it might not have any orbit at all unless it was close enough to one of the stars so that star's gravity clearly dominated the planet's orbit. It is more likely that ' … untold numbers of worlds have … fallen into their suns or been flung out of their systems to become "floaters" that wander in eternal darkness.' (97)

Unfortunately, for those who prefer to imagine a universe, which is teeming with life, most stars are either binary or triple. Some estimates have placed the portion at ninety percent. (98) Examples of binary stars within the near portion of the Milky Way Galaxy are; Eta Cassiopedia, Procyon, Sirius, (brightest star in the sky) Ross 614, Gl 166, Gl 702, Gl 783 and Groomsbridge 34. Examples of triples are; Alpha Centura A and B along with Proxima Centura, (the nearest star to the solar system) Gl 166, Gl 1245, Gl 570, Gl 663 A and B along with Gl 664 and EZ Aquarii. (99)

All of the stars in the vicinity of our Sun are multiples. In particular, the closest star to the Sun is a triple. There is no point whatsoever in considering a planet orbiting in this type of situation as a possible candidate for life support. As a result of all of this observation and scientific comment the probability factor seems to be in the range of 0.1 to 0.2. If we use 0.15 the number of possible life-supporting candidate stars in the galaxy is reduced from about 5,000,000 to = 750,000.

8.3.4 The Dead Star Factor

Occasionally astronomers refer to a star as having a dead partner. It is never exactly clear what this means. There was a time when heavenly bodies were readily placed into categories. There were stars, planets and moons. Identifying which category to use was never a problem because this is the way that the solar system is arranged. The situation is not quite so identifiable for other assemblies of stars and objects that orbit around them. Many are so strange as to confirm the biologist J. B. S. Haldane's famous remark that "The universe is not only queerer than we suppose, but queerer than we can suppose." (92) A dead star would seem to be an object that was once glowing as a star and then it cooled off and stopped glowing – at least in the visible spectrum. It would not seem appropriate to simply call it a planet because planets are not hot – not that hot anyway. A probability will not be suggested for this factor but if any dead stars exist it only reduces the number of possibilities of finding a suitable host star.

In any event if we review the factors from the above categories we have; Probability = 100,000,000 x 0.05 x 0.15 x 2 other factors = <750,000. In other words there is a probability that less than 750,000 stars in the Milky Way Galaxy might be appropriate to support a life-bearing planet.

8.3.5 The Void Solar Proximity Zone

The next factor to be considered is the region immediately round the candidate star. In this case immediate will refer to the region out to the habitable zone and beyond for a distance of about another four or five times the distance from the star to the habitable zone. For example, if the thermally habitable zone was 100 million miles from the host star, the distance out to about 500 million miles must be virtually void of planets of any significant size. The reasoning for this is that large planets within this zone would have a detrimental effect on the circularity of a potentially-habitable planet's orbit. An orbit that is close to circular enables the heat received from the star to be constant. A potentially-habitable planet absolutely must have constant heat input. The surface temperature of any planet that is trying to support life must be about one-half way between the freezing point of water and the body temperature of animals. (The body temperature of animals is very close to the upper temperature at which seeds can germinate.) Even seemingly small deviations from this are seen by many scientists as detrimental. The gravitational influence of even one large nearby planet on a small planet like the Earth would not just cause its orbit to be non-circular but could even fling a small planet right out of the system. 'Even if an Earth-type planet existed in this system (i.e. a bright star called Upsilon Andromedae in the constellation Andromeda c/w three orbiting gas giants) it might risk being obliterated ... ' (106)

Recently, a report was filed that Earth 2.0 has been found. ... It's sun is about half the size of our own, and it completes its orbit in 130 days. It has four neighbouring planets that are closer to the star but Kepler 186f is the only one located in the Goldilock's zone, with conditions allowing for liquid water – and therefore life – on the surface. (114) The size of the other four planets was not given but concern is immediately raised that they might be large and therefore prevent the planet of interest from having a stable circular orbit. Also the size of the star indicates that it is a red dwarf and hence not an appropriate candidate for a life-supporting planet anyway.

Jupiter is approximately 500 million miles from the Sun but Jupiter with its enormous mass is understood to be a stabilizing factor for the orbit of the Earth. We probably owe our existence to Jupiter, whose massive gravity has stabilized the orbits of the other large planets in the Solar System. If Jupiter's orbit was as eccentric as some extra-solar planets, Earth would never have had a chance to have a stable climate. (113) At the same time it is not really a stabilizing factor for objects in its vicinity such as asteroids, as it is credited with modifying their orbits and sometimes causing them to move further into the region near the Earth. It would supposedly also disturb any other object that drifted too close. The best arrangement for monsters such as Jupiter is to keep them at a safe distance to avoid trouble altogether.

By mid 2016 more than 1800 exo-planets had been discovered. This number will only become greater during the coming years. For the time being we could, from a scientific viewpoint, simply state that there is a one chance in 1800 of finding a solar system that meets the 'void-proximity-zone' requirement. The 'one' is our own solar system and the 1800 is all of the other discoveries to date. Therefore scientifically we multiply the previous result by 1/1800 to arrive at approximately 500 possible host stars for a life-supporting planet.

8.3.6   The Sun Size Factor

A host star must be in a certain size range. 'Too small' means that the star is too young. 'Too large' means that the star is too old. Stars are understood to increase in size as time goes by so an old one would probably not last long enough to allow life to develop or if it had developed it would probably have already died out. (i.e. Some commentators suggest that intelligent life would annihilate itself within a few centuries of obtaining the atomic bomb. (68)) 'Too small' would suggest that there hasn't been enough time for life to develop. For example, our Earth is thought by some to be 4.5 billion years old and that human life has only just arrived. Therefore any star much smaller (i.e. younger) would probably not be a good place to investigate because any small life forms would not yet be ready to show themselves. All of this type of reasoning derives from the thinking accompanying the Theory of Evolution which must be accepted if one is thinking in terms of billions of years. Of course if one does not think in these terms, shorter 'temperature windows' would be acceptable.

The complicating factor accompanying such extremely long-term thinking relates to the theories of how a star operates. Nuclear fusion is the prime suspect and with nuclear fusion a star would be expected to steadily increase in both size and brightness (i.e. heat output) as time goes by. Our Sun, for example, is thought to have been about 25% cooler 4.5 billion years ago which begs the question of it being a good source of heat at that time because it would not have been hot enough to have kept the Earth from being frozen solid. If the Earth was ever frozen and covered by a layer of ice, the albedo or reflectivity factor would have been very high. Most of the incoming heat from the Sun would have been reflected away and very little would remain as a source of heat for life. How then would it ever thaw out? There isn't even enough heat from the Sun at the present time to keep the surface temperature of the Earth above freezing if it wasn't for the greenhouse effect. We need both the current input of heat from the Sun as well as the greenhouse gas inventory (to retain some of this heat) just to keep the surface temperature of the Earth where it is at the present time with very little deviation from this point being allowable. In fact without the greenhouse factor - which would not be fully in place until the Earth actually thawed out and released some water vapour into the air - the temperature of the Earth would still not be above the freezing point and would not be for another two or three billion years. This complication really does present a conundrum for 'long-agers' from which it is difficult to exit.

In any event certain commentators suggest that the size of any star that is to be a possible host for a life-supporting planet must be in the range between 0.83 and 1.2 solar masses in order for an orbiting planet to avoid either a runaway greenhouse effect or a permanent ice age, as Mars would be having if it had more water. (332) This immediately eliminates red dwarfs (which had to be eliminated in any event) as well as blue giants (which are understood to produce too much ultraviolet light) as well as a large percentage of 'sun-like stars'. This reduces the number of possible stars to a fraction of the number remaining from the previous restrictions. While a probability factor cannot be applied to this situation with any degree of certainty it is unlikely that even 10% of the remaining stars would qualify. If this were the case the number of possible stars would now be reduced to less than one hundred.

8.3.7   The Solar Stability Factor

Stars are not necessarily stable. Sometimes they explode as novas. Ideas have been advanced to try to explain these explosions and some of them apply to multiple star systems which this discussion has already covered. Also as stars get old they are thought to become unstable but 'old' is a relative term which is difficult to define with any precision. Further, it is doubtful if all types of unstable solar behavior is really understood. Some probability might be attached to this factor but it will not be attempted here. In any case it would reduce the number of possible candidate stars even further.

8.3.8   Stable Solar Mass

In order to be a suitable candidate for a life-supporting planet, a star must have a stable mass. Unfortunately solar masses are not necessarily stable. Stars have an enormous amount of mass. That is, they consist of a great amount of material. The amount of material in any given solar mass determines the orbit of every object that has been captured by it and every object that continues to endlessly orbit around it. In fact, it is the orbital features of an orbiting object that are used to determine the mass of the host star. Therefore, if a star has a planet in orbit around it, the length of that orbit and the time it takes to make a single revolution can be used to determine the star's mass. The same technique can be used to determine the mass of a planet. If a planet has a moon, the characteristics of the moon's orbit can be used to determine the mass of the host planet. Determining mass in the absence of an orbiting object can also be done but it is more difficult. However, if a star has an orbiting planet, the mass of the host star can be immediately calculated.

A problem arises if the host star does not have a stable mass. If its mass should increase, for example, the orbit of a planet would change and in fact it would spiral inward into a smaller orbit. This would not be satisfactory for any planet that was the host for either animal or plant life of any kind. While achieving the proper orbit is most improbable in the first place because there is always such a small margin for deviation it is a wonder that even the Earth can stay in the very narrow habitable zone that is so essential for life because the mass-stability of the Sun ,while unknown is probably increasing.

The orbit of the Earth is referred to as being a Goldilock's orbit. It is 'just right'. It is slightly elliptical and this works synergistically with the surface material of the Earth to provide the best possible conditions for temperature stability in exactly the right range. Orbital deviation is not permissible. In fact it has been determined that even a 5% deviation would spell disaster and the surface temperature of the Earth would be outside of the permissible range for life to exist here at all. (100) What would happen then if the mass of the Sun was increasing and causing the Earth to spiral inward closer to it?

Unfortunately the mass of the Sun is increasing and increases every day. This is because the mass of the Sun is so very great that it continually attracts more and more material into itself. One staggering example is the comet that went into the Sun in 1979. This comet was a monster and various reports suggest that it was larger than the Earth. (101) That might not represent very much mass in comparison to the enormous mass of the Sun but what would be the result if this type of activity continued to happen over one billion years? However, it does continue to happen with comets falling into the Sun on a regular basis. Other material also falls into the Sun and it is a wonder that even more is not pulled in as well. The gravity of the Sun is enormous and very difficult to resist. This makes it hard to send satellites out of the solar system because they must have enough energy to overcome the pull of the Sun's gravity. It would be even worse for any spaceship returning from an interstellar journey because it would have to resist the Sun's gravity all of the way in through the solar system or it too would be pulled in. If a returning spaceship were to enter the solar system at some small fraction of the speed of light it would have to apply the brakes full time just to avoid going right into the Sun at full speed.

The Sun also loses mass. Every time that a solar flare bursts out, some mass is ejected outwards. Products of hydrogen fusion could also be leaving the Sun and this too would deplete its mass. Over the course of a year none of these activities would matter but over the very long time frames associated with most of the theories of the universe they would matter a lot. Could the Sun retain a constant mass and keep a planet in a stable nearly-circular orbit for billions of years? While this type of question can never be settled for certain it does raise a doubt about long- term stability. The Earth cannot be allowed to spiral inwards closer to the Sun by even a small amount or it would overheat. Similarly it could never have been in any orbit further away from the Sun because it would have been frozen solid and would still be frozen solid. If it has been gradually spiraling inwards for 4.5 billion years it must have been further away in that distant past. The Sun would have been dimmer at that time and the double factor of being further away and having a dimmer Sun would have ensured total long-term freeze-up. Since this has not happened the long-term view is cast into doubt. Neither the mass of the Sun nor the orbit of the Earth must have ever changed over the long time period or life on the Earth would never have been possible. The same criteria must apply for all other possible candidate life-supporting stars in the universe. A probability factor will not be applied to this situation because of the sparseness of data but the criteria can plainly be seen to be important.

8.3.9   Stable Earth Mass

While the mass of any sun must have long-term stability in order to be considered as a possible host heat-source for a life-bearing planet, the planet must also have long-term mass stability for basically the same reason. Unfortunately in both cases large objects like stars and planets have a very strong gravity field. Strong gravity fields attract matter. Over the short term this would not be a problem but over the long term it can plainly be seen that it would be a problem. Very simply, the orbit would not be stable and repeatable but would be constantly getting smaller as the planet spiraled into its sun. Either the increasing mass of the sun or the increasing mass of the planet would cause this to happen and long-term stability would not be possible.

8.3.10   Review

The last three factors mentioned have not had a probability factor applied to them because it isn't really possible to identify such a factor from the available information. For consistency however applying a conservative 'for the time being' factor would be reasonable. If we allowed a factor of one in one hundred the number of possible candidate stars is reduced to just a few.

The greatest problem associated with trying to identify a reasonable overall probability factor is the complete lack of agreement among the astronomers and other commentators on the basic factors. There really isn't any agreement on how many stars there are in a galaxy, nor the actual number that are in the galactic habitable zone. Neither is there any solid understanding of how many stars might be Sun-like. Many scientists eliminate red dwarf stars as possible host stars for a life-supporting planet but some still insist on observing red dwarfs for any indication that there might be planets nearby in spite of the recognized tidal problem. The chances of life existing on exoplanets in the traditional in the thermally-habitable zone of Red Dwarfs is pretty bleak due to tidal effects. For another Earth you need to find a second Sun. (330)

There does seem to be considerable agreement that brown dwarf stars would not be suitable but there is little agreement on whether brown dwarfs are stars or planets. In many cases the assumption that stars and their associated planets formed from a collapsing cloud of space dust is used to try and explain the observed lack of consistency in both brown dwarf as well as other formations. Such explanations are repeatedly used in spite of the fact that the very first exo-planets that were identified ruled out the idea. (326) The difficulty in getting a collapsing dust cloud to ignite has also been pointed out but this is also commonly ignored when comments about brown dwarfs are made. (328) Also, in order to be Sun-like, a star must be singular. However occasionally someone will insist that there is a possibility that a multiple-star, under just the right circumstances, could accommodate a life-supporting planet. Such opinions indicate a very strong conviction that life exists out there which is reflective of the conviction years ago that Mars was inhabited.

The great disparity in deciding what is true and what is speculation leads one to recognize that real understanding is hard to find. Never-the-less it is clear that suitable stars for life-support will be in extremely short supply. Where the universe is concerned 'hard' data is rare but strong convictions are plentiful. In spite of these realities the brief discussion offered above points out the great improbability and virtually utter hopelessness of ever concluding that there is any significant number of stars that might be suitable to support life on some far-away planet.

8.4   A Planet-Moon System

While we have really been looking for a planet all along, it is now plainly obvious that looking for a planet prior to identifying a suitable star would have been a waste of time. While this has occasionally happened, the two factors cannot be separated. In fact three factors including star, planet and moon must be simultaneously addressed or the quest will not be fruitful. In spite of this necessity there are a few factors which can be separated out for an initial evaluation.

8.4.1   Mass and Size

A planet can be neither too large nor too small. While there is some leeway with this factor when the planetary discoveries to date are reviewed the leeway is not really very great. The Earth is just large enough to hold an atmosphere. While it cannot hold hydrogen which raises a doubt concerning our long-term water supply, it can hold both oxygen and nitrogen without difficulty. If the diameter of the Earth was only about one-half of its present value (Assuming similar density) the mass would be reduced to about 10% of its present value. The Earth's force of gravity would be reduced (to about the same as Mars) and escape velocity would be lower. It would be impossible to hold a substantial atmosphere under such conditions. On the other hand if the Earth had a diameter that was twice as large as it is at present, its mass would have increased about eight times and the force of gravity would be approximately twice as great. In this case an oxygen-nitrogen atmosphere could be retained without difficulty (as well as several much less desirable gases like ammonia) but it would be more difficult to evaporate water. Astronomers prefer that a planet be in this size range and preferably well within this outer boundary. Of the first 1800 extra-solar planets discovered, only five meet these criteria when considered together with the thermally-habitable zone criteria. (334)

8.4.2   The Habitable Zone

Whenever a far-away planet is discovered, the first question that is usually raised is; Is this planet in the habitable zone? There simply isn't any reason to proceed further with any investigation if a newly-discovered planet is outside of the habitable zone. (In this case we are thinking in terms of the thermally-habitable zone. There is also a structurally-habitable zone.) A planet absolutely must be in a location where it receives an adequate amount of heat but not too much. 'If they (i.e. astronomers) manage to discover a rocky planet roughly the size of the Earth orbiting in the habitable zone – not too close to the star so that the planet's water has been baked away, nor so far out that it has frozen into ice - they will have found what biologists believe to be a promising abode for life.' (163) In the case of the Earth even though it has numerous mechanisms to regulate and distribute heat, its habitable zone with respect to the Sun is very narrow. Just as there is a galactic habitable zone (which is at a very particular distance from the center of the galaxy (146)), so too there is a very particular location for a planet to be in order to receive adequate, but not too much, heat from its host star.

The habitable zone in the solar system is exactly where the Earth is located. This zone is very narrow and it is understood that if we were only 5% closer to the Sun we would seriously overheat. (89) Overheating is easily done simply because the acceptable average surface temperature for the Earth is about half way between the freezing point of water and the body temperature of animals and it cannot be allowed to deviate from this level by more than a few degrees. While the temperature of the Earth is currently at the right level there is serious concern that it is rising due to too much greenhouse gas in the atmosphere. The rise anticipated over the next one hundred years is only a few degrees but if it became more than this, wide-spread disaster would be expected to follow. Even a cursory survey around the near universe will confirm that a temperature change of that magnitude is really quite small. This is of very little comfort however if disaster really does accompany such a small increase. While the habitable zone is really very narrow to begin with, it is well understood that all of our other temperature regulation factors must also be fully functioning because simply being in the habitable zone might not be sufficient to keep the Earth habitable. While being 5% closer to the Sun is considered disaster being a similar distance further away would also be disaster as the Earth would be much too cold. This requirement is partially recognized in the following comment. An astrophysicist named Michael Hart concluded that Earth would have been uninhabitable had it been just 1% farther from the Sun or 5% closer to the Sun. Since then the figures have since been made a little more generous – 5% nearer and 15% farther. (100) While the commentator in this case clearly recognized the narrowness of the habitable zone in general, being 15% further away would result in the Earth receiving about 25% less heat. This would not be acceptable because the Earth would simply freeze up. (50) Other commentators have also recognized the narrowness of the thermally habitable zone. Recent studies show that Earth just barely qualifies as a suitable abode for life. If the planet Earth had been placed in an orbit only 5% closer to the Sun, a runaway greenhouse effect could by now have turned the planet into a hothouse. On the other hand, if the Earth was only about 1% further away we would now have a 'snowball Earth'. (104) While glacier formation requires heat and therefore would not happen if the Earth chilled, the idea that it would become ice-covered and hence uninhabitable, is well recognized by such comments. (We recall that being covered with ice is not an ice age. See chapter 4.)

Drifting out of the habitable temperature range could happen more readily than one might at first think because of the various viscous cycles that would cut in to exaggerate any initial change. For example if the Earth should suffer a chill on a world-wide basis (for any conceivable reason) some significant fraction of our current inventory of water vapour would be lost. Since water vapour is our most influential greenhouse gas, the temperature would fall even further due to this loss. (50) In fact it would spiral down until any further loss of water vapour would not result in any further drop in temperature. Unfortunately, the final temperature would be about -25C. (50) This is the main reason that any outward deviation of the Earth's orbit be limited to only a very few percentage points.

A planet's distance from its host star is the most obvious and easiest factor to identify but ellipticity of orbit is intimately connected. The orbit must be very close to circular to ensure that the strict distance limits are not violated. While the orbit of the Earth is slightly elliptical, this works together with the Earth's axial tilt as well as both the distribution and type of surface material to provide an optimum heating control arrangement. For example, when the Earth is closer to the Sun the southern hemisphere is having summer. This is most convenient because when it is summer in the southern hemisphere the large ocean surfaces are facing the Sun more directly. The water is therefore able to absorb the extra heat and release it later thereby helping to regulate the temperature in a most advantageous manner. Also, when it is summer in the northern hemisphere, the Earth is slightly further away from the Sun which means that the larger land areas will not tend to overheat. Land is much more readily warmed than water but land cannot store the heat for later use nearly as well. If the northern hemisphere was closer to the Sun (during its summer season) instead it would overheat which would restrict the overall habitability of the Earth. It is difficult to imagine a more favorable arrangement than the one that the Earth has and it is one that is certainly improbable.

Any planet that provides an environment where temperature is not regulated in a manner similar to the way that it is done on the Earth will not be acceptable as a life-supporting location and the search should simply move on.

Occasionally the habitable zone for our solar system is claimed to be so wide that it includes Mars as well as most of the space in towards Venus. (334) However if the Earth were located much closer to Venus it would be receiving much more solar heat than it does now. The Earth's barely-allowable limit for temperature increase (from the 5% orbital variation mentioned above) is only about 10% whereas in closer to Venus' orbit the increase would be more like 100%. Clearly this would not be acceptable. Earth's current average temperature of about +15C would rise until it was well above the temperature at which seeds can germinate as well as the body temperature of animals. The currently-inhabitable areas would all be over-heated and even near the polar areas it would be too warm. Just to make matters worse, even if some areas near the poles were within an acceptable temperature range they still wouldn't be able to support life because there isn't enough light at those latitudes to grow plants. (90) Consequently any possibility of survival for either plant or animal life over virtually the entire world would disappear.

At the other extreme it is well understood that Mars is so cold that the temperature seldom rises above the freezing point of water. 'With its distant orbit – 50% further from the Sun than the Earth – and slim atmospheric blanket, Mars experiences frigid weather conditions. Surface temperatures typically average about -60C (-76F) at the equator and can dip to -123C near the poles. Only the midday Sun at tropical latitudes is warm enough to thaw ice on occasion. But any liquid water formed this way would evaporate almost instantly because of the low atmospheric pressure.' (112) Also its distance from the Sun varies so much that the variation in heat received is plus or minus about 20%. (69) Consequently neither plant life nor animal life would be possible. For this reason alone Mars is not habitable and never will be colonized.

It is increasingly clear that maintaining an appropriate temperature to enable life to thrive on Earth is going to be a serious challenge over the next millennium. So trying to include either Marsor Venus in the habitable zone of the Sun seems most dubious.

When remote exo-planets are declared to be in their respective star's habitable zone one would naturally think that the astronomers are thinking in terms of habitable for animals and humans. Unfortunately this is not the case. In spite of the fact that all forms of life are based on carbon and would appear and function the same way that animals do on Earth, 'habitable' is extended to include 'extremophile forms of Earth-like life.' (148) While this is valid to a certain degree one must be careful to recall that what appears as extreme to a human being isn't extreme in the broader sense. The worms that live near the hot vents on the bottom of the ocean are an example of 'extremophile'. One end of these worms is in the cold water of the bottom of the ocean (about 4C) and the other end is in the hot water flowing up from beneath the ocean floor. Such an arrangement would not be the least bit satisfactory for any of the common forms of animal life and therefore to call it extreme is reasonable. However, even in cases like this, the creature must have protection from ultraviolet light as well as the other types of deadly radiation that emit from stars. They must also have a supply of nutrients even though the amount might be very small. In other words the complexity of the supportive environment, even for extremophiles, means that the vast majority of far-away planets must be eliminated from any 'habitable' list.

When considered together with the size restrictions discussed above, only five of the numerous extra-solar planets discovered to date meet the plus or minus 5% criteria.

While there is clearly a thermally-habitable zone there is also a structurally-habitable zone. This second necessity for habitability is very seldom recognized as a habitable criteria but it is just as important as the thermal factor. The surface of any life-supporting planet must be structurally stable and reliable. How could any buildings be constructed or roads built if the crust of the Earth was in constant turmoil? Even on Earth there are places where roads, buildings or bridges can never be placed. One of these areas is the Amazon River valley in South America. The Amazon is probably the most powerful river on the Earth when both the flow rate and the volume of water are considered. Along the shores in certain areas towns exist but they must be located with full recognition of the variability of the Amazon's flow volume. Both the depth and width of the river can change dramatically within a few hours. On one occasion a report was received that at sundown the river appeared to be less than one mile across. At sunrise the water level had risen twenty-five feet and the width had increased to about two miles. With such dramatic variability and with the general lack of structural stability of the shore material, there will never be a proposal to build a bridge across the Amazon. Another example of an area with structural limitations is a fault line where the land on one side is moving with respect to the other side. Would anyone ever build any building across a fault line like this?

In order for a planet to be structurally habitable, it must be so remote from its host star that the pull of gravity of the star will not continually distort it's surface material. This means that the star must be hot so the planet can stay well back and still receive enough heat. These criteria exclude 95% of stars as possible candidates to support life because they are just not hot enough. To satisfy the structural criteria a planet must be remote enough from its star so that the pull of gravity of the star is reduced to a level which allows for structural stability of the planet's surface. Only about 5% of the stars in the galaxy are considered hot enough. (147) This criteria excludes red dwarfs because a planet would have to be quite close to a red dwarf to receive enough heat and this would put it too close for structural stability and hence structural habitability.

In spite of these restrictions red dwarf stars are continuously presented as possible candidates to support life. At the same time it is admitted that these planets, while in the thermally-habitable zone of their host stars, have suffered from the gravitational pull of these same stars to the degree that they would be tidally, permanently distorted. This means that even if they had rotated at one time, they would not be rotating any more and one side would be continuously locked in the direction of the star. An example of this is given by Gliese 581c. 'It is tidally locked (always faces the parent star with the same face) so if life had a chance to emerge, the best hope of survival would be the 'twilight zone.' (148) Unfortunately this type of hope is no hope at all. When one side of a planet is always facing a star, that side will be much too hot. At the same time the other side will be much too cold. In the twilight zone the surface temperature might be in the habitable range but this does not necessarily make the planet habitable. If there had ever been an atmosphere any volatiles like water would boil on the hot side and migrate to the cold side. On the cold side they would freeze solid and thereby become locked out of circulation altogether. This would also apply to carbon dioxide which, if any had been present, would become locked up as ice on the cold side as well. (325) Therefore just being warm enough in the twilight zone would not make a planet habitable and red dwarfs should not be considered as possible candidates for a life-supporting planet.

8.4.3 The Axial Tilt (discussed in an earlier section but valid here as well)

The Earth has an axial tilt. This enables the Earth to have seasons and seasons are not just convenient, they are necessary. If there wasn't any axial tilt and the Sun shone directly down on the equator all year the equatorial regions would over-heat. While the heat from the Sun is currently near the limit of what human beings can tolerate, if the Sun just kept shining straight down all year on the equator, both plant and animal life would be seriously jeopardized. At the same time areas remote from the equator including the mid-latitudes and higher would be too cold. Winters in the mid-latitudes would be just as severe as they currently are north of the Arctic Circle. South of the mid-latitudes spring would be late and fall would be early. The growing season would always be short and frequently plagued by unfavourable weather. Even as survival near the equator would be doubtful and survival above the mid-latitudes would be doubtful, survival in the remaining areas would not be much better. Hence the overall habitability of the Earth would be dramatically reduced. With respect to the search for a suitable life-supporting planet near a far-away star, habitability is the factor of primary interest and if a planet isn't habitable then it will be of no further interest with respect to the search for extraterrestrial life. Just being in the thermally-habitable zone would not guaranty habitability because while the incoming heat must be adequate it must also be distributed around the planet.

A reduction of the thermally-habitable zone would happen on Earth if the axial tilt was reduced to zero even though the planet might remain well within the plus or minus 5% limits mentioned above. In such an event, the heat of the Sun would not be properly distributed and the equatorial regions would over-heat while the temperate regions would be too cold. Any possibility of life existing would only be in a narrow band between the equator and the mid-latitudes but even this would be dubious because the atmospheric circulation patterns would continually bring cold air down from the continuously-frozen regions to the north. To say the least, life would be tenuous even with all of the other life-support factors that the Earth has in place. This means that any hope of survival in those narrow twilight zones of red dwarf stars really is virtually impossible. However, this reality does not suppress the enthusiasm for the possibility of life in such a place as stated by one astronomer. 'Personally, given the ubiquity and propensity for life to flourish wherever it can, I would say, my own personal feeling is that the chances of life on this planet are 100%.' (149)

While this type of unsubstantiated enthusiasm is disappointing, it was even more disappointing when the very existence of 'this planet' was cast into doubt and then discounted altogether. 'The M dwarf GJ is believed to host four planets including one (GJ 581d) near the habitable zone that could possibly support liquid water on its surface if it is a rocky planet. The detection of another habitable-zone planet – Gj 581g – is disputed as its significance depends on the eccentricity assumed for d. Analyzing stellar activity using the Ha line we measure a stellar rotation period of 102 + or – 2 days and a correlation of Ha modulation with radial velocity. Correcting for activity greatly diminishes the signal of GJ 581d while significantly boosting the signals of the other known super-Earth planets. (This means that) GJ 581d does not exist but is an artifact of stellar activity which, when completely corrected, causes the false detection of planet g.' (327) In other words the chances of life were 100% even though the host star was a red dwarf and the planet itself did not exist!

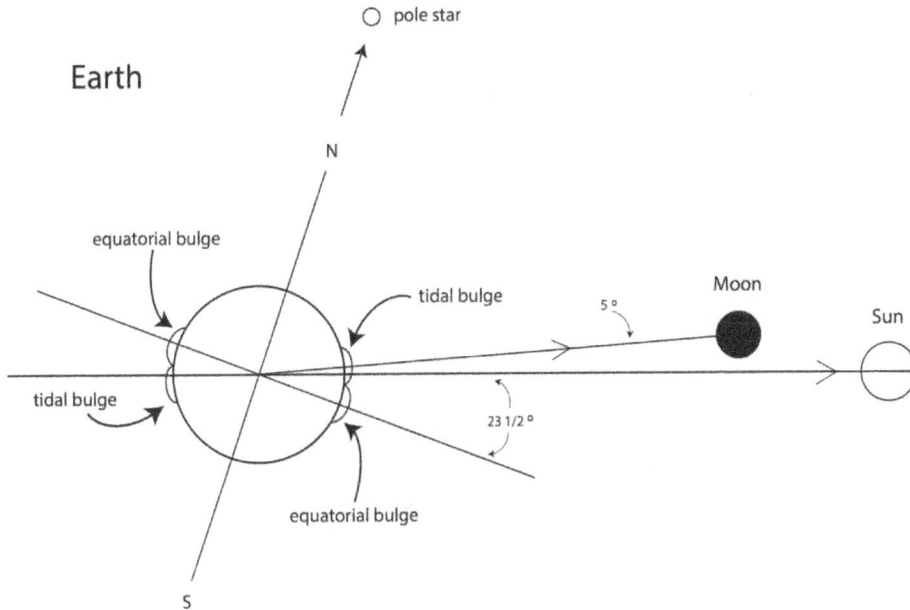

The Axial Tilt   23 ½ degrees

It is the Earth's axial tilt that enables so much of the Earth to be habitable. The magnitude of the axial tilt is 23 ½ degrees. This is, in fact, the optimum angle enabling the habitable area of the Earth to be maximized. With any other angle, less area would be hospitable to life and the overall habitability of the Earth would be reduced. The axial tilt) has been caused by and is maintained by interaction between the Earth and the Moon. In other words in order to have an axial tilt there must be a suitable moon.

Maintenance of the Earth's axial tilt is active (or dynamic) with the maintenance functions being able to overcome and deal with gravitational disturbances that occasionally come our way. While significant disruptions are expectedly rare, when they happen they must be accommodated or the Earth would simply drift away into some other orientation which would certainly be devastating for all forms of life. For example in 1979 a giant comet (Howard-Koomen-Michels 1979X1) came into the inner solar system and went right into the Sun. (101) It came in from the far side of the Sun which was a great relief for the Earth because it was so large it would have disturbed our orbit as well as our axial tilt. The head of this comet was larger than the Earth which, for a comet, is unusually large. (101) Even if a comet that size passed within one million miles of the Earth we would be in danger. If it passed within ten million miles it might have disturbed our angle of inclination but the stability mechanism which we have in place would probably have been able to deal with an upset such as this and re-establish the angle once again. It is an active control system and it operates through the gravity of the Moon acting on two of the Earth's bulges. The particular features of the Earth that provide the required bulges are the equatorial bulge and the tidal bulge. (Such features would have to be found on a remote planet in order for it to have an appropriate axial tilt as well.)

8.4.3.1 The Earth – a Fluid Body

In order to provide the necessary bulges, a planet must be a fluid body covered by a relatively thin crust. This would necessitate that the interior be hot because otherwise the interior would be solid. While rotation is also necessary, rotating too fast or too slow will not be acceptable. The objective is to have an equatorial bulge and such a bulge would not be useful if it was either too large or too small. The required bulge would be part of the axial stability function. A proper spin rate and a fluid interior are absolutely necessary. It is a virtual certainty that without all that magma swirling around beneath us we wouldn't be here now. (105) The molten interior and the spin rate working with several other factors as well are necessary to generate the appropriate axial tilt.

The Earth is constructed as a large fluid/semi-fluid sphere with a comparatively very thin crust. The flexibility provided by this arrangement enables bulges to form if the appropriate forces are applied. If the Earth was a solid body, bulges would not be able to form. However bulges can form on the Earth due to either gravity pulling on the Earth or by the rotation of the Earth and both types of bulges are involved in establishing and maintaining the Earth's axial tilt. The pull of the Moon's gravity causes the tidal bulge to form. The rotation of the Earth causes the equatorial bulge to form.

8.4.3.2 The Equatorial Bulge

The equatorial bulge exists because the Earth is a fluid body and because it is rotating. As a fluid body the Earth is slightly malleable so when it rotates its incredible mass causes it to bulge slightly at the equator. Actually this bulge is significant and causes the distance from sea level to the center of the Earth to be different by several miles between the equator and the poles. On a large object like the Earth this bulge is not noticeable particularly because it is very spread out. Alternately it can be stated that the poles are slightly flattened. (To the inhabitants of the Earth none of this is the least bit noticeable and does not interfere with any activity on the Earth.) The lack of symmetry provides something for the Moon to pull on. This happens with any bulge that the Earth might have had because the pull of the Moon's gravity on the protruding bulge on one side is greater than it is on the protruding bulge on the other side. If the Earth was a perfect sphere instead there would not be any particular preference for the Moon to pull on. While the pull of gravity of the Moon is greater on one side of the Earth than it is on the other side, since the plane of the equator is inclined by the axial tilt (i.e. 23 ½ degrees) the pull on the equatorial bulge causes a torque (i.e. a twisting force) to develop which tries to pull the Earth back upright and thereby eliminate the axial tilt altogether. At the same time it will be understood that the more that the Earth tips over thereby increasing the axial tilt, the greater the restoring torque provided by the Moon to bring it back up again. (within certain limits)

A similar situation exists between the Earth and the Moon with respect to the reality that one side of the Moon always faces the Earth. This is caused by the center of gravity of the Moon being offset from its geological center. This enables the gravity of the Earth to differentially pull on the material of the Moon and keep one side facing the Earth all the time. In effect this causes the Moon to rotate on its axis only once for every circuit it makes around the Earth. As the Moon orbits the Earth, the side that is facing the Earth tries to just keep moving straight ahead without rotating but as soon as the misalignment reaches a few degrees the pull of the Earth causes the near side to once again swing around and become aligned with the Earth. This drifting and correcting goes on repeatedly. The Moon is consequently said to librate (93) with one component of the vibration attributable to an 1178 impact. (107) This means that it rocks back and forth slightly so that technically more than 180 degrees of its surface is visible to us on the Earth. The libration will continue indefinitely because there isn't any damping factor to reduce it. Similarly the gravity of the Moon tugs on the part of the Earth's equatorial bulge that is closest to it a little more than it tugs on the part on the Earth's far side. Since there is an angle of inclination in place and the equatorial bulge is below a line drawn from the Earth to the Moon, this tugging is in the direction of trying to pull the Earth back upright to reduce the angle. The equatorial bulge provides the material that enables the Moon to reduce the angle of inclination. The tidal bulge does the opposite.

8.4.3.3   The Tidal Bulge

The second bulge that the gravity of the Moon operates on is the tidal bulge. Once again the flexibility of the Earth's crust comes into practical use. The Moon pulls directly on the crust of the Earth and because it is slightly flexible it will respond by rising up slightly towards the Moon. While this bulge is not extremely large it is quite well defined and has the shape of a mound almost directly in line with the Moon. Since the Earth is continuously turning, the mound reaches maximum elevation slightly after the location that is directly under the Moon. The tidal mound therefore leads the Moon slightly and the tidal bulge keeps circling the Earth repeatedly. While there is a mound almost directly under the Moon there is also another one on the opposite side of the Earth.

The second part of the tidal bulge is the water in the ocean. This is the most familiar portion of the bulge and is well known to all mariners. The water bulge or water tide is quite measureable but it is also localized to a region almost directly under the Moon. The combination of crustal bulge and water bulge are the elements that the Moon pulls on and in this case the pulling is towards increasing the axial tilt. It is generating an overturning force which, if left unchecked, would continue to increase the axial tilt until the North Pole was facing the Sun. In other words it would be a disaster. Without the compensating equatorial bulge, the tidal bulge would tip the Earth over and the bulge would spiral from the mid-latitudes northward until it reached the high latitudes and its torque-generating ability was no longer effective.

8.4.3.4   Axial Tilt Stability

The axial tilt is dynamically stable. This means that upsetting forces can be accommodated and corrective action can be taken. This adaptive, dynamically-stable function is not arbitrary, it is necessary. If the axial tilt were to somehow increase slightly, the equatorial bulge would be further below the line between the Earth and the Moon. Consequently the pull of the Moon on it would generate a slightly greater restoring force which would reduce it. Then as it was reduced the restoring force would be reduced. This would enable the torque generated by the tidal bulge to be more effective towards increasing the angle once again. The two forces are exactly balanced (for the present rotation speed of the Earth) when the axial tilt is 23 ½ degrees and this seems to be the optimum angle for the most widespread habitability of the Earth. With any other angle, habitability would be reduced. (Unfortunately since the rotation of the Earth is slowing down, the equatorial bulge will be reduced and the habitability of the Earth will be reduced with it.)

Mars has two moons but neither of them is very large and neither of them is effective at generating and maintaining an axial tilt. Therefore the axial tilt of Mars will wonder and is expected to drift until one of the Martian poles temporarily faces the Sun. (70) Any similar arrangement for the Earth would be a disaster. None of the other planets in the solar system have any better arrangement. In fact the other planets in the Solar System do not have any moons that are able to provide an axial tilt stability function at all.

While the Earth's axial tilt is very close to optimum, it is transient and does not have long term stability. As discussed in chapter 3, the energy loss due to tidal movement is causing most of the rotational energy of the Earth to be lost by moving the tidal water. This is causing the rotation of the Earth to slow down at an alarming rate which will likely result in the Earth tipping over within another 20,000 or 30,000 years. Therefore the axial tilt, which is so necessary for life on the Earth, is a narrow temporal 'window of life'. Could such a narrow 'window of life' exist elsewhere?

8.4.3.5   Lunar Criteria

The Earth has an unusual moon in that it has a diameter that is a significant fraction of the diameter of the Earth. The other moons of the solar system are much smaller when compared to their host planets. This is significant for several reasons with the Earth-Moon system commonly referred to as a double-planet system. Orbital stability for the Earth is thereby enhanced. The Moon's gravity keeps the Earth at the right angle to provide the type of stability necessary for the maintenance of life. (102) Even as the Earth has an almost-circular orbit around the Sun so the Moon has an almost-circular orbit around the Earth. The Moon's distance from the Earth varies from about 252,710 miles to about 221,463 miles. This means that it only varies by 15,623 miles from the average which is less than 7% (71) and that the pull of gravity of the Moon does not vary significantly.

The Moon provides two significant stability functions for the Earth. The first is the pulling force that causes the axial tilt to be generated and maintained. Secondly, the Moon provides an orbital stability function without which the Earth might drift out of its precise orbit. These functions provided by the Moon mean that an Earth without the Moon would not be habitable and neither would a planet in some far away solar system.

If the Earth did not have a moon, the pull of gravity of the Sun would be the only force involved in operating on the Earth's bulges. The Sun's gravity or tide-generating capability is dramatically less than the force provided by the Moon. On Earth, it is the combination of both the gravity of the Moon and the gravity of the Sun that generates the tides. Without the Moon the tides would only be about one-quarter as high but the equatorial bulge would be the same so the pull of the Sun would reduce the angle of inclination to practically zero. Therefore it would be possible for the axial tilt) to disappear altogether if the gravitational pull only came from the Sun.

The offset of the pull of the Moon from the pull of the Sun is what guarantees that there will be an axial tilt in the first place. While this offset is just barely over 5 degrees (103) it is that difference that guarantees that an axial tilt will exist. Otherwise a situation can be visualized where the equatorial bulge and the tidal bulge exactly line up which means that no axial tilt) would occur.

If there wasn't any axial tilt at all, The Sun would shine straight down on the equator continuously. Now suppose that the Moon is installed at the 5 degree offset from the direction of the Sun. The Moon would immediately pull up a tidal bulge that was offset from the equator by about 5 degrees and the Earth would start to tip over generating an axial tilt. Then it would continue to tip over until the axial tilt reached 23 ½ degrees and there it would settle out and stabilize. This means that the Moon could have generated an axial tilt 'starting from scratch' if it had not previously existed. It also means that the angle is dynamically stable and if it was somehow upset slightly, it would just become re-established again. A more beneficial arrangement is hard to imagine. Unfortunately this stability will disappear as the rotation of the Earth slows down.

8.4.3.6   Axial Tilt Probability

While a candidate planet in some far away solar system must meet some very specific requirements, it must also have a moon which also meets some very specific requirements and the two bodies must have a synergistic relationship to generate and maintain an appropriate axial tilt or they will not be useful as a life-supporting system. As mentioned earlier only five of the approximately eighteen hundred exo-planets discovered so far simultaneously meet both size and thermally-habitable requirements simultaneously. But how many of these will have a moon of the right size and distance together with an arrangement of planetary material to enable a tidal bulge to be generated? Also, will the planet be spinning at the right rate to provide an appropriate equatorial bulge? The probability of such criteria being met is obviously very very small.

The number of possible candidate stars is only a few from the previous discussion and that number must be multiplied by the probability decided here. This means that it will be virtually impossible to find a suitable arrangement in some far-away solar system even before the 'windows of life' of the next section are included. The belief that there is life throughout the galaxy is obviously not rooted in reality and so must relegated to some other realm.

8.4.4   Planetary Crust

The structure of the crust of the Earth is more complicated than it first appears. The continents are made from one type of material and the rest of the Earth is made from another type of material. The continental material is very slightly lighter than the other material and this enables it to 'float' on top. The amount of continental material is only a very small part of 1% of the other material and this relatively small amount means that continents will only exist over a small percentage of the Earth's surface – that is if the continental material is clumped up as it currently is found. Why is it clumped up at all? It certainly wouldn't be if it has ever been liquid. In that case it would have spread out evenly over all of the other material and any ocean water that existed would also have been spread out. (Also neither gold nor lead would be found in the crust because they would have sunk to the center of the Earth.)

With the present arrangement the relatively miniscule amount of continental crust is mostly clumped into the northern hemisphere. This arrangement works together with the slightly elliptical orbit of the Earth to provide an optimum temperature regulation system. With any other arrangement, the habitability of the Earth would have been reduced.

These various relationships illustrate that if there is to be life on a far-away planet there must be an appropriate relationship between the various types of solid and liquid surface material. Otherwise even if the planet is within a proper thermally-habitable zone it might not really be habitable because not only is the appropriate amount of energy needed it must be managed properly or habitability will be seriously curtailed. What is the probability that a 'rocky' planet will be found that has the proper temperature regulation systems in place including not only an axial tilt and the necessary amounts of different kinds of solid crustal material but also an ocean of the appropriate size and in the best-possible location?

8.4.5 Review

In order to be considered as a possible host for life, a planet must be of a very particular size and weight (mass). It must be located at a very particular distance from the star which it orbits and the allowable variation in these criteria is very small. It must have a moon which must also be of a very particular size and distance and have a nearly circular orbit which will enable an optimum angle of inclination to be generated and maintained. The planet must rotate at a very particular rate and be flexible enough to enable an equatorial bulge to develop. A tidal bulge is also necessary as well as a massive ocean located where it will work synergistically with a slight ellipticity in the planet's orbit to provide optimum temperature regulation. The relationship between the Moon's distance and mass and the magnitude of the tide must be such that the Moon will not recede from the planet too quickly and cause the planet to become uninhabitable – even if it is right in the middle of a thermally-habitable zone. The planet must have a very small amount of crustal material that is just slightly lighter than all of the rest of the planet's material and clumped up so that it projects above any ocean that exists. Overall these are very difficult criteria to meet and trying to apply an appropriate probability to them is equally difficult. It would be optimistic to suggest that there would be one chance in a million (or maybe even a billion) that any such arrangement might be found.

8.5 Planetary 'Windows of Life'

8.5.1 Magnetic Field

The Earth has a magnetic field. While the magnetic field is completely invisible it is absolutely necessary for our survival on the Earth. The magnetic field blocks and captures radiation coming to the Earth from space. The

Van Allen Radiation Belt is thereby formed and this is a region of space high above the surface of the Earth where the incoming radiation is trapped and prevented from reaching the surface of the Earth where numerous species of animal life thrive. Radiation is a deadly enemy for all forms of animal life as it interferes with the reproduction process at the cellular level. Interference can result in cancerous cells forming so if the magnetic field was lost a great many more animals and humans would develop cancer. While the protection being provided by the magnetic field seems reasonable at the present time, measurements indicate that the strength of the magnetic field is dropping rapidly ... about 15% in 170 years. (124) It is also true that the incidence of cancer formation is on the rise. These two developments could be connected. While there has been noticeable advances in the treatment of cancer the actual incidence of new cancers is increasing. The decrease in magnetic field strength could be directly connected to the increase in cancer formation. If this is the case, the outlook for the inhabitants of the Earth looks pretty bleak. At the present rate of decrease the magnetic field will be gone within a few thousand years. (124) As that time approaches, the incidence of cancer formation would be expected to accelerate and there is a high probability that almost every human being will be involved.

A weak magnetic field would not offer protection from incoming radiation – particularly from our own Sun – but a very strong magnetic field would bring with it another set of hazards. Before 20,000 BC the currents needed to produce the field would have overheated the Earth. If you go back even further, say, to a million years, there would be so much heating associated with these currents that the whole Earth would be destroyed by the heat energy. The strength of the magnetic field therefore indicates that the Earth cannot be very old, and not more than 20,000 years at the absolute maximum. (318)

If any planet in a far-away solar system is to be a possible home for animal life, it must have a magnetic field. There must be protection from the harmful radiation produced by that planet's own sun. In other words the presence of a sun near any planet in its vicinity will guaranty that it will be constantly bombarded by radiation - totally jeopardizing the possibility that any life can exist. In order to offset this hazard a magnetic field must be in place. What are the chances that this will be the case and what are the chances that the strength of that field will be in the necessary range to provide protection but not be so strong that the planet would be over-heated and effectively pushed out of its thermally-habitable zone even though the distance from its star would indicate otherwise?

8.5.2   Ozone

When the ultraviolet light from the Sun acts upon atmospheric oxygen, ozone is formed. The formation of ozone thereby absorbs the ultraviolet light and keeps it from proceeding on toward the surface of the Earth and contacting animal life. Ultraviolet light is very-high-energy light. Every photon of it is energetic and this means that it is harmful to delicate tissue like retinas and skin. An overdose of ultraviolet light will burn exposed skin and damage the eyes. A good layer of ozone high above the Earth is our first type of protection which means that if the ozone is either depleted or lost, greater incidence of skin cancer will result. More ozone would be better whereas less ozone would spell trouble.

Certain substances destroy ozone and a great amount of effort has been applied during the last few years to interrupt this destruction and enable the protective characteristic of ozone to continue. Ozone can be considered a 'window of life'. It is, like the magnetic field, a one-sided window because more is beneficial whereas less is detrimental.

8.5.3   Wind

Light winds are quite pleasant while heavy winds are not welcome because of the destruction that they cause. Actually winds are necessary as they bring the needed moisture and remove unwanted atmospheric contaminants. Heavy winds are destructive of property and trees as well as being very dangerous to people. What if we lived on a planet which had heavy winds all the time like the planet Jupiter? Survival would be seriously curtailed or even impossible.

On Earth the winds are one of the main means that equatorial heat is transferred to regions that receive less heat. They are a means of spreading the heat from the Sun so that more of the Earth is habitable. If there wasn't any wind, tropical areas would overheat and the mid-latitudes would be colder. Wind might therefore be recognized as a 'window of life' because it enhances the habitability of the Earth.

8.5.4   Carbon Dioxide (CO2)

Carbon dioxide forms a very small portion of our atmosphere but is a major participant in most of life's processes. It is the carbon in carbon dioxide that provides the atomic building blocks so that plants can grow. Trees, as well as grass and all of the plants that are used for food are made from the carbon in carbon dioxide. However there is a limit to the quantity of carbon dioxide that is acceptable. If there was too much, some of the oxygen in the air would be displaced and this would make breathing more difficult for all types of animals. While 'more' enhances the ability of plants to grow, 'too much' inhibits animals from breathing. Too little is also a problem. A reduced amount of carbon dioxide would result in stunted plant growth. Animals could tolerate a reduction but plant life would not be vigorous. This makes carbon dioxide a two-sided window and significant deviation from the present level would not be beneficial.

Carbon dioxide is also involved with temperature regulation. The temperature of the Earth must be regulated. In fact the range of temperature that we can tolerate is really very small. The temperature regulation characteristic of carbon dioxide is manifest through the greenhouse gas inventory. Greenhouse gases have the characteristic of absorbing and re-radiating heat from the surface of the Earth right back to the surface of the Earth. Without this factor operating in our atmosphere, the temperature around the world would deviate much more. It is the combination of the incoming heat from the Sun and the heat absorption and reflection factor of the greenhouse gases that keeps the Earth's surface temperature in the habitable zone. If our greenhouse gases were lost, the Earth would not be habitable. On the other hand if the greenhouse gas inventory increased too much, the Earth would over-heat and not be habitable for that reason either. In fact, there is serious concern at the present time that our carbon dioxide inventory is increasing too much and that this is resulting in a world-wide increase in average temperature. Even a few degrees is anticipated to be a disaster. Carbon dioxide is therefore a 'window of life' and a two-sided window at that. We cannot tolerate too little and neither can we tolerate too much.

8.5.5   The pH of Water

Water is very seldom found in a pure state but commonly includes other ingredients some of which might be beneficial for animal life and some that might not be beneficial. Just because we have discovered a body of water does not mean that we should expect that it will be beneficial to any form of life. Water is very easily contaminated and once this happens it is virtually impossible to render it pure again. Animals drink water. An animal body requires water and in fact water constitutes a large part of most animal bodies. Since the water in an animal body is being continually lost through respiration and other means, more water must be taken in periodically for the animal to continue functioning. However the water that is taken in to replenish the volume that was lost must not compromise any function of the body.

Water is said to have a pH value. This is a measure of the water's acidity or alkalinity. A pH of 7 is neutral. Lower than 7 is acidic and higher than 7 is basic. Even if a planet has oceans the water must have a neutral pH to enable cells to grow. (115) Water with a neutral pH would be safe to drink as long as no other impurities were present. While slight variations in pH are allowable, most types of contaminant are not allowable at all.

8.5.6   Ocean Purity

Ocean water is salty and the ocean is a source of food. It could not be a source of food if the saltiness was closer to saturation. The Dead Sea is also salty but it is not a source of food. The salts of the Dead Sea have long since gone right to the saturation level which means that absolutely no more salt can be admitted. As more comes in from the Jordan River an equivalent amount must precipitate out. At such a high level of saltiness animal life cannot survive.

The saltiness of the ocean is only about 10% of the saltiness of the Dead Sea and while numerous types of creatures thrive in the ocean there are many types of animals which could not live with such a high level of salt in the water. With respect to animal life, the saltiness of the ocean might be considered a one-sided window. While no salt would be acceptable, too much salt is totally unacceptable. Unfortunately the status quo will not be maintained. Rivers continue to bring more salt to the ocean which means that over time the general saltiness of the ocean will increase. While some of the salts that are brought can be used by some of the small creatures of the sea there is a question about whether or not this type of salt removal can indefinitely prevent the general saltiness from increasing as time goes by.

As a significant source of food for the entire world, the ocean would lose this characteristic long before any of the salts of the ocean reached saturation. While dry land also produces vast quantities of food the loss of the ocean as a source of food would seriously curtail the ability of the world to support the vast numbers of people that it does at the present time. While salt is a contaminant that would curtail the use of the ocean as a source of food there is an unlimited number of other contamination possibilities that would achieve the same results. One of these is plastic. The ocean has become seriously contaminated with plastic which takes a long time to break down but when it does it gets into the food chain at the very bottom. This has already been happening for some time and the situation will only get worse as more plastic breaks down. As a source of food, the ocean could be removed from the list of possibilities by this contaminant alone. While saltiness might be considered a one-sided window, an unending list of other contaminants could also be considered in the same light. If a far away planet was discovered that had a large ocean (which would be so necessary as a component in the generation of the axial tilt) which was contaminated and unavailable as a source of food, would such a planet be a suitable home for animal life?

8.5.7   Planetary Crust, Thermal

The Earth is basically a massive sphere of hot molten low-viscosity material covered by a relatively thin solid crust. Since it is quite hot in the interior of the Earth the heat that slowly comes up to the surface is helpful in modifying the temperature on the surface. This fact is not widely appreciated but if one descends into the Earth (e.g. goes down into a mine) it will readily be appreciated that there is a lot of heat not very far below the surface. This slightly-modifying component of surface temperature would not appear at all if the crust of the Earth were much thicker. In that case surface temperature would vary more and surface weather patterns would be more extreme. This temperature modifier does not really show up very dramatically but it is noticeable in a 'deep' mine (one mile deep) and under the glaciers of Antarctica. There the temperature is very close to freezing and liquid water exists even though just two miles above on top of the glacier the temperature might be -100F.

The thickness of the crust of any planet is expected to increase as time goes by as internal heat is dissipated. More and more of the molten material will solidify until sometime in the distant future the entire sphere would be solid material. How much will the surface temperature vary when that happens?

On the other hand and of greater importance would be the situation if the crust was much thinner as it supposedly was in the distant past. In this more readily appreciated case, the surface temperature would be higher, volcanoes would be closer to the surface and sub-surface water would be hot. The oceans would be warmer. The tidal bulge in the crust would be greater implying that the angle of inclination be greater. However, the equatorial bulge might also be greater implying that the angle of inclination would be less. How would these uncertainties affect habitability? The best possible situation seems to be to have a crust that is about the same thickness as the one that the Earth currently has because the Earth's current angle of inclination is at a value which maximizes the habitability of the Earth. If a planet similar to the Earth were found it too should have a crust very similar to one we have in order that the stability factors dependant on the crust be the same as they are on Earth at the present time.

Mines are occasionally sunk to a considerable depth into the Earth and one of the difficulties of working in a deep mine is the temperature. One of the deepest mines ever operated was a nickel mine near Sudbury Canada.

This mine eventually went down for more than one mile but even at the one mile depth the temperature was too hot for the workers without having cool air piped in. A similar temperature on the surface of the Earth would inhibit both plant and animal life from being able to survive. The ground would just dry out and water would never stay in any reservoir. If the surface of the Earth was at the temperature of a deep mine, the Earth would not be habitable. These factors illustrate that the thickness of the crust and the idea that it formed from a molten state by cooling, are important ideas with respect to the length of time that the Earth could have been habitable.

The same type of thinking must be applied to planets in distant places. Surface temperature is partially determined by the thickness of a crust and this is of great importance in deciding if a candidate planet has a reasonable chance of being a life-supporter. How thick would a crust need to be to enable a planet to have a surface temperature in the habitable range? What is the probability that a planet near some far away star will have a crust of the appropriate thickness?

Time, in this case, is the relevant factor. If the crust of a mostly-molten planet thickens with time and if there is a range of thickness which enables life to thrive on the surface, in order to be a candidate for life-support, that planet must be identified within the time-window where the crust thickness is appropriate. The same could be said for the Earth. Life could only exist here after the crust had reached a certain thickness. Possibly it will only be able to continue existing here until the crust thickness reaches some maximum because a very thick crust would conduct less heat from the accompanying smaller molten interior. Supposedly this would make only a minor difference to the average surface temperature but – as pointed out earlier – differences as small as just a few degrees are expected to be significant. If this is true, there is a definite time-window allowing life to exist here. If greater thickness is irrelevant it is a one-sided time window but we would still not expect life to exist until the crust thickness exceeded the critical minimum. Will another planet with a molten interior and a crust with the appropriate thickness ever be found?

8.5.8 Planetary Crust, Structural

If the crust of a planet is to have any hope of supporting life it must be structurally stable. Even minor earthquakes cannot be happening continuously all over the Earth. Material from the interior cannot be repeatedly oozing out onto the surface. The crust cannot be littered with active slip joints. How could any roads, bridges or buildings be constructed if the stability of the ground wasn't very highly dependable? However, it has occasionally been declared that a newly-discovered planet near a red dwarf star is in the habitable zone. However which habitable zone is intended? Greater clarification would result if it was recognized that while the planet appears to be in the thermally-habitable zone, it wouldn't necessarily be in the structurally-habitable zone. There really isn't very much credibility in talking about a planet being habitable just because it might, on the average, be warm enough, while at the same time the tidal forces caused by its host star are so great that the surface would not be usable!

Any planet that was close enough to a red dwarf star to be in its thermally-habitable zone would be too close to have structural stability. The pull of gravity of the star on such a planet would raise a tidal bulge which would be excessively large. If such a planet had been rotating at one time it would have soon slowed down and would not be rotating any more. An excessive tidal bulge would have slowed any rotation to zero within a very few years. Up until rotation ceased, the tidal bulge would have continued to roll around the planet and totally disrupt the crust. These crustal tides can render the thermally habitable zone around low-mass stars uninhabitable. This makes the chances for life existing on exoplanets in the traditional habitable zone of low-mass stars (the most numerous stars in the galaxy) very bleak. If you are looking for a second Earth, you first need find a second Sun. (91) This basically means that the candidate star must be hot like our Sun so that the thermally-habitable zone can be far enough away from the star so that a planet would also be in the structurally-habitable zone.

In order for there to be any hope of being habitable, a planet can only exist in association with a hot star. This reduces the number of candidate stars to about one in twenty. (147) When a star is hot, the thermally-habitable zone can be far enough from the star so that a planet can also be in the structurally-habitable zone which means that any bulges caused by tidal effects will be of very modest size. This could never happen with a planet near a red dwarf (i.e. a low-mass star) and there is therefore no hope whatsoever of any such planet being habitable even though the exact opposite has been declared by over-zealous commentators. (149)

The Earth is located at such a distance from the Sun that the pull of gravity of the Sun is not strong enough to raise anything more than a very modest tidal bulge. The Moon actually raises a tidal bulge that is more than two times as great as the one raised by the Sun making the influence of the Moon much greater with respect to allowing the crust of the Earth to be structurally stable enough to allow the Earth to be habitable. When the Sun and the Moon line up in the sky their combined gravity is at its maximum but it still is not enough to compromise the crust of the Earth into structural failure and thereby prevent the possibility of the Earth being habitable. However this might not always have been the case.

The Moon is receding from the Earth. It is receding both theoretically and measurably. This is caused by the tides and has been discussed in chapter 3 above. The fact that the Moon is receding from the Earth means that it was closer to the Earth in the distant past which in this case is not that distant. Calculating from the current recession rate and factoring in energy loss due to tidal action and the general non-linearity of the recession pathway, the Moon would have been in actual contact with the Earth much less than one million years ago. The implication of this conclusion is that much less than one million years ago the tides caused by the Moon would have been totally overwhelming and would have made the crust of the Earth totally unstable.

Also, because those same tides have caused the Earth to spin more and more slowly, it would have been turning much faster at that remote time. Combining these two factors means that there would have been monster tides every few hours and that the crust of the Earth would have been in continuous turmoil. That is, the Earth would not have been habitable from a structural viewpoint and it would only have become marginally habitable much more recently. It also means that the upper portion of the crust should have been fractured into innumerable pieces by all of the turmoil but it isn't and the relative recency of the nearness of the Moon to the Earth means that the Earth would not have been structurally-habitable throughout most of those declared long ages anyway.

These are most unfortunate pieces of evidence for those who understand that the Earth is really very old because the lack of evidence of excessive fracturing means that it cannot be very old.

If any planet in the galaxy is to be considered as a possible location for life to thrive, the crust of that planet absolutely must have structural stability.

8.5.9 Oxygen

Oxygen is one of the basic and necessary components of the atmosphere of any planet that is to provide a home for life – both plant and animal. While it is an absolutely necessary ingredient, only a certain percentage of an atmosphere can be oxygen because of oxygen's reactive characteristic. It will readily combine with numerous other elements and while many of the combination possibilities are necessary for our survival many of them could get out of hand if an atmosphere had too much oxygen in it. Fire is one manifestation of oxygen reacting with something else. If the atmosphere consisted mostly of oxygen, fires would break out quite readily and cause general planet-wide instability. Any minor spark would initiate a fire with the result that both plant and animal life would be continuously threatened. There was a situation several years ago during the Moon exploration program where such a catastrophe took place. In those days the atmosphere of the crew quarters of the Apollo space capsules was 100% oxygen. One day a fire broke out even before the machine left the launching pad. Before anybody could take remedial action, all of the crew members died. The same thing would happen on Earth if the atmosphere was mostly oxygen. While oxygen is necessary as a component of any life-supporting atmosphere, it must be diluted. In the Earth's atmosphere this is done with nitrogen. Nitrogen is the ideal gas for this purpose for several reasons. While it can react with oxygen it is one of those reactions that is difficult to achieve. Nature does it occasionally but the circumstances must be exactly right or the reaction will not proceed. Secondly the weight of nitrogen is very close to the weight of oxygen. This is most convenient because the two gases are consequently able to freely mix without there being any significant tendency for the nitrogen, which is slightly heavier, to settle out. Carbon dioxide also mixes freely throughout the atmosphere but it has a much greater tendency to settle out. If a parcel of air was contained where there was no circulation or heat input, the carbon dioxide would tend to settle to the bottom. This would be deadly for any animal life caught in the situation because there is a definite limit to the amount of carbon dioxide that an animal can tolerate before succumbing. Clearly, an animal cannot survive by breathing nitrogen any better than it can by breathing carbon dioxide but the mixing ability of atmospheric nitrogen is so great that it will always stay completely mixed with atmospheric oxygen. If, on some far-away planet, some other non-reactive gas was involved that had any tendency to settle out, animal life would be impossible. It follows immediately that the atmosphere of any potential life-supporting planet must be nearly identical to the atmosphere on Earth which appears to have a very-close-to-ideal mixture of gases. Also is there a protective ozone layer or is the surface scorched by harmful ultraviolet rays? (115)

8.5.10   Carbon Dioxide (CO2) Again

Carbon Dioxide is in the atmosphere of the Earth and it is important to recognize its value with respect to the possibility of life on other planets. There are several reasons why CO2 is absolutely necessary in order to have life – both plant and animal - and any candidate planet must have CO2 in its atmosphere in an amount that is within certain very strict limits. On Earth it also fulfills the function of being a greenhouse gas, without which temperature regulation would be seriously diminished and quite possibly be absent altogether. We need CO2.

All plants are constructed from carbon – the carbon that they extract from the CO2 in the atmosphere. If any plant is to even exist and carry out structural repairs there must be CO2 in the atmosphere. This is true for the grass on the lawn, for the flowers in the garden and for the trees in the forest. All of these things are constructed from carbon taken from the CO2 in the air. While this is true at the present time it was also true long ago regarding the plants which formed the great coal beds which are being mined at the present time.

This important factor has been discussed elsewhere (i.e. 3,3,7,6 The Interrupted Carbon Cycle) but it will be reviewed here with respect for the importance of carbon dioxide. Coal is carbon which came from plants and the remains of numerous different types of plants has been detected in coal. There is a lot of coal. This means that there is a lot of carbon stored in the great coal storehouse which is really a great carbon storehouse. Some of the plants which have been identified include; ' ... tulip tree, magnolia, sequoia, poplar, willow maple, birch, chestnut, alder, beach, elm, palm, fig, cypress, oak, ... and plum and many other species.' (116) (It will immediately be noted that these are not swamp trees making it clear that coal did not form in swamps.) While this is revealing on its own, the vast quantity of coal opens another avenue of enquiry. The quantity of coal that is still in the ground is exceedingly large and when the amount of carbon that is tied up in this coal is compared to the amount of carbon that is in the current biosphere (i.e. the total of all living things – both plant and animal) it is postulated that there is possibly 50 times as much carbon in the coal as there is in the biosphere. (117) All of this carbon has been trapped away from the carbon cycle and has not been available to recirculate since it was trapped.

Normally carbon circulates through the biosphere. Animals eat plants and the carbon combines with oxygen in the animal body to produce CO2 which is expelled back into the atmosphere. Plants can also rot or burn and both of these processes produce CO2 which is released into the atmosphere. While these are the three main ways that carbon gets into the atmosphere, any process that combines carbon with oxygen and forms CO2 would also release CO2 into the atmosphere. In recognition of this, the carbon which formed the coal-bed plants must have come from some source other than the metabolization process of animals. And neither did it come from the burning or rotting of the plants that are forming the coal because the plants in the coal have neither rotted nor burned – they are still there in the coal. However, it must have come from the air because that is the source of all plant-forming carbon. But how did it get into the air?

Three possibilities present themselves;

1. The first possibility is that the required CO_2 was formed by a burning process, which used primeval or virgin carbon as a source. This burning process introduced the virgin carbon into the atmosphere and into the carbon cycle at just the right rate to enable the trees and other plants, which would later form the coal beds, to grow. In order to be internally consistent, it must be recognized that a vast amount of virgin carbon was required. In fact exactly the same amount of ancient virgin carbon was required as is presently found in the coal beds as well as that which has been removed since coal mining began. This type of arrangement is recognized as an artificial construct because nature has been conveniently arranged to bring about a result, which isn't otherwise credible.

2. Is it possible that there could have been enough carbon dioxide in the air at some ancient time to enable the coal-forming plants to develop simply by depleting this CO_2? The amount of carbon dioxide which is in the atmosphere at the present time is more than 380 parts per million (118) which means that 380 out of every million molecules in our atmosphere at the present time are carbon dioxide molecules. If all of the carbon in this carbon dioxide were assembled together to make coal, about $6 \times 10(11)$ metric tons of coal would result. Current estimates of the world's coal reserves are $12 \times 10(12)$ metric tons. (119) The current atmospheric carbon inventory is therefore equivalent to $6/120 \times 100$ or about 4.6% of the world's coal reserves. Therefore, if, prior to formation of the coal bed plants, the amount of atmospheric CO_2 had been about 25 times as great as it is now, the coal bed plants could have grown and depleted that higher level of CO_2 down closer to its present level. Therefore, one possibility to explain the great amount of carbon required to form the coal-bed plants is for an ancient CO_2-rich atmosphere to have been depleted of its CO_2 burden down to an atmosphere with much less CO_2.

3. Prior to the formation of the coal beds there was an ancient biosphere which was 50 times more extensive than the one we have at the present time. Forests, swamps and meadowlands were filled with an abundance of all kinds of plants. In addition a greater area of the Earth was involved including the high arctic lands, Antarctica and areas now below sea level. Then suddenly this ancient biosphere was annihilated and its carbon is now found in coal. This explanation basically shoves the question back because now we must ask where the carbon came from to form this massive assembly of ancient plants. Did they just suddenly appear? Were they created?

In summary, there appear to be three possibilities for the formation of the coal-forming plants.

1. The ancient biosphere was formed from CO_2 which was produced by an unknown, virgin-carbon burning process.
2. The ancient biosphere (with all of the coal-bed plants in it) was formed by depleting an even more ancient atmosphere of its CO_2 down from a level approximately 25 times as high as the present level of 380 ppm.
3. A massive ancient biosphere was created and the plants from it became available to form coal.

However, all of these possibilities come with attachments. If the ancient atmosphere had 25 times as much CO_2, the average surface temperature of the world, due to the greenhouse gas effect, would have been much higher and the trees would not have been able to grow properly because they would have sweat too much. Also, the surface temperature would have been above the body temperature for most types of animal life as well as too high for the trees themselves.

While the third possibility is totally unacceptable to many people, with both the first and second, it would have been necessary that none of the plants which grew during the extended times required, were burned, eaten or decayed because this would have released their carbon back into the atmosphere and it would not have been available to contribute to the coal beds. There were no forest fires caused by lightning (which currently strikes the Earth several thousand times every day). Also it was a rot-free forest wherein no significant quantity of material was eaten.

Therefore it can be seen that trying to come up with an explanation for the presence of CO_2 in the atmosphere of the Earth is not a straight forward task. However, since CO_2 is absolutely necessary for both plant and animal life to exist, one wonders how an appropriate arrangement could have been produced on a far-away planet. The probability of this happening is obviously extremely low.

8.5.11   Temperature Stability

Life on Earth is only possible if the average surface temperature is held very close to where it is at the present time. A deviation either way of even a few degrees is anticipated as disaster! Currently the Earth – with all of its temperature regulation factors – is heating up. (306) This is being attributed to the increase in CO_2, one of the most influential greenhouse gases. However it is occurring at a time when the Sun is cooling off! What if the Sun reversed this cooling off trend and started heating up again? How long would it take for the average surface temperature of the Earth to rise to a level that would make the Earth totally uninhabitable (which would probably only require an increase of less than ten degrees C.)? How did the Sun maintain a steady heat output for 4.5 billion years and how could any star be expected to maintain a steady heat output for very long periods of time without which life of any kind on a far-away planet would not have any chance of existing?

8.5.12   Earth's Axial Tilt

As pointed out above, the Earth has an axial tilt. Currently, this tilt is 23.4 degrees and is expected to wander slightly between 23.5 degrees and 23.1 degrees due to the slight variations in the Earth's orbit and the Moon's distance. Such minor variations are of no concern to us and will not affect the habitability of the Earth at all. However the calculations involved in making this prediction did not recognize the fact that the Earth's rotational speed is reducing at a very rapid rate.

The Moon maintains the axial tilt by tugging on two of the Earth's bulges – the equatorial bulge and the tidal bulge. While it is the distance to the Moon that determines the tidal bulge it is the rotation speed of the Earth that determines the equatorial bulge. Centrifugal force is what causes the equatorial bulge and centrifugal force is a square law. This simply means that minor variations have an exaggerated result. Therefore the Earth will not have to slow down very much before the equatorial bulge will reduce substantially. Then there will be nothing for the Moon to pull on to offset the tidal bulge. The Earth will consequently 'tip over.' When this happens the Sun will be seen to rise very high until it is north of the mid-latitudes and then it will fall until it is completely out of sight at the northern mid-latitudes. There will be extreme variations in the Earth's climate and the Earth will not be habitable! The rapid reduction in the Earth's spin indicates that the 'window of life' provided by the axial tilt of the Earth is, at the most, a few tens of thousands of years in either direction from the present time! If such a narrow exists on the Earth, what is the probability that it exists anyplace else in the entire universe?

8.5.13   Minimum Necessities

There is a minimum set of conditions required to enable any form of animal life to exist. There must be an atmosphere which contains oxygen in a breathable form. There must be a supply of food. There must be a temperature-controlled environment. It would be possible to survive a raging sub-zero blizzard if you were in a small log cabin with a wood stove and plenty of wood. With a supply of food survival could be sustained for some time and this type of thing has happened repeatedly. However, sooner or later the survival criteria would shift somewhat because other body functions would demand attention. In any event the minimum setup would quite clearly only be adequate for a short period of time before the bigger picture would present itself. One cannot just stay in a log cabin indefinitely. At first the wood supply must be replenished and the food supply must be restocked but before too long other questions will come up and the initially-simple arrangement will not be adequate.

The proposed mission to Mars enables such factors to be seen more clearly. In the Mars case, it is quite clear that the mission will not be sustainable. While the minimum necessities could be accommodated for a short period, they cannot be sustained indefinitely and pretty soon the food supply, the oxygen supply, the water supply or the source of heat will run out. But possibly prior to that, one might appreciate a shower and a change of clothes. It will probably be difficult to stand up and walk around because of the prolonged period of weightlessness during the time it took to get there. Sooner or later, but probably sooner, the environment on Mars would terminate the mission. The environment on Mars includes an unending flow of ultraviolent light which is deadly for all forms of animal life. Protection is possible if one stays in the shade but staying in the shade will not help when there is a major disruption on the Sun. During such times great bursts of radiation are emitted and would even wipe out life on the Earth except for our protective magnetic field. There isn't any such protection on either the Moon or Mars, The manned mission to the Moon in the 1970's barely escaped without incident. It too could have ended with the astronauts being caught exposed during a solar storm. The universe is really a very hostile place and without the numerous protective mechanisms in place that we have here on the surface of the Earth, life would not be sustainable for very long. The minimum necessities would only work for a minimum time and for a particular location. If life is too be sustained on a far-away planet for an indefinite period of time, the entire list of support factors must be in operation or any type of life that developed there or any that came from the Earth - would only have a brief history before being gone forever.

8.5.14   Windows of Life Review

Several 'Window of Life' have been mentioned. It will be obvious to the serious reader that the above list is not exhaustive. In fact, depending on one's professional discipline, the list can readily be extended considerably. It would not be an exaggeration to suggest that the total number of 'windows' would actually number several thousand. The more that the operation of the human body is studied and understood, the more it becomes apparent that life is really a very complicated matter. How an appropriate probability can be applied to the 'minimum necessities scenario' is even more difficult than it was for the planet-moon system but for argument's sake suggesting that it would not be better than one in several million would be appropriate.

8.6   Summary and Review

8.6.1   The Star

In pursuit of the possibility of life elsewhere in the galaxy, it is quite appropriate to begin with the task of identifying a suitable star. This would initially seem to be a relatively easy task because there are so many possibilities. Previous estimates of the number of stars in the Milky Way Galaxy were about one hundred billion. More recent estimates place this number considerably higher at three hundred billion. Surely with so many possibilities there must be a suitable star out there someplace. Unfortunately, this very large number shrinks dramatically when it is understood that only stars in the Galactic Habitable Zone can be possible candidates. This is a thin slice of the galaxy far from the central region where the stars are rotating around the center of the galaxy at the same rate as the arm of the galaxy at that distance. This location is called the co-rotation radius and is not really very wide in galactic terms. These criteria narrow the possible number of stars of interest to a very small fraction of the total that are in the galaxy and pulls the number down to 1% or even less. In reality, by far the great bulk of stars are located closer to the center. A life-supporting planet could not exist there because of the intensity of the radiation and the supernovas. For these reasons it is a most hazardous place to be. Even if the number of stars in the Galactic Habitable Zone is only 1% of the total this still leaves three billion stars to investigate. Star density is much reduced that far from the center so the more realistic number is probably more in the order of 0.1% than 1% which would leave three hundred million stars as possible targets for the search.

This number immediately shrinks to one/twentieth of these possibilities because of the brightness requirement. Most stars are not hot enough to allow a potential planet to keep back at a safe distance and still be in the thermally-habitable zone. Where 'cool' stars (i.e. red dwarfs) are concerned, if a planet was close enough to one of them to be in their thermally-habitable zone it would be out of the structurally-habitable zone.

Next, there is the singularity requirement. Temperature stability is paramount if a planet is to be a candidate for life and a circular orbit is a minimum necessity for temperature stability. Unfortunately, most stars exist as multiples and there is no possibility that a multiple-star system would allow an otherwise-habitable planet to have an endlessly repeatable circular orbit. Therefore all multiples must be ruled out. Since the ratio of multiples to singles isn't any better than one in five the number of possible star candidates is reduced to between twenty million and two million. (1,000,000,000 x 0.1 x 0.2 = 20,000,000 or 100,000,000 x 0.1 x 0.2 = 2,000,000)

The host star must be stable. All stars are not stable but the ratio of stable to unstable stars is not known. Stability in this case is required for extended periods of time. Even our own Sun might not be stable for long periods of time and the fact that observations have only been made for a few thousand years does not enable the stability question to be firmly settled. Appealing to distant stars as proof of long-term stability isn't valid and because great distances are involved it isn't even possible to state for certain that any of those distant objects even continues to exist. After all, if it takes light 100,000 years to reach the Earth how can anyone state that they know what is happening to that star at the present time. When we are observing distant objects it is clear that whatever we are seeing, happened a long time ago and that we have no way of knowing what is happening out there now.

The candidate star must not pour out vast quantities of radiation. The radiation that our Sun ejects during periods of major sunspot activity is serious enough and would be devastating to life on Earth except for the magnetic field of the Earth, which unfortunately is diminishing. If sunspot activity was more energetic than it appears to have been, life on Earth might not have been possible even with a magnetic field. Unfortunately, the Earth's magnetic field is expected to drop to insignificant within another few thousand years which means that cancer will be rampant and life on Earth will be virtually impossible due to this factor alone.

8.6.2 The Planet

There are also rigorous requirements for any potential life-supporting planet and a host of factors must be met before there would be any hope at all that it might be acceptable. One of these is size. If a planet is a little too small it will not be able to hold an atmosphere. Earth is just barely large enough and even it cannot hold hydrogen which keeps streaming away by the ton every day. (74) Neither can it be too large. About twice the diameter of the Earth is considered the upper limit and even this is being quite generous. At such a size the force of gravity would be much greater and water would be reluctant to evaporate. All forms of life would be quite stocky which would work all right for some of the creatures on Earth but certainly not for all. In particular all flying creatures would require major design changes.

A star must not have any large companions near its thermally-habitable zone. While they cannot be in between the thermally-habitable zone and the star, neither can they be outwards of the thermally-habitable zone either. Like the solar system, a large planet would have to be as far out as the planet Jupiter in our solar system or their influence on the orbit of a relatively small planet like an Earth that was located in the thermally-habitable zone, would be devastating – possibly even ejecting it from the system altogether. (73) Unfortunately in the search for an acceptable planet in a far away system, all but two of the more than 1800 planets discovered so far are in the unacceptably-large category and are located within the zone defined above. This would be devastating to the orbit of any small Earth-like planet in this region. If we apply a probability to this factor it would be something like $2/1000 = 0.002$ and the overall probability becomes $0.002 \times 20,000,000 = 40,000$ or $0.002 \times 2,000,000 = 4$. The possible number of life-supporting stars has therefore dropped into the 4 to 40,000 range.

In order to be habitable, a planet must absolutely be in the thermally-habitable zone. This is a very particular distance from the host star and the distance cannot be varied – even on a seasonal basis. A little too close and overheating would result. A little too far away and freeze-up would happen out of which it would be very difficult to come. An orbital variation of even a few percentage points from the center of the habitable zone would make a planet uninhabitable.

We cannot help but notice that all of the planets, moons and asteroids in the solar system have been pummelled repeatedly by other objects. Why would it be any different for a far-away planet? Would far-away planets also get pummelled and how would any form of animal life survive such pummelling any more than it could on Earth? How did animal life on Earth survive it?

8.6.3   The Planet-Moon System

The angle of inclination of the Earth must be duplicated by an exo-planet and this requires that there be a moon. Since it is the pull of gravity of the Moon on the Earth that causes the angle of inclination, it follows immediately that the candidate planet must have a moon. The mass and distance of the moon together with the tidal bulge that must be produced on the planet's surface must work with an equatorial bulge caused by the planet's rotation. All of this necessitates that the planet be a fluid ball encased in a thin crust and have a massive water ocean. It also necessitates that the moon be of a very particular size and at an appropriate distance. Also the moon's orbit must be close to the plane of the planet's orbit around its host star. Of course the moon must not be too close to its planet or it will render it to be structurally uninhabitable and if it is too far away its pull of gravity on the planet's crust will not be sufficient. All of these necessary characteristics have a fairly narrow window. Further, there is a definite time limit to the arrangement of only a few tens of thousands of years. If a far-away planet is to support life, a similar development would be required and if it did, there is no reason to believe that it would exist any longer!

As with the Earth, an exo-planet must have the various and numerous 'windows of life' - some of which have been discussed with the understanding that the items mentioned were only representative of a much larger list in reality.

The basic conclusion is that the Earth-Moon system with which we are so familiar must be duplicated by any far-away system before there is any possibility of there being life there at all! From a probability viewpoint it is really quite obvious that this is most unlikely.

It is recognized that numerous far-away planets have been 'discovered'. With very rare exception (307) all of these 'discoveries' are the product of a computer program which is operating on data from a star and what is shown by the media as a discovered planet is actually an artist's conception. If the surface of an exo-planet could actually be observed, it would be a lot easier to decide if there was any hope that life could exist there.

8.6.4   Long Term Stability

If life has ever or ever will develop on a far-away planet it would presumably have done so by natural processes. That is, no miracles allowed. Any such development would have taken time to happen. No one would argue with such a simple and seemingly self-evident assumption. However in order for this to have been the case, long-term stability would be absolutely necessary. Unfortunately long-term stability is rare. It does not exist here in our Solar System and terrible and destructive events are often observed throughout the universe. First of all, within our Solar System we have the stability of the Sun to think about. No one knows if it has long term stability and since it has not been observed for very long to suggest that it does is simply conjecture. In fact, for the past forty years the output of the Sun is observed to be decreasing. Also, we observe that the Sun is continually receiving more mass. The gravity of the Sun is enormous and if any object in space does not have an appropriate orbit, it could easily be drawn into the Sun. The massive comet Howard-Kooman-Michels 1979X1 was one such example. The head of this comet was reportedly about the same size as the Earth (101) Other comet material is also observed to go into the Sun on a regular basis. Comets are relatively easy to spot but what about other types of material like asteroids. It will probably never be known how much material is being continually added to the Sun but for long-term planetary stability, the Sun must not increase it's mass at all. It is the mass of the Sun that determines the orbits of the planets so any change in that mass on either a plus or minus basis is not welcome news. With a continual increase in solar mass the planets will spiral into the Sun and with a decrease they will spiral outward. Even a minimal change in either direction would put an otherwise habitable planet out of the habitable zone and since that zone is very narrow even a small change would not be welcome.

Mass stability of any potential life-supporting planet is also necessary. If any planet attracts more material to itself, its orbit around its host star will not be stable and it will slowly spiral into that star. This raises considerable uncertainty about whether or not a possible life-supporting planet might ever exist because long-term orbital stability is an absolute necessity and both the host star and the potential life-supporting planet must have mass stability in order for this to happen.

If the Sun and the stars are powered by nuclear fusion there is the problem of the Faint Young Sun Paradox which basically recognizes that for most of the supposed 4.5 billion years of the Earth's existence it would not have been in the thermally-habitable zone of the Sun because the Sun would have been much dimmer. In fact it would only have come into the thermally-habitable zone quite recently – that is within the last few hundred million years. This would hardly have been enough time for long-term processes such as evolution to really get going. All of life's processes require long-term thermal stability but where the source of heat is based on nuclear processes this is not possible.

As mentioned immediately above, the Earth necessarily has an angle of inclination which is determined in part by the distance to the Moon. Unfortunately the Moon is steadily receding from the Earth so long-term stability is not in the offing. In fact, from this factor alone the Earth will probably not be habitable within another 100 million years and much much less than this when the energy loss associated with tidal action is included.

8.6.5 Conclusion

It has been shown that the number of stars in the Milky-Way galaxy that could be possible supporters of life is not very great and will not be more than a very few at most. Then when the requirements for a life-supporting planet are factored in, it becomes quite clear that what we have here in our own solar system is really quite special and it is not very likely that something similar will be found elsewhere – particularly when the choice of stars is so limited. The final conclusion is that it is exceedingly, exceedingly unlikely that there is any life elsewhere in the universe and mathematically the probability is virtually zero. The fact that so many factors must be lined up for life to exist at all is amazing enough but when we factor in that they do not have long-term stability (just a few tens of thousands of years at the most) one is left with a serious doubt that they could ever exist ,or ever have existed, elsewhere. Consequently, the conviction that extra-solar life exists, is more firmly planted in the fairy-tale category than in any other.

9.0 Instant Creation

9.1 The Big Bang Theory

Part of the way through the 20th century The Big Bang Theory appeared. The basic idea declares that all of the material of the universe suddenly came into being and instantly commenced to expand outward from its initial location. Actually it would be better to just say outward because there wasn't any universe to be located in immediately prior to the appearance of the material. The appearance of the material was exactly the same as the appearance of the universe and the expansion of the material was the expansion of the universe. The universe did not exist outside of where the material was located even though it might have been moving at very high speed away from the initial location in all directions. This is a concept that is difficult to grasp because we always think of the universe being everywhere with the material (i.e. the stars, clouds of space dust, planets, comets and everything else) moving around in it. However the 'it' and the universe are the same thing. After moving outward, the material became very spread out and, in comparison to the starting situation, it became very thinly spaced as well. Currently, it is certainly very thin with stars being separated by exceedingly great distances. Part of the idea is that the universe is still expanding. However it is not expanding into a void region because there is no such thing. If the universe has a boundary (if it is expanding from an initial starting place) that boundary would be the boundary of reality as we understand it. There wouldn't be anything past the boundary because the boundary of the universe would be the boundary of reality. To suggest that there is something beyond the boundary has no meaning at all.

The idea of expansion is just as difficult to grasp. To suggest expansion is to suggest some particular distance. The idea of expansion in this case however involves the expansion of space. There is obviously space between the stars and space between the galaxies but there is also space within and between atoms. However if space is actually expanding, all of it must be expanding so the 'space' within and between atoms must also be expanding. How then could one measure anything to determine the amount of expansion? If space is expanding, the yardstick used to make the measurement must also be expanding and the idea of expansion therefore runs into serious measurement and confirmation difficulty.

What would be happening to the speed of light if the universe is expanding? The speed of light is understood to be a constant independent of the movement of any material. One might therefore suspect that with an expanding universe, it would take longer for light to travel from one place to another within it. However this would not necessarily be the case because the means of measurement would also be changing so the time taken for light to travel between two locations might always be the same. Such uncertainty makes it difficult to determine if expansion is actually happening.

The 'Red Shift' is commonly appealed to as evidence of expansion with the greater red shifts of more remote objects indicative of them having both increasingly greater speed as well as greater distance. However, if their speed is continuously increasing, it means that all of the material of the universe is being continuously accelerated but acceleration requires continuous and ongoing force. Any initial impulse applied at the beginning could not supply a continuous and ongoing force - with that force allegedly becoming greater and greater with increasing distance and 'time'. In addition, the nature and source of the required force has never been identified!

The 'Red Shift' also has observational difficulties because galaxies are observed that have numerous different 'Red Shifts; within their boundaries whereas there should not be so much variation. In other cases, structures which appear to be connected together visually, have dramatically different 'Red Shifts'. Such observations are suggestive that the 'Red Shift' isn't an appropriate indicator of expansion at all but actually an indicator of something else altogether – possibly involving Einstein's Theory of Relativity wherein gravity has an effect on the movement of light or possibly because each photon of light energy has lost a bit of energy on its way towards the Earth..

Background radiation has also come up for discussion with respect to the Big Bang Theory. The idea here seems to be that if any irregularities can be observed in the spectrum of the background radiation, it would be indicative of 'clumping' in the early universe (i.e. galaxy formation) and therefore supportive of the Big Bang Theory. However this too is not conclusive because the background radiation could simply indicate that the universe is warming up. It is very curious on this point that the 'warmth' might simply be the evidence of all of the energy that the stars have been emitting into the universe for years and years. A situation like this would be comparable to a small stove in the middle of a large closed-in stadium. If we could keep the stove going long enough, the stadium should warm up a little. With this approach, the observed 'warmth' of the background radiation could indicate a universe that is only a few thousand years old!

Our system of dimensions is intimately connected to these ideas as well. We have the three spacial dimensions and use them to describe where something is with respect to something else. Time, the fourth dimension is an integral aspect of the idea as well. However, time beyond a finite universe has no meaning at all. Neither does it have any meaning prior to the appearance of the material initially. As the material appeared so also did time come into being. Otherwise there was no time. Therefore there has never been a 'time' when the universe did not exist because 'time' started just as the material appeared and commenced to expand. Time and space are only of importance with respect to the universe as we see it. If there are other universes they would expectedly also have some parameter like time, but since we know nothing of any such relationships there is absolutely nothing that can be said about time or space apart from our universe. Time and space are notions integral to the universe and do not necessarily apply to any other situations whatsoever. In summary, some of the ideas involved with the Big Bang Theory are indeterminate or simply inconclusive. However, the basic idea of the Big Bang Theory is completely supportive of the notion of Instant Creation.

9.2   LaPlace Nebular Hypothesis of Solar System Formation

This theory has dominated the thinking of many ever since it was proposed and because it requires long periods of time it appears to contradict the ideas of both recent and instant creation – at least with respect to the Milky Way galaxy. It has run into insurmountable difficulties with the discovery of planets orbiting very close to their host stars in far-away places but in spite of this, a replacement has not been offered.

Numerous other problems with the whole idea of long time frames for the existence of the Earth have been pointed out in chapter 3 above but the difficulties with the Nebular Hypothesis will be discussed here and there will necessarily be some overlap in these discussions.

'The Sun, the planets, and associated debris such as comets and asteroids all formed from the gas and dust residing in a flat, spinning disk known as the solar nebula. The protosun, at the center of the nebula, pulled in so much material that eventually the pressure and temperature in its interior rose to the point where thermonuclear reactions could begin. Meanwhile, chunks of rock and ice called planetesimals duplicated the process on a smaller scale, sweeping up enough material to form planetary cores ... These cores, in turn, gravitationally attracted the nebula's abundant hydrogen and helium gases, building up the huge gas giant worlds of the outer solar system. Then the Sun turned on and blew away the leftover nebula material, leaving the solar system as we see it today.' (505) In other words, a large cloud of space dust and gas collapsed. Most of the gas collapsed into the center and formed the Sun while the rest formed the planets. As it collapsed, it began to rotate and this accounts for the fact that the Sun rotates and that the planets orbit around the Sun.

Unfortunately, this popular science version is contradicted by authorities such as 'G. R. Burbidge, a recognized authority on the "evolution of elements" in stars: The problem simply is that the condensation of a star from interstellar material would violate our understanding of the laws of nature. (506)

9.2.1   Gas Cloud Collapse

Included in these 'laws of nature' is the observation that gas clouds in free space do not collapse at all unless they are already very small. Usually they expand. An explanation for the initial formation of stars is missing because stars will not spontaneously form in space since the dominant outward gas force will forbid collapse. Gas clouds always dissipate outward.` (507) Even if the cloud of gas is relatively cool, it is still expected to expand. The outward push due to the thermal movement of the molecules, even at 100 Kelvin, is greater than the gravitational pull inward. (508) Therefore in order to make this theory work, the gas cloud must have been forced to collapse down to its critical size, which is the size from which it could continue to collapse due to its own gravity. The agent, which has been specified in the literature for this task, is another gas cloud. (305) 'The scheme that is invoked here is simply this: surround the cloud you wish to compress with a hotter cloud, so that the molecules at the surface of the inner cloud will be bombarded by the faster moving molecules of the outer cloud and pushed inward.' (305) This outer, hotter gas cloud would then put pressure on the inner gas cloud (i.e. the future solar system) to collapse it. The inner gas cloud must be condensed down to a critical size (called the Jean's length, after Sir James Jeans who initially developed the mathematics) by some other mechanism before gravitational collapse will work. (246)

In order for the outer gas cloud to put pressure on the inner gas cloud, it too must have already been down to its critical size. Otherwise it would have been expanding out into space and could not have put any pressure on anything (What caused the mystery gas cloud to start collapsing?). Somehow after the inner gas cloud was down to the right size, the outer gas cloud disappeared. (Why didn't the outer mystery gas cloud continue to collapse also?)

Therefore, it is clear that three artificially-arranged conditions are necessary for the theory to work. An outer gas cloud must be present, it must already be at a certain critical size and temperature and it must disappear after the inner gas cloud is down to its critical size and a solar system is well on its way to being formed.

9.2.2   Angular Momentum

Momentum is the idea that when something is moving, it will keep on moving unless it is somehow forced to stop. For example, if a stone rolls down a hill, it will not stop right at the bottom. It will keep right on rolling past the bottom. Similarly, if a wheel is turning and the turning force is removed, the wheel will keep on turning until something stops it. The wheel is understood to have angular momentum or rotational momentum, whereas the stone has linear momentum.

The main problem with the idea that a gas cloud can collapse and form a star is that the angular momentum of the finished product(s) does not match the angular momentum of the original cloud. According to observation, interstellar clouds have up to 100,000 times as much angular momentum as their progeny stars. Any theory of star formation must therefore describe how a cloud disposes of the extra angular momentum before it collapses to form a star. (444)

In addition, the Sun does not have angular momentum, which is in proportion to its size. While the Sun has most of the material of the solar system (i.e. 99% (509)), it has hardly any of the rotational factor of the solar system. (i.e. It only has 1%) In order for the Sun to have angular momentum, which is in proportion to its size, it would have to be spinning one hundred times faster. It is not rotating nearly fast enough to agree with the theory. The outer planets including Jupiter, Saturn, Uranus and Neptune have most of the angular momentum of the solar system while the Sun has most of the material. Therefore, the declaration that a large gas cloud collapsed to form the solar system and rotated as it collapsed does not coincide with observations of the solar system and is in direct contradiction with the principles of well-known physics.

9.2.3   Deep Time Problems (Faint Young Sun Paradox)

It is well recognized that the Earth must have a very stable temperature because if it wandered from its present orbit or if the energy output of the Sun changed, the thermal 'window of life' on Earth would close. With these necessities in mind, it is of interest to note that the Earth is declared to have been formed 4.5 billion years ago subsequent to the collapse of a massive gas cloud. (505) During the extended period of time since the Earth was formed, the Sun was shining on the Earth and life is declared to have somehow developed. At the same time, theories of solar function tell us that the output of the Sun would increase over a 4.5 billion year time-frame by a factor of 25%. Over the supposed 4.5 billion year history of the Sun its luminosity has increased about 25 percent. This would result in an ice-covered Earth. This problem is called the Faint Young Sun Paradox, and contrasts sharply with the presumptions surrounding the geological history of the Earth. (33) (This idea was discussed in chapter 3 entitled; The 4.5 Billion Year-Old Earth and is repeated here because of its relevance to the time required for creation.)

Since we appear to be at just the right temperature now, it must have been a lot colder on the Earth through much of the previous 4.5 billion years up until just a short time ago. How then did life develop? Twenty-five percent less energy is equivalent to being about 12% farther from the Sun. This is well past the 5%, which is recognized as being the limit for life on Earth. (104) It may therefore be concluded that either the theory of solar system formation is invalid or that the theory of solar function is invalid or both of them are invalid and opens the possibility that the Earth might not be very old at all.

Current understanding of the warming effects of greenhouse gases further invalidates the notion that the Earth is several billion years old. The present average temperature on the surface of the Earth is about 15C. (Please refer to Climate and the Oceans by Vallis p. 165) This is due to incoming solar energy AND the characteristic of the greenhouse gases (mostly water vapor and CO_2) to retain some of this energy to keep the surface temperature at its present level. The energy from the Sun by itself is currently not sufficient to keep the surface temperature at the proper level. This means that if the Earth had been frozen solid at some time in its past, the Sun would not be hot enough yet to thaw it out because water vapor would not appear to assist with warming until some ice thawed. It would actually require another one or two billion years into the future before the Sun would be hot enough on its own to raise the surface temperature of the Earth until it was above the freezing point. Once this happened, water vapor would appear in the atmosphere further helping to raise the temperature. However by then the Sun would be considerably hotter than it is at present and together with the heating affect of the added water vapor, the temperature would continue rising until it was too high for life to exist. The transition from 'starting to melt' to 'too hot' would only take a few years – not nearly long enough for life to develop by any natural process.

9.2.4   Retrograde Rotation

If all of the planets of the solar system formed as a result of a collapsing gas cloud, they would be expected to be rotating in the same direction. However they are not. Venus has so-called retrograde rotation because it turns in the opposite direction to the other planets. Also three of Jupiter's moons and one of Saturn's moons have retrograde rotation. 'A rotating nebula could not produce satellites revolving in two directions.' (510) The large moon, Triton, of Neptune, has a retrograde orbit. (511) The theory cannot explain exceptions such as these. In addition to a planet with retrograde rotation, there is a planet with an axis which points toward the Sun every time it comes around. Uranus does not spin with an axis basically pointed at the pole star but with an axis that is close to the plane of its orbit around the Sun. Also the moons of Uranus revolve in a plane almost perpendicular to the orbital plane of the planet. (68) This is not good news for the Nebular Hypothesis of Solar System Formation because it does not fit into the expectation that rotation should be the same way for all of the planets. By implication it casts a serious doubt on planetary-formation ideas that require long periods of time.

9.2.5   Planetary Rotation Periods

The planets do not rotate with the same speed and the deviations from any suggested average are enormous. The rotational speeds of the planets are: Mercury 58.6 days, Venus 243 days, Earth 1 day, Mars 1.03 days, Jupiter 0.41 days, Saturn 0.45 days, Uranus 0.72 days, Neptune 0.67 days and Pluto 6.39 days. (512) The Nebular Hypothesis for Solar System Formation should have addressed this problem. Since it hasn't, it is incomplete and its validity is questionable.

9.2.6   Gravity Insufficient

Gravity is the basis for the formation of the Sun at the center of the rotating gas cloud as well as for the formation of the various planets. However, if the central majority of gas separated to form the Sun, there would not have been sufficient self-gravity in the remaining portions of gas to cause them to coalesce to form the planets. Even if several gas clouds had broken away they would not have balled into globes. (513) This problem is compounded with respect to the moons. If there wasn't enough gravity to form the planets, there certainly wasn't enough to form the moons.

9.2.7   Patterns Nonexistent

Neither Mercury nor Venus has a moon. Earth has a very large moon. Mars has two extremely small moons. Jupiter has sixteen moons. Saturn has twenty-one moons, Uranus has fifteen moons. Neptune has eight moons. Pluto has one moon. Mercury is quite small. Venus is larger and Earth is larger still. Jupiter is very large but Mars which is located between Jupiter and Earth, is quite small. Between Mars and Jupiter there very likely was another planet which exploded to form the asteroid belt. Saturn is smaller than Jupiter while Uranus and Neptune are similar in size. 'No orbit is an exact circle; there is no regularity in the eccentric shapes of the planetary orbits; each elliptical curve verges in a different direction.' (514) 'The celestial harmony is composed of bodies different in size, different in form, different in velocity of rotation, with differently directed axes of rotation, with different directions of rotation, with differently composed atmospheres or without atmospheres, with a varying number of moons or without moons, and with satellites revolving in either direction.' (515) It appears to be by chance or some other reason that our planet has oceans, atmosphere, oxygen and one moon and that it is located between Venus and Mars with a nearly circular orbit but the Laplace Nebular Hypothesis of Solar System Formation offers no explanation for any of these circumstances.

9.2.8   Metal Distribution

If an Earth-forming gas cloud should contract, it is reasonable to expect that heavier elements would settle toward the core. While it is commonly held that the core of the Earth may be iron, there is no explanation why iron is found throughout the crust right to the very surface. Even a violent explosion cannot explain this because the iron is well mixed with other elements over untold thousands of square miles of the Earth's surface. The theories explaining gas cloud collapse are unable to explain iron distribution throughout the crust of the Earth.

Further, gold and lead, our heaviest metals, are found in the crust of the Earth far away from the center. If there had ever been a molten phase, these elements would now be at the center of the Earth. (see 9,4,2 below)

9.2.9   Multiple Star Systems

Most of the stars in the sky do not occur as single entities but as doubles or triples. (516) In some cases, two stars orbit around each other and a third one orbits further away around the first two. If these stars formed at nearly the same time, they should be nearly the same age. This however, is apparently not the case. There is a class of binaries known as semi-detached systems, in which the stars almost touch each other. These systems have a large secondary of lower mass, and a small primary of higher mass. The larger secondary would appear to be at a later stage of evolution than the primary but this presents a problem for computer models because they predict that the more massive star should have evolved more quickly than the less massive one. So why is the star of lower mass, larger than the other? (36) It is common for binary star systems to include a white dwarf. Such stars are about the size of the Earth and about as massive as the Sun, while the companion is much larger and not nearly as massive. (517) The Laplace Nebular Hypothesis of Solar System Formation cannot account for these discrepancies.

9.2.10   Asteroid Shape

An integral aspect of the Nebular Hypothesis is that small objects like asteroids would never be found with a spherical shape. A planet's mass and gravity must have been large enough to have pulled it into a sphere. Asteroids and comets do not possess enough mass for this to occur, and thus they are often seen to have irregular shapes. (180) Unfortunately for this theory, the asteroid 2005 YU55, which passed the Earth in November 2011, is only 400 meters in diameter but it clearly has a spherical shape. In fact it has been referred to as a round mini-world. (518)

Numerous asteroids appear as nearly-perfect spheres; from the largest right down to the very smallest and many are irregularly shaped. All of the shapes can be explained by temperature. If a particular mass was hot enough it would have been able to obtain a spherical shape before it cooled down and its shape became fixed. If it wasn't quite hot enough, it's limited self-gravity would not have been able to bring it into a spherical shape before it cooled down too much. In this case, the shape would remain irregular but it would still have a soft rounded appearance. Also, any impact marks would have a soft appearance – some of which could be quite deep. On the other hand, if the asteroid material had been hard at the time of impact, any impact marks would be shallow and more sharply defined. Hard impact marks are seldom found on asteroids. This indicates that they must be very young. Otherwise, one would expect more hard impact marks than there are soft ones because the time over which soft marks could have been made would only have been the cooling-down time right after the asteroid had been formed (from a previously-existing warm body). Expectedly, the cooling-down period would not have lasted very long.

9.2.11   Ignition Temperature Problem

Anything that is warm cools off. Whenever a substance is heated so that it's temperature rises above absolute zero, heat will be lost to space. In fact, the hotter an object becomes, the faster the heat will radiate away. Warm objects do not stay warm. If a collapsing cloud of gas started to heat up why wouldn't it just cool down again? When this reality is factored in, it isn't clear that a gas/dust cloud would heat up at all! An even greater dilemma was identified more recently regarding how hot such a cloud could possibly become. Of course, the temperature of interest is the ignition point for nuclear fusion. This temperature is very high and it is very questionable whether a collapsing cloud of gas and dust could ever reach such a temperature. In fact one knowledgeable commentator pronounced that the temperature would not have risen higher than about 1,000,000 C which is too low for fusion to begin. This was confirmed recently by a leading astronomer at the NATO Advanced Study Institute. There A. G. W. Cameron presented a paper in which he calculated that if the Sun had contracted from a gas cloud, and material fell inward to form a core with the present density, the temperature would have only reached one million degrees which is far too low for nuclear reactions to begin. Thus if the Sun had formed according to an evolutionary process, nuclear burning would never have begun. (328) So the theory was under serious attack even prior to the discovery of extra-solar planets.

9.2.12   The Death Knell

The Nebular Hypothesis was held to be the explanation of how the solar system formed up until other solar systems were discovered. Somehow, the Nebular Hypothesis anticipated that rocky planets would form closest to the Sun and gaseous planets would form much further out. Since this was indeed the case with the Solar System, the hypothesis seemed to be correct. Then, other solar systems were discovered. These other systems are all constructed contrary to how a solar system should be constructed as they have large gaseous planets near their host star. In fact, this type of arrangement enabled them to be discovered. It is, needless to say, difficult to 'see' a planet near a faraway star. The light from a star is so bright when compared to a non-glowing object like a planet, that visual detection is very difficult. However, if a star has a planet nearby and if it is big enough, it should put a slight wobble into the host star's pathway. Then it would not be necessary to actually see the planet but only the wobble. The Sun has a wobble. The Sun and Jupiter both orbit around a common center which is located 50,000 km above the surface of the Sun. (144) Therefore, if the Sun were observed from far away, its wobble would enable an observer – if he had the patience to wait for Jupiter to complete one or two orbits around the Sun - to deduce that Jupiter existed.

In the 1990's the first extra-solar system was discovered and in 1999 a system was found that included three massive planets much closer to their host star than Jupiter is to the Sun. (345) This was, of course, an exciting discovery and prompted considerable speculation but did nothing to help the Nebular Hypothesis. In fact it contributed to its demise.

The presence of very large objects, in some cases, very near their host star, effectively disabled the Nebular Hypothesis. Unfortunately, its replacement is slow in arriving as the following illustrates. 'The presence of such huge bodies so close to their stars challenged prevailing theory. How could gas giants form so close to their suns? Could they have formed elsewhere? If so, how did they get to their present locations? Are their orbits stable? And what does all this say about our solar system? Is it a freak whose apparent orderliness is illusory? The discovery of more than a dozen extra-solar planets (the number has steadily risen since these comments were offered) has forced a serious rethinking of many details of the solar nebular theory and, in particular, the subtleties of orbital dynamics. One cornerstone of the standard theory has been that the planets first formed at or near their present locations relative to the Sun. But the news from afar, combined with more sophisticated computer modeling techniques, has suggested a more complicated and literally chaotic scenario for planetary formation.' (326) In other words, the old theory is dead and there isn't a new one.

The LaPlace Theory of Solar System Formation is clearly not credible. Perhaps some credibility should be given to other ideas like 'instant creation'.

9.3   The Plasma Theory of Solar System Formation

While The Big Bang Theory has gained widespread recognition as a valid theory there is another approach that seems to offer a better explanation for numerous measurements and observations. However we must all be cautious about readily accepting abstract ideas such as these because the bottom line and the basic truth of the matter is that sitting here on the Earth and making declarations about what really happened long ago or is currently going on in places that are very far away is an exercise that can never be finally resolved. In the meantime listening to various ideas seems reasonable. This comes at a time when several observations and discoveries right here in the solar system have upset some long-held opinions. One such opinion held that it would be impossible for two objects to share exactly the same orbit. It has now been discovered that this type of thing happens repeatedly throughout the solar system. Even the Earth is involved in this phenomenon as the Earth has a Trojan asteroid just as Jupiter has numerous Trojan asteroids. (Trojan objects are those that follow exactly the same orbit as the host planet.)

The rings of Saturn also get into the act as they include two small moons that travel around in the same orbit and even exchange positions with one leading and then following the other as they orbit around the planet. While this shouldn't be happening, it is, and sometimes strongly-held opinions must give way to observation.

Several observations have been made throughout space indicating that the universe might be driven by electromagnetic phenomenon instead of by gravity. One of these relates to how the stars move. In the Milky Way Galaxy it is observed that the stars orbit the center of the galaxy at the same speed no matter how far they are from the center. This has long been known by promoters of the Big Bang Theory and accommodated by declaring that there must be material and energy out there that cannot be observed. Hence Dark Matter was invented. Unfortunately 95% more matter and energy is required to get the theory to work. Alternately, the theory might be 95% wrong.

On the other hand, if movement of material throughout the galaxy is being determined by electromagnetic phenomena the observed movement is expected. 'The simulations solved another long-standing mystery – the 'flat' rotation curves of the galaxies. If the speed of gas rotating around the galactic center is plotted against its distance from the center, the curve first rises rapidly but then levels off. However, if the disk-shaped galaxy is held together by gravity alone, the speed should fall steadily as distance (from the center) increases. As in the Solar System, the outer planets move more slowly than planets close to the Sun. Astrophysicists had long seen this (i.e. the strange rotation speeds of the stars about the center of the galaxy) as evidence of a halo of gravitating dark matter surrounding the visible galaxy: within such a sphere, a flat rotation curve would be possible, though by no means necessary. But ... the flat rotation curve emerges quite naturally in a galaxy wholly governed by electromagnetic fields.' (527) This means that the strange rotation patterns could be explained by forces other than gravity.

The same observations upset the idea that a black hole exists. 'But scientists at the University of Arizona ... measured the velocities of stars within a few light-years of the center of our galaxy and found the velocities twenty times slower than the plasma velocities in the same area. Since the stars must respond to any gravitational force, their low velocities show that no black hole exists.' (528)

Just to further upset certain long-held ideas, in a plasma universe, when the solar system formed, the Earth-forming material would have collapsed into a planet a few days before the Sun would have formed and turned on. Unfortunately, ideas like this are a little too close to the Genesis account of creation which promoters of the plasma universe idea refer to as 'myth' and associate with the Big Bang Theory thereby attempting to discredit both of them as being in the realm of myth rather than science. (530) On the other hand, if the Earth did receive its final shape prior to the Sun being formed, it is clear that some aspects of the plasma universe idea are supportive of Instant Creation just as some aspects of the Big Bang Theory are supportive.

One would expect that scientists measure everything and determine logically what is happening. This is clearly not happening and ideas are retained long after observation indicates that they are not valid. Such retention could explain why there is such a wide diversity of opinion among scientists whereas the lay person looking in from the outside expects a more consistent picture to emerge - but there is no consistency. Fortunately, where the idea of instant creation is concerned, there is an abundance of understandable evidence available to all of us and some of this evidence is discussed in the following sections.

9.4   The Crust of the Earth

9.4.1   Radiohalos

Radiohalos are found in the rock that forms the crust of the Earth. Of course most of the material that forms the crust of the Earth is rock of one type or another. Occasionally when a piece of rock is broken open, a series of very small coloured concentric circles are exposed and have been given the name 'radiohalos' because they are understood to have been formed by radioactive process. Apparently these very small circles were noticed more than 100 years ago but at that time there wasn't any explanation. More recently an explanation has been identified which involves the 'decay' of polonium down through a series of steps until lead is formed. Lead, of course, is not radioactive and never becomes anything but lead. It will always be lead. It is the final product formed from polonium after the decay process is complete.

It is understood that radioactive elements do not actually decay. They simply transform from one type of material to another. The time for these transformations to take place is not the same for all types of material and neither is any one of the transformations really ever complete. This is where the idea of a 'half-life' comes in. It is the nature of all radioactive transformations that after a period of time called the half-life, one-half of the original amount of material will still be left. This means that if we started with ten pounds of material, after one half-life there would still be five pounds left. It is further the nature of this type of process that after another half-life has passed, there would only be two and one-half pounds left. No matter how much material there was at the beginning of the time period of interest, after one half-life there will only be one-half of it left.

All radioactive materials have a half-life and there are never two of them with exactly the same half-life. Carbon14 is radioactive with a half-life of about 5,700 years. On the other hand the materials involved in the decay of polonium have half-lives between fractions of a second and many thousands of years. There is simply no consistency based on weight or any other characteristic of the material. For example, carbon14 is a gas whereas uranium is a metal. The point of immediate interest however, is that some of the materials which were involved in the formation of the radiohalos found in rock have extremely short half-lives.

The circles that are usually found in rock are indicative of polonium218 (1/2 life = 3 minutes), polonium214 (1/2 life = 164 microseconds), polonium210 (1/2 life = 138 days). When these circles are found in rock they require nearly instantaneous formation of the rocks with the creation of the polonium atoms. (519) In other words, they require 'Instant Creation'.

9.4.2 Heavy Elements

Continental crust material is slightly lighter than the rest of the material of the Earth so it basically 'floats' on top of the rest. From the viewpoint of modern geophysics, the continental platforms are understood to float on the underlying material. (529) Also we observe that continental crust (i.e. the granitic continental platforms) is clumped together so that the continents actually appear above water. Even when it is arranged into these clumps, it just barely projects above the ocean. While the average depth of the ocean is about 2 ½ miles (250) the average elevation of the continents above sea level is only about one-fifth as much. In reality we barely have our heads above water. It has been declared and appears to be commonly accepted that the Earth and the other planets formed from a collapsing dust cloud and went through a period of time when they were completely molten. Then, over a period involving hundreds of millions of years, a crust formed. One commentator offered that it would have taken a billion years for the Earth to have cooled enough for a crust to form. (128) However, if the material which now forms the continents had ever been in a molten state, it would have spread out evenly over all of the rest of the material. No suggestion has ever been offered as to why it would have clumped together as it is currently found. If it had been molten it would have spread out and above-water visible continents would not exist at all. Why would anyone expect otherwise? Alternatively, there never was a time when the continental material was molten. It must have been formed just the way that it is presently found or it would be spread out in a thin layer over all of the rest of the material.

This problem is further compounded by the presence throughout the crust – even at the very top – of heavy elements like gold and lead. How can this be explained? Both of these elements are about twice as heavy as the other material where they are found so if the crust had ever been molten, they would have sunk to the very center of the Earth. This reality of physics has recently been recognized with respect to the structure of certain asteroids. In fact it has been explicitly declared that certain asteroids must have experienced a molten state because their heavy elements are not found on the surface. Many planetary bodies in our solar system must have melted significantly, allowing denser materials to sink to their centers and in some cases melting was dramatic. The researchers determined that both asteroids (i.e. Vesta and another one unnamed) experienced widespread melting with more than 50 percent of each object being liquid. (340) Unfortunately, there is no suggestion concerning why or how they melted! (Please refer to The Asteroid Theory of the Ice Age for more discussion on this topic.)

An explanation for how heavy metals have become distributed throughout the crust of the Earth has never been offered. The presence of iron in the shell of the Earth or the migration of heavy metals from the core to the shell has never been explained. For these metals to have left the core, they must have been ejected by explosions, and in order to remain spread through the crust, the explosions must have been followed immediately by cooling. (537)

The other problem immediately associated with this one is trying to explain how iron still exists as iron. If in the beginning, the planet was a hot conglomerate of elements, as the nebular as well as the tidal theories assume, then the iron of the globe should have become oxidized and combined with all available oxygen. But for some unknown reason this did not take place. Therefore the presence of oxygen in the atmosphere of the Earth is unexplained. (537)

In recognition that heavy material would sink, the crust of the Earth could never have experienced a molten state because all of the heavy metals found in the crust would have sunk to the very center of the Earth - just as massive quantities of iron are thought to have done to form the core of the Earth. This recognition from basic physics does not bode well for any idea that the Earth went through a cooling-down period which would have required hundreds of millions of years. Instead, it indicates that the relatively-light crust material was present in a solid state from the very beginning and was therefore able to support the heavy metals found therein. 'Instant creation' is the logical explanation.

It is proper to recognize at this point that there is a considerable body of current opinion that has abandoned the idea of a molten Earth and opted instead for one that was formed from solid chunks of material instead. While this might appear to partially satisfy the problems with a molten phase for the Earth it does not address the fact that the entire Earth is currently molten except for a very thin layer on top. If it formed from solid chunks, it should still be chunky and irregular. It is a good thing that it isn't because our atmosphere does not extend upward very far and if the surface was seriously irregular much of the surface might be above it. All of this only further indicates the problems that arise in trying to explain origin of the Earth and underscores that ideas such as 'instant creation' are also well supported by observation and hence just as valid or even more so than other ideas and should not be set aside too soon or too easily.

9.5   Coal Formation

9.5.1   Radiohalos

There is evidence which indicates that the trees and other plants that formed coal were themselves formed almost instantly. (i.e. numerous different types of plants have been identified in coal including; ferns, cycads, sassafras, laural, tulip tree, magnolia, cinnamon, sequoia, poplar, willow, maple, birch, chestnut, alder, beech, elm, palm, fig, cypress, oak, rose, plum, almond, murtle, acacia, and many other species. (116)) This evidence has been recovered from the study of very tiny, concentric, coloured circles found in both rock and coal as mentioned in 8,4,2 above. Also as mentioned previously, the circles that are usually found in rock are indicative of polonium218 (1/2 life = 3 minutes), polonium214 (1/2 life = 164 microseconds), polonium210 (1/2 life = 138 days). When these circles are found in rock '… they require nearly instantaneous crystallization of the rocks simultaneously with the synthesis or creation of the polonium atoms.' (519) However, the first two of these circles are not found in coal. This indicates that the material that would become coal, had not yet formed around the polonium and that it did not form for at least five half-lives (or about 15 minutes) after the polonium218 started to decay. However, polonium210 is included indicating that the material had enclosed the polonium within days of its formation. (i.e. well within 138 days, the half-life of polonium210)

Therefore it can be concluded that the radiohalos in coal indicate that the trees and other plants that formed the coal were themselves formed after the polonium214 and polonium218 had decayed but before the polonium210 had decayed. Therefore the coal-forming plants were formed between about 15 minutes and several days after the original polonium sample started to decay. This means that they would have been formed within a few days of the initial formation of the crust of the Earth along with the polonium that it contained. This is direct evidence of 'instant creation'.

9.5.2   Unexplainable Coal

The problem in trying to explain the existence of coal and CO2 was applicable to the discussion of the age of the Earth as well as to the possibility of life on far-away planets. It is just as relevant to the present discussion of instant creation so for completeness and at the risk of being repetitive, it will be repeated here.

As coal beds are formed, the carbon cycle is interrupted. We understand that all of the plants, which became part of the coal beds, were formed from carbon obtained from atmospheric carbon dioxide. However as the coal-bed carbon accumulates, it is effectively trapped and is no longer available to circulate as part of the carbon cycle. It is being trapped off into a great carbon storehouse. In fact, if such a situation were allowed to continue, more and more carbon would become unavailable and the great carbon cycle would have less and less carbon in circulation. Since both plant and animal life need the carbon to keep circulating, life would consequently become less and less viable. This process could actually have led to a carbon-starvation death for the Earth. In recognition of how much carbon is presently stored in the coal beds of the Earth, in comparison to the amount in the biosphere, it is a wonder that this didn't happen. Indeed it might have happened, except that the Industrial Revolution reintroduced great quantities of carbon back into the atmosphere. If there had not been an industrial revolution and vast quantities of coal had not been burned, carbon dioxide levels would have kept dropping and would be much lower today than they were prior to the Industrial Revolution. If this had happened, life at the present time would be much less viable.

Now we are in a position to recognize the great problem which exists in trying to explain the coal beds. The carbon from the atmosphere must have formed the plants for the coal beds, but the carbon in those particular plants has not been allowed to circulate back into the atmosphere. It is still in the coal. Therefore, the carbon, which is in the coal beds, has been diverted from the carbon cycle, and has become trapped out of circulation. It is therefore appropriate to ask where the carbon in the coal came from in the first place. It is obvious that it was not exhaled by any animals, which had eaten plants, because the carbon from the plants, which formed the coal-beds is still in the coal, which hasn't been eaten at all. Neither did it come from any plants, which were burned, because if they had been burned, their carbon would have combined with atmospheric oxygen to form CO_2 and would consequently not be in the coal beds either. The same carbon cannot be in two places at the same time.

Once coal is formed, its carbon is tucked away in the great coal storehouse and it is out of circulation. Therefore, all of the carbon in these coal formations has become completely unavailable for carbon dioxide formation and the possible production of more plants. Hence it is appropriate to seek an explanation for the source of the carbon dioxide, which supplied the carbon for the great coal deposits of the world in the first place.

The coal beds do contain a very great amount of carbon. Various estimates have been made and compared to the amount which exists in the biosphere (the total of all living things). The coal beds might contain 50 times as much carbon as the biosphere. (269) All of this carbon has been trapped away from the carbon cycle and has not been available to recirculate since it was trapped and because of this trapping, the carbon, which formed the coal bed plants, came from some source other than the metabolization process of animals. Neither did it come from the burning or rotting of plants. Of course, it came from the air because that is the source of all plant-forming carbon, but how did it get into the air if animals had not breathed it out?

Three possibilities present themselves.

1. The first possibility is that the required CO2 was formed by a burning process, which used primeval or virgin carbon as a source. This burning process introduced the virgin carbon into the carbon cycle at just the right rate

to enable the trees and other plants, which were forming the coal beds, to grow. In order to be internally consistent, it must be recognized that a vast amount of virgin carbon was required. In fact, exactly the same amount of ancient virgin carbon was required as is presently found in the coal beds. However, this type of arrangement is recognized as an 'artificial construct' because nature has been conveniently arranged to bring about a result, which isn't otherwise credible.

2. It is possible that there could have been enough carbon dioxide in the air at some ancient time to enable the coal-forming plants to develop simply by depleting this CO_2. The amount of carbon dioxide which is in the atmosphere at the present time is more than 380 parts per million (38) which means that 380 out of every million molecules in our atmosphere at the present time are carbon dioxide molecules. If all of the carbon in this carbon dioxide were assembled together to make coal, about $6 \times 10(11)$ metric tons of coal would result. Current estimates of the world's coal reserves are $15 \times 10(12)$ metric tons. (520) Atmospheric carbon is therefore equivalent to $6/150 \times 100$ or about 4% of the world's coal reserves. Therefore, if, prior to formation of the coal beds, the amount of atmospheric CO_2 had been about 25 times as great as it is now, the coal beds could have been formed by growing plants and depleting this higher level of CO_2 down to near its present level. Therefore, one possibility for coal formation is that an ancient CO_2-rich atmosphere could have been depleted to an atmosphere with much less CO_2 as the coal-forming plants grew.

3. Prior to the formation of the coal beds, the ancient biosphere was 50 times more extensive than it is at the present time. Forests, swamps and meadowlands were filled with an abundance of all kinds of plants. In addition, a greater area of the Earth was involved including the high arctic lands, Antarctica and areas now below sea level. Then suddenly this ancient biosphere was annihilated and its carbon is now found in coal. This explanation basically shoves the question back because now we must ask where the carbon came from to form this massive assembly of ancient plants. Did they just suddenly appear? Were they instantly created?

In summary, there appear to be three possibilities for the formation of the coal-forming plants.
1. The ancient biosphere was formed from CO_2 which was produced by an unknown, virgin-carbon burning process.
2. The ancient biosphere was formed by depleting even more ancient atmospheric CO_2 from a level approximately 25 times as high as the present level of 380 ppm.
3. A massive ancient biosphere was 'instantly created' and the plants from it were available to form coal.

However, all of these possibilities come with attachments. If the ancient atmosphere had 25 times as much CO_2, the average temperature of the world would have been much higher (i.e. CO_2 is a greenhouse gas) and the trees would not have been able to grow properly because it would have been too hot. Also, the temperature would have been above the body temperature for most types of animal life.

While the third possibility is totally unacceptable to many people, with both the first and second possibilities, it would have been necessary that none of the plants which grew during the extended times required, were burned, eaten or decayed. They had to be preserved as plants or they could not have been used to form the coal. There were no forest fires caused by lightning (which currently strikes the Earth several thousand times every day). Also it was a rot-free forest. The most obvious conclusion is that the coal-forming plants were 'instantly created'.

9.5.3 The Carboniferous Period

The Carboniferous Period is declared to have been a time when the Earth was covered by massive forests of trees and numerous other plants. All of these plants were large and extremely healthy and apparently covered most of the land. There is some evidence that numerous plants lived in water as well and floated as great vegetation mats. Then all of this material was destroyed and is now found in the Earth as layers of coal. A time-frame has been associated with the Carboniferous Period as beginning about 350 million years ago and lasting about 80 million years. (531) Layers of coal are not found during other times in this declared extensive history of the Earth and one is prompted to ask why there wasn't any forests growing during these other times as well. Was the Earth basically forest-free for the vast majority of geological time? How could any life have existed if there were no forests?

It is a further curiosity that included in the layers of material in the Carboniferous Period formations are both sandstone and limestone. These are both types of sedimentary rock which means that the material for them was placed by water. Actually it is also apparent that the layers of coal themselves were placed by water movement as seams of coal are often split with sedimentary material in between. This means that most of the material throughout the Carboniferous Period layers was placed by water movement. It is therefore apparent that some great catastrophe was in progress which completely begs the question of time altogether. If a great world-wide catastrophe was happening, all of those layers could have been placed in a matter of days and millions of years would not be necessary.

In support of the uncertainty of time, 'Erratic Boulders are often encased in coal.' (425) However, with conventional thinking Erratic Boulders were to have been moved by ice during the Great Ice Age which was, by comparison with the Carboniferous Period, a very recent event. One cannot have it both ways as the Ice Age involves thousands of years while the Carboniferous Period Involves millions of years. Either, one of these ideas is wrong or, they are both wrong. The evidence indicates that the Great Ice Age happened quite recently effectively dismissing the idea that there ever has been millions of years which brings the idea of 'instant creation' further into the realm of feasible.

9.6 Polystrate Fossils

A fossil is a stone that has the exact shape of something that was once alive. That something could have been either part of a plant or part of an animal. The fossilization process requires that the object be buried completely. Then minerals in the soil replace the various parts of the object until the original material is all gone and has been replaced by stony material. The time required for fossilization will depend on how quickly the surrounding soil can provide the replacement material but it seems that it would not be very long or else the dead object would simply decay and disappear. After all, this is what normally happens when something is buried. In many cases, even after one year very little of the original material can be found. Rapidity of fossilization is particularly evident in the case of the Petrified Forest of north-eastern Arizona. The logs that have been preserved by fossilization are massive. That fossilization occurred quickly is manifest by the exquisite detail of these petrified (i.e. fossilized) logs. If it had not happened quickly, the logs would have started to degenerate. As it is, very fine detail has been preserved and without the petrification process having been rapid, there is no other explanation.

Polystrate refers to the involvement of more than one layer of the Earth. Poly simply means many and strate is another form of stratum or simply layer. A polystrate fossil is therefore a fossil that projects up through several layers.

As with the layers involved in the Carboniferous Period, the layers involved where polystrate fossils are found are thought to have required many millions of years to accumulate. (532) However this cannot be known but only declared. The times involved with the Carboniferous Period are questionable enough but suggesting that great spans of time were involved with polystrate fossils is preposterous. Exposed wood cannot be exposed for a million years while soil and other burial material slowly accumulates around it. The wood would simply rot and do so in much less than a million years. In fact it would be well on its way to rotting within a very small fraction of a million years and more likely within just a few years.

This reality relating to the polystrate fossils raises a doubt about the idea that immense periods of time were involved just as the sedimentary rock layers of the carboniferous period could not have taken millions of years to accumulate. If a world-wide catastrophe was underway, things would happen quickly and great amounts of time would not be involved.

9.7   The Theory of Evolution

9.7.1   Structural Similarity

One of the planks in the platform of the proponents of the Theory of Evolution is that since various animals have similar structures there must be some ancestral relationship between them. This idea is given the name 'converging evolution'. On the surface it might appear to have some validity but with a little more scrutiny it is obvious that it doesn't.

A situation like this can be compared to automobiles. If one is shopping for a car they will altogether likely pay a visit to various places where cars are on display and being offered for sale. If the customer isn't sure which car they wish to buy, they will probably visit several dealerships and gather information on the various features and benefits of one car compared to another. Suppose that the potential customer visits a dozen car lots, makes some notes and then returns home to look over the information.

It will come as no surprise that the following features were found on all of the cars reviewed. It will be found that all of the cars had;

  A. a windshield
  B. a steering wheel
  C. a seat for the driver
  D. rear view mirrors
  E. a back window
  F. a door for the driver
  G. other doors for the passengers
  H. front lights
  I. rear lights
  J. an engine
  K. four wheels
  L. turn signals
  M. a grill
  N. fenders

as well as numerous other features which might have appeared quite similar and which provided a similar function.

Or consider a person who is looking for a house to buy. In all likelihood they will travel around in an area where there are numerous houses and they might also attend at a real estate agent's office to review photographs of houses that are being offered for sale. Then they might return home and review the information that they have gathered. When they do this, they will find that all of the houses had several features which appeared similar.

  A. a front door
  B. a foundation
  C. a roof
  D. a front window
  E. an electrical power supply
  F. a water supply
  G. a kitchen sink
  H. a bathroom
  I. a closet
  J. several rooms
  K. flooring
  L. lights
  M. a heating system
  N. a waste elimination system

None of these similarities will be considered odd by any home buyer who would, in reality, be able to extend the list quite readily.

There appears to be a principle involved in both of these examples namely; SIMILARITY OF FUNCTION NECESSITATES SIMILARITY OF STRUCTURE.

When we consider the animal kingdom a very similar situation presents itself. All animals need to be able to move around - both to defend themselves and to obtain food. This means that they must have some means of locomotion such as legs or wings or fins. All animals require an energy supply. They must therefore take in nutrient. This necessitates a means of reception for the food and a means of processing it so that it can be useful. Some portion of the food that is taken in will have no use so must be eliminated. An elimination means is therefore necessary. Once the food is broken down it must be distributed around the entire body in order to get to where it can actually be used. A circulation system is therefore necessary. In order for the food to be useful and provide heat to keep the animal warm the food must be chemically combined with oxygen during a chemical reaction. This reaction must be very tightly controlled of course because we do not want it to occur explosively but be spread out over the entire day. After all, an animal cannot be active for a few minutes and then dormant the rest of the time. Since the carbon in the food must be combined with oxygen to get the appropriate reaction, the oxygen must be available at the proper location or it will not be useful at all. Therefore it too must be distributed around the entire body in a manner that enables it to come into proximity with the carbon that was in the food. The circulation system must service every last cell in the entire body so it must be arranged so that this can happen. Also since movement of the circulating fluid is required there must be a pump which will provide the power to accomplish the movement. Ironically, the pump must also have a power supply and it would be most ingenious if the pump could supply itself with the required ingredients. Incidentally this is exactly what the circulating pump does in every animal body.

As the animal moves around, means to detect its surroundings is essential and a combination of detectors would be best. Sight would appear paramount but smell and hearing would also be very useful.

The more similar the movement is between various animals the more similar they would appear. If, for example, both of them must move through water then they both must have fins. Or if they both move through the air then they both must have wings. If movement across the surface of the Earth is required, legs would appear advantageous and if the required movement required speed long thin legs would be best.

As with the previous examples from cars and houses, the animal kingdom indicates the same principle; SIMILARITY OF FUNCTION NECESSITATES SIMILARITY OF STRUCTURE

With such necessities in mind is it any wonder that numerous different kinds of animals have very similar structures. While cars must transport passengers and houses must shelter their occupants animals must find their food and nurture their young. In all cases, we should expect numerous similarities to exist but such a development wouldn't necessarily mean that they were 'related' any more than cars or houses are 'related'. Perhaps they are just carrying out similar functions. Possibly the Creator recognized these necessities.

9.7.2   Animal Ancestry

Quite often when The Theory of Evolution is discussed, it is declared that a certain type of animal descended from some other type of animal. Occasionally diagrams are included to illustrate the point wherein animal B descended from animal A and animal C descended from animal B and so on. In some cases the illustrations will be in full colour and quite persuasive and include a dozen stages or steps of evolutionary development. All of these diagrams are based on similarity of appearance and function.

However from a scientific point of view it must be recognized that the genetic distance between any two of the types in the usual sequences shown, is enormous. To illustrate this point consider the elephant and the Woolly Mammoth. These two animals appear very similar – in fact much more similar than many of the animals or animal categories that are shown in evolutionary diagrams. For example, they both have trunks. They both have large round legs with feet that are not much larger in diameter than the legs. They both have tusks. Their bodies are very similarly shaped and include a tail with a small collection of hair at the end. It would be tempting to think that the elephant might have descended from the Woolly Mammoth. Basically the Woolly Mammoth is larger than the elephant but if we could devise some way to make the elephant larger, the matchup would appear quite striking.

However, the DNA of both elephants and Woolly Mammoths has been mapped. From that analysis it is understood that there are about 50,000,000 differences in the DNA between these two very similarly-appearing creatures. That really is a lot of differences and if one is to declare that one animal descended from the other, an explanation for this great genetic difference would be a reasonable expectation. However an explanation has never been offered. In fact it has been declared that they are two distinct species. This is a reasonable conclusion – well supported by the scientific evidence.

Unfortunately, when one examines any of the evolutionary diagrams, it is immediately evident that the genetic distance from any one of the species shown to those shown on either side, is enormous. It will usually be much greater than the 50,000,000 for the elephant-Woolly Mammoth case. Unfortunately, explanations for these incredible genetic distances, is never provided. It is instead a leap of faith that is not justified. In that manner the basic scientific approach is just abandoned and replaced by declarations. This is exactly the approach taken by Richard Dawkins in his book entitled 'The Greatest Show on Earth' while trying to explain the extreme improbability of life developing. I must stress that the details of my story are pure fiction because neither I nor anybody else can tell you the details of how Evolution happened. (549)

There are a few cases where an ancestor for some particular animal has been identified and it will be immediately obvious because of appearance that the genetic make-up will be very similar. The most obvious of these examples is the wolf-dog situation. Some breeds of dog appear, in fact, very similar to the timber wolf so suggesting that one was bred from the other will not strain the imagination. By the same token all breeds of dogs are dogs and only appear distinctive because of very determined efforts by breeders to obtain some particular characteristic. Such a procedure cannot be extended to nature however, because there is no element in nature which has ever extended a focused and determined effort to achieve some particular result. In fact, in nature any offspring that cannot hold its own among the others will not even live long enough to breed. The Theory of Evolution stresses that everything happened by random chance. There was no guiding intellect involved. The stresses of survival will not favor any development that cannot fully and equitably participate in the necessary survival activities. In particular the miniature dogs that have been so carefully and intentionally bred would not have any chance of survival without continuous protection. In fact, if dogs are just left alone to breed randomly, within a few generations there will not be any miniature animals left.

Wolves, evidently, help this function along by only allowing the strongest pair in any particular group, to mate. African lions carry out a similar procedure as any male that overcomes an incumbent male will immediately destroy all of the incumbents cubs and proceed to breed his own right away. In this manner any male that was not competitively-strong would never be able to breed and propagate his own genetic short-comings. While such breeding activities have been going on for thousands of years, there is no sign of any super-wolves or super-lions showing up. In both cases, the DNA is being preserved without any expectation that a large block of it will suddenly be transformed enabling a super-animal to appear.

Unfortunately for adherents to the Theory of Evolution, genetically-miniscule steps like wolf-to-dog cannot be extended any further without invoking unsupportable conjecture. It is scientifically more prudent to just recognize what nature is telling us than to try to make it support some hypothesis for which there is no evidence.

Some examples of the type of declarations that have been made are; frogs descending from fish, land animals descending from amphibians and birds descending from dinosaurs. These are all unsupportable declarations and in the dinosaur-to-bird case we note that proto-avis, an ancient bird, was on the scene much earlier than many of the dinosaurs! 'Proto-avis is so much 'older' than Archaeopteryx, that it must have existed long before all the dinosaurs that are supposed to have been its ancestors. This indicates that birds existed in the strata 'before' their dinosaur ancestors. This completely upsets the theory of bird evolution. (533) Incidentally, one wonders why birds (which supposedly were evolving from the dinosaurs) were able to survive the cataclysm that wiped out every last dinosaur, flyers, walkers and swimmers, large and small. How could birds, which are mostly land-based creatures, survive, when the dinosaurs, including both the very large and the very small, couldn't? If one is to offer such serious claims as these, some explanation would be in order but none has ever been offered.

The further problem is the 'start-up' problem. How did it all begin? An explanation for the beginning has never been offered! If one is to offer a theory such as The Theory of Evolution, then it would be reasonable to expect an explanation for start-up. How can anyone offer a theory without an explanation? This problem was well recognized by Francis Crick, one of the co-discoverers of the DNA double helix when he expressed frustration at the almost impossible task of giving a numerical value to the probability of what seems a such an unlikely sequence of events. It actually appears that an honest man, armed with all of the knowledge available to us now, could only state that the origin of life must be a miracle because of the conditions which would have had to have been satisfied to get it going. There is simply too much speculation running after too few facts. (524) One of the 'facts' that must be faced is the unthinkable quantity of information contained in every DNA molecule. In fact there is enough information in a single human cell to store the Encyclopoedia Britannica three or four times over. (551)

The problem relates directly to the basic claim of the theory which is that all the living forms in the world have arisen from a single source which itself came from an inorganic form. (491) Trying to explain how it all began has posed a seemingly insurmountable problem so much so that a prize has now been offered for a reasonable explanation. There is currently a million-dollar prize for anyone who proposes a chemically-plausible naturalistic origin-of-life scenario. The Origin of Life Prize will be awarded for proposing a 'highly plausible mechanism' for the spontaneous rise of genetic instructions in nature sufficient to give rise to life. (466)

No one has claimed the prize. This includes Richard Dawkins, the current chief promoter of the Theory of Evolution. One is left wondering why he wouldn't file a claim when he is so convinced of his position that he has written several books including 'The Greatest Show on Earth' to try to explain everything.

The enormous amount of 'information' (i.e. approximately 3 billion groups of miniscule atomic assemblies) in an animal's DNA has proven impossible to explain and the enormous DNA distances between any one type of animal and the next in any proposed evolutionary sequence has not been explained at all. The Theory of Evolution is therefore woefully incomplete. This leaves the question of animal ancestry unanswered because there isn't any animal on the Earth for which a series of ancestors can be identified through a sequence of changes in DNA. Therefore as things stand at the present time, animals have no ancestors. Without any ancestors it can only be concluded that all of the animals of the world have been CREATED INSTANTLY in the forms that we find them in at the present time.

9.7.3   Time

In order for animals to evolve time is required. In fact the 'necessary' time or more properly 'hypothesized time' is very well recognized. The time that has elapsed since the common ancestor of all of today's animals walked the Earth is about two million centuries. ... The time that has elapsed since our fish ancestors crawled out of the water onto the land is about three and one-half million centuries. (552) Of course no support is given for these comments leaving one to wonder if they have just been made up!

As an aside we recall that it was the ancient Greeks who first, as far as we know, offered that humans descended from fish! One of the first naturalistic evolutionary theories was proposed by Thales (640 – 546 B. C.) of Miletus, a city in the province of Ionia on the coast near Greece. Thales was also evidently the first person to advance the idea that life originated in water. ... One of Thales most famous students, Anaximander (611 – 547 BC) developed the idea even further. Anaximander taught that life originated in the sea ... He even concluded that humans evolved from fish or fish-like creatures. (554) No support for these ideas are offered either!

Without vast amounts of time The Theory of Evolution is simply unthinkable and the requirement for vast amounts of time has never been in dispute but simply accepted because it was available! However this is an assumption without support as numerous evidences will testify.

9.7.3.1   Axial Stability

When the atomic clock came on line several years ago, it was finally possible to obtain an accurate value for the length of a day. It turns out that the length of the day is getting longer at an alarming rate. In fact, since the atomic clock became active in the 1970's about 500 leap-seconds have been added to the length of the day. This means that a day will be twice as long in approximately 120,000 years! The reason that the Earth is decelerating and the day is getting longer is because of the tides. As the Moon pulls up the tides (because the Earth rotates) the tidal bulge is continually circling the Earth. Tidal water is moving in and out of bays and estuaries, around islands and through straights. Moving all of this tidal water requires a very great amount of energy (i.e. The water movement in the Bay of Fundy alone has been estimated at 500 billion tons. (550)) This energy is being supplied by the rotational energy of the Earth. Consequently the Earth slows down. However this in turn means that the equatorial bulge is getting smaller. When it gets critically smaller, it will not be able to counter-balance the tidal bulge. When this happens the pull of the Moon on the tidal bulge will tip the Earth over making it uninhabitable. The time until this happens will be measureable in tens of thousands of years – not millions. When we look into the past, within a similar amount of time (with a faster-rotating Earth and a larger equatorial bulge), it is clear that the Earth would have been sitting too upright to have been habitable. Both the spin rate of the Earth and the axial tilt of the Earth must be within a very narrow range for habitability. This is one of several time factors indicating that evolution could never have happened! There wasn't time for it to have happened.

9.7.3.2   The Moon

The Moon is heavily involved with the question of time. As discussed above, the Moon is hot inside, repeatedly ejecting clouds of gas and often showing lighted areas. Also, there is no dust on the Moon! All of these factors are suggestive that the Moon cannot be very old. The question of even greater interest with respect to the Earth and time is the fact that the Moon is receding from the Earth. This too has been discussed above and indicates that this factor alone will cause the Earth to stop rotating even if tidal energy loss were non-existent. While the time frame in this case is several hundreds of millions of years, a non-rotating Earth will be an uninhabitable Earth. Also, in this case, as with an Earth that is being slowed by tidal energy loss, looking back in time it is clear that the Earth would have been rotating much faster in the past and would have been much closer to the Earth raising continent-crossing tides. Within a similar time frame (i.e. several hundreds of millions of years) a day would have been a mere six hours long with over 1200 days in a year. (553) A rapidly-rotating Earth with monster tides every few hours is an uninhabitable Earth.

9.7.3.3   Asteroid Impacts

Numerous life-destroying asteroids have hit the Earth. One commentator offered that there have been 14 mass extinctions. (27) This is clearly conservative because even 'small' asteroids would destroy most of the life on the Earth. It is thought that impacts of objects larger than one kilometre are large enough to cause global consequences such as climate change that would kill most inhabitants of the Earth. (525) In all probability the Earth has been hit thousands of times by objects that large (418 + other listings) and claiming that every last species had a remnant that survived isn't credible. While large animals would preferentially have been killed, how would small ones deal with speeding (i.e. 1000 mph) monster continent-crossing tsunamis followed by world-wide darkness lasting for months and the asteroid winter that would accompany it. In fact, a survival scenario has never been offered. With respect to the Theory of Evolution a survival hypothesis is absolutely necessary. Otherwise it is necessary to explain how a diversity of animals evolved in time for the next wipe-out which would have happened on the average every few tens of thousands of years. The surviving cohort from one wipe-out had to get ready for the next wipe-out every time. Finally, the cohort that survived the last wipe-out had to have been the source for the current diversity of animal life on the Earth. While such a development is not the least bit credible, neither has there ever been an explanation for how this might have happened. Without an explanation the proponents of the Theory of Evolution are operating in a vacuum.

9.7.3.4   C14 Startup

A discussion of C14 startup has been included in section 3,3,8 above. Therein it was pointed out that the production of C14 has only just begun. It has only begun during the last few thousand years. This has implications for The Theory of Evolution because it points to an Earth that might not be very old at all. How could anything evolve if the Earth is only a few thousand years old?

9.7.3.5   The Ice Age

The Ice Age has been discussed extensively herein as well as in The Window of Life and The Asteroid Theory of the Ice Age. The Earth could barely have tolerated even a single Ice Age because of not only the direct disruption of the conditions which are directly required for animal survival but also for the necessity to maintain the life-enabling Greenhouse Effect. In fact in spite of the extensive interest in the great Ice Age there has never been recognition that the Greenhouse Effect must be maintained at all costs. Only the Asteroid Theory of the Ice Age made this recognition. The implications for The Theory of Evolution relate to the times involved. At the very most the entire ice age time-frame could only have involved a few hundreds of years because the Greenhouse Effect had to be retained and restored. With a prolonged ice Age this would not have been possible. Millions of years were not in the offing.

9.7.3.6 The non-Fluid Earth

As discussed in section 9,0 above the Earth has never experienced a fluid state. If it had ever been entirely fluid all of the material that was heavier than rock would have slipped through and sunk to the center of the Earth. Instead we find it right up in the crust of the Earth – in fact right on the very top. If the Earth never had a fluid stage then how old is it? The same reasoning has been recently applied to certain asteroids which are declared to have been fluid because they do not have any heavy elements near the surface. (535) ' The researchers determined that both asteroids (i.e. Vesta and another unnamed asteroid) experienced widespread melting with more than 50% of each object being fluid.' (340) This evidence completely destroys The LaPlace Theory of Solar System Formation as well as the necessity for the Earth to cool down enough to be habitable. The crust of the Earth never had to solidify because it was never fluid. All of this indicates Instant Creation and that there has never been time for anything to evolve. This evidence demolishes The Theory of Evolution.

9.7.3.7 Non-existent Time

The time that would have been required for evolution to have been possible has never existed. The evidence from Nature clearly indicates that very little time has actually been involved since the beginning of the habitability of the Earth. Proponents of the Theory of Evolution have always insisted that since there have been vast amounts of time since the Earth became habitable there has clearly been plenty of time for evolution to have happened. Unfortunately for that point of view Nature indicates the exact opposite. The necessary great amounts of time have never existed.

9.8 Review

While the Big Bang Theory has been offered as a scientific theory and quite widely accepted as such, it really isn't scientific at all because it ignores basic physics. For example, an explanation for the ongoing force that is causing the increasing rate of expansion of the universe has never been identified. Worse, the very idea defies our understanding of basic physics. Science is all about making measurements and then trying to make some sense out of them with an explanation. If a measurement cannot be made at least the discussion should recognize aspects of reality that have been well tested and verified. In spite of such shortcomings, the Big Bang Theory is based on the idea that the universe appeared instantly and it thereby supports the idea of 'instant creation'.

The LaPlace Nebular Theory of Solar System Formation has run onto stony ground and should be discarded completely. In fact, the very first extra-solar planets that were discovered totally undermined it and the only recourse would be to discard it. From both a theoretical and observational point of view it is not valid. It is basically a Fairytale for Adults. On the other hand, the Plasma Theory of Solar System Formation has not been discredited even though it has faced fierce opposition. Since it totally undermines the LaPlace Theory it is no wonder that it has upset a large number of people. Inherent in the Plasma Theory is the assertion that the Earth took shape slightly before the Sun began to shine. The ideas involved in the Plasma Theory are completely supportive of the notion of 'instant creation'. It is no small wonder that people have tried repeatedly to discredit it.

The crust of the Earth is readily observable and its contents can be measured. The presence of heavy metals in the crust tells us that the crust had to have formed instantaneously in order to have supported these heavy metals as soon as they appeared. Otherwise why wouldn't they currently be near the center of the Earth instead of right up here on the very top of the crust? This conclusion is fully supported by the radio-halos found within the crust which also tell us that the material of the crust formed instantly. It formed so fast that it was able to capture the first decay product of polonium. This means that the crust material had to be there at the same time as the polonium appeared and the time involved was only thousands of a second. This is a basic observation from nature and offers a scientifically-valid support for the notion of 'instant creation'. Incidentally it also means that the crust never had a liquid phase but was cool and solid enough to support heavy metals right from the very beginning.

Coal also provides support for 'instant creation'. In fact coal does this in two different ways. First, coal also contains radio-halos but in this case they indicate that the trees that formed the coal were themselves formed instantly within a few days of the formation of the crust of the Earth.

Secondly, coal provides support for 'instant creation' because there is far too much coal to be explained any other way. Coal was, of course, made from trees. Of this there is no doubt. The trees would normally have been formed from CO_2 in the air but there isn't any explanation of where such a vast amount of CO_2 could have come from except by 'instant creation'. Technically there is a possibility that the coal-forming CO_2 was itself formed by a virgin-carbon burning process at just the right rate to enable the trees to form and for animal life to exist. However this option is not very convincing as a totally artificial setup must be in place for it to have happened. This leaves 'instant creation' as the most reasonable explanation.

Animal DNA further supports the 'instant creation' explanation because animals do not have ancestors. If they have evolved over long periods of time there expectedly would be plenty of ancestors but since none have been found, the pendulum swings towards 'instant creation'. Reinforcing the absence of ancestors factor is the total lack of any explanation for how the animal kingdom ever got started. A start-up scenario has not been identified even though a prize has been offered to anyone who can provide one. 'Instant creation' fills the gap so there isn't any need for an explanation anyway. Scientific evidence supports the notion of 'instant creation' and arguments countering the evidence have not been successful. 'Instant creation' is the logical explanation and hence it is not a fairytale.

10.0   Summary

Eight ideas have been discussed. The four that are usually held to be fairytales have been shown to be valid and the four that are widely held to be valid, have been shown to be fairytales!

While the vast majority of people have never thought about the length of the year as possibly having been shorter at one time, whenever most scientists comment on the matter at all, they dismiss it as not being valid. However the evidence indicates otherwise. There is no doubt whatsoever that the Earth has been struck numerous times by very large asteroids every one of which would have changed the Earth's orbit by a certain amount. Is it possible that they could have changed it enough to change the length of the year? The evidence indicates that it is possible! The entire Earth was much warmer at one time which could logically have been caused by greater heat input. The atmosphere was also different because the temperature of the Earth was nearly the same from pole to pole. Records from ancient time tell us that everything was warm. If anyone is to dismiss the matter it would be reasonable to deal with all of that evidence and show that some other explanation has greater validity. The literature is void of any such offerings.

For the last several years the Earth has been declared to be 4.5 billion years old. While this is a very bold assertion it is not well supported by anything other than more assertions. Scientific integrity demands more than this. It is not valid to just pile one assertion on top of another. On the other hand the evidence is strongly biased towards an Earth that is not nearly that old and one which has a very bleak near future. The Sun is not very helpful as a support for an old Earth because the output of the Sun is not constant. Theoretically, the Sun is warming up which is nothing short of disaster for the 'old-habitable-Earth' idea, because the Earth would have been 25% colder in the distant past and not be warm enough to thaw out yet if that had been the case. Neither could the Sun possibly be gravitationally constant for an extended period of billions of years because it is constantly gathering more material into itself. This means nothing other than that all of the planets are spiralling into it! How could that be a satisfactory long-term condition for life on the Earth?

If mountains are a recent development it means that for billions of years the continental crust material was only barely sticking out of the water. Why didn't it erode? It erodes now. Even granite cannot be exposed for even a few hundreds of years without eroding. Further, even a billion years ago the Earth would have been rotating much faster and the Moon would have been much closer. This is a bad combination because high-speed monster tides would have appeared every couple of hours and repeatedly washed over the land. Even now the land is – on the average – only hundreds feet above sea level. Why didn't the land all wash away long ago?

And why hasn't the ocean become filled with silt? Incoming material from space arrives quite regularly every day and (together with even a modest amount of eroded material from the land) should have filled the ocean basins enough for the ocean to completely cover the land by now. The floor of the ocean does not have nearly enough material on it to explain long periods of time and some in parts of the ocean the floor is rock-bare!

The Moon is receding from the Earth and being pulled into an increasingly-higher orbit by the peak of the tidal bulge. This is accompanied by a reduction in the rotation speed of the Earth. If all of the rotational energy of the Earth went into lifting the Moon, it would ultimately result in it being about 50% further away than it is now. However this is not going to happen because most of the rotational energy actually goes into friction – in this case the friction involved in moving the water of the ocean around obstacles and up and down estuaries and so on as the tide comes and goes. The net result is that the rotational speed of the Earth is being reduced much more rapidly than it would be if it simply lifted the Moon. In fact, the length of a day will become about one minute longer over the lifetime of a person. The ongoing reduction in the rotational speed will result in a smaller equatorial bulge and since the bulge is needed for axial stability, the axial tilt of the Earth will increase. This will happen quite slowly at first but later the Earth will simply tip over and will no longer be habitable. At current rates this could happen well within another 100,000 years. The opposite situation would have been in place within a similar time frame in the past where the Earth would have been sitting upright and been uninhabitable. These realities mean that any declared periods of time involving millions of years are totally unrealistic and should be set aside.

Understandably it is hard to accept that a habitable Earth is only a short-term scenario but that is what science tells us repeatedly.

The Astronomical Theory of the Ice Age has been repeatedly refuted but continues to hold the dominant position among scientists. This is possibly the case because they have not identified an appropriate replacement. Unfortunately, even though a valid replacement is now available, there will not be any mass reconsideration. The Asteroid Theory of the Ice Age on the other hand recognizes that the greenhouse effect cannot be violated except with the result that the Earth could never again be suitable for human existence. It would have required a balance of heat and cold to retain the greenhouse effect. The Asteroid Theory of the Ice Age recognizes both of these factors. Being supported by both theory and evidence it is therefore a valid theory of the Earth.

Several years ago Beringia was conjured up as an excellent place for untold thousands of animals to thrive right through the Great Ice Age. This is blatant nonsense. There has never been an animal that could survive in even one hundred feet of snow let alone ten thousand feet of snow. They were not only declared to survive right through the Great Ice Age but when it was over they conveniently buried themselves along with great boulders and trees and the other material which forms part of the 'muck' of Alaska as well as other places in the north. Only when everything was buried did the ground freeze solid. Either this was a most amazing achievement or the entire story is just a fairy-tale.

Keeping the Earth in a condition to be able to support human life seems to be an increasingly greater challenge. This is even on a planet that has everything imaginable to carry out the task. Is there any possibility that there could be another planet far away that is similarly up to the task? The argument usually starts with the recognition that there are a large number of stars in the galaxy so there should be at least a few places where life could thrive. However even a preliminary investigation shows that the initially large number is reduced to a very small number after only a few factors are addressed. The necessity of a Moon is never mentioned. We have an appropriately-sized and appropriately-nearby Moon but even then it will not provide the necessary support for very many more years before it will actually contribute to the demise of the Earth's habitability - a condition from which there will be no escape. The notion that the Earth has been here for 4.5 billion years does not really deal with any of this scientific evidence at all but is basically only a declaration. What could possibly be better on a far-away planet than it is right here on the Earth at the present time but where even here the near future looks pretty dismal?

The Earth has been pummelled by numerous large asteroids every one of which would have rendered the Earth uninhabitable. This does seem ironic but a single isolated impact would produce an asteroid winter from which there would be no recovery. On the other hand, a shower of asteroids would have provided the conditions for an ice age as well as the conditions required to preserve the greenhouse effect, without which life on Earth would not have been possible. It is difficult to explain how any animal life could have withstood repeated wipe-outs by several dozen asteroids arriving every few million years while all of those animals were struggling to regroup from the previous impact. Animals cannot stand being washed right off of the continents and swept around the Earth by giant water-flows. How could any animal - large or small - stand being swept across a continent at several hundred miles per hour only to be swept back again by another water flow going in the other direction. Surviving an asteroid shower does not sound like much of a picnic but dealing with a permanent asteroid winter (which would have developed following every single major hit) would not have been possible at all. On the other hand, dealing with a short-term ice age would have been just barely possible.

The idea that the Earth was created instantly is not totally foreign to the scientific community because it is an integral part of the Big Bang Theory (which does have numerous critics). The evidence from the radio-haloes points to 'instant creation'. Also the total absence of any ancestor for any animal on the entire Earth indicates the same thing. The evidence from coal is similar. No one has ever explained where the carbon in the great coal beds came from. If there was an explanation, it would have showed up by now and nobody has ever offered an explanation for why heavy metals like gold and lead are found up on the very top of the crust of the Earth. Why didn't they sink to the center of the Earth? The rocks where they are found must have been there as soon as the heavy metals appeared. Whether we like an idea or not is not justification to ignore it! The evidence should always be the guide.

Eight ideas – four are fairytales and four are not. Will they ever be recognized appropriately?

Appendixes

1. Energy Considerations
2. Asteroid-Crater Size
3. Bibliography
4. References
5. Energy table
6. Illustrations and Photos
7. Index

Appendix 1 - Energy Considerations

In order to lengthen the duration of an Earth year from 360 days to 365 ¼ days the Earth must be lifted further from the Sun. This would require energy which would have to be supplied from an outside source. The amount of energy can be determined from the speed that the Earth has in its present orbit and the speed that it would have had in a 360-day year orbit. The total energy is required and is given by the formula for objects that orbit the Sun.

$$\text{Total energy} = -1/2 \, m \times v(2)$$

The mass of the Earth is about $6 \times 10(24)$ kg.
The speed of the Earth in its present orbit is approx. 29.6 m/s $\times 10(3)$
The speed that the Earth would have had in a 360-day year orbit would have been
approx. 29.72 m/s $\times 10(3)$

Appendix 2 - Asteroid-Crater Size

Suppose that an asteroid that was about 5 miles in diameter (8 km) impacted the Earth at 75,000 mph. (33,300 m/s) Also, suppose that the crust of the Earth was 20 miles thick. In order not to punch right through the crust, this asteroid must be brought to a dead stop within 20 miles (which is only 4 diameters of the asteroid). The force required to stop this asteroid would be;

$$F = m \times a$$

Where $m = 962 \times 10(12)$ kg and $a = 33,300$ m/s (2)

Therefore $F = 320 \times 10(17)$ newtons.

The cylindrical crustal plug that must be pushed ahead of the asteroid in order to allow it entry would have a mass of about;

$$M \text{ (plug)} = 5.76 \times 10(15)$$

At the surface of the Earth this represents a force/weight of about;

$$F(\text{plug}) = 5.76 \times 10(15) \times 9.8 \text{ m/sec}(2) \quad \text{or about } F(\text{plug}) = 53.5 \times 10(15) \text{ newtons}$$

Appendix 3 - Bibliography

	Source	Abbreviation
1.	A Short History of Nearly Everything, By Bill Bryson, Anchor Canada, a division of Random House of Canada Limited	Short
2.	American Petroleum Geologist Bulletin 56 No 2 1972,	Am Pet1
3.	An Ice Age Caused by the Genesis Flood by Michael J. Oard, Institute for Creation Research, P.O. Box 2667, El Cajon, California 92021	Ice Age
4.	Apocalypse When? Cosmic Catastrophe and the Fate of the Universe by Frank Close William Morrow and Company Inc., New York	Comets
5.	ASHRAE Handbook Fundamentals, American Society of Heating Refrigeration and Air-Conditioning Engineers Inc., 1791 Tullie Circle, NE Atlanta GA 30329	Fun 1981
6.	By Design, By Jonathan Sarfati PhD, Creation Book Publishers, Atlanta Georgia	By Design
7.	Canada from Space, By Brian Banks, Camden House Publishing, Suite 100, 25 Shepherd Ave. West, North York, ON M2N 6S7	Canada
8.	Cassell's Atlas of Evolution, By Andromeda, Weidenfield & Nicolsen, London UK	A of E
9.	Chemistry by James V. Quagliano Prentice – Hall Inc., Englewood Cliffs, New Jersey, USA	Chem
10.	Climate Wars, by Gwynne Dyer, Random House Canada	Climate
11.	College Physics by Weber, White and Manning, McGraw-Hill Book Company, New York NY	College
12.	Comets and Asteroids and Future Cosmological Catastrophes compiled by Glen W. Chapman, www.2s2.com/chapmanresearch	Cosmic
13.	Creation Matters, Creation Research Society, P.O. Box 8263, St. Joseph MO 64508-8263 USA	CM
14.	Creation Research Society Quarterly, 6801 N. Hwy 89, Chino Valley AZ 86323	CRSQ
15.	Design and Origins in Astronomy, By George Mulfinger, Jr., Creation Research Society Books	Design
16.	Earth Impact Database, www.unb.ca/passc/ImpactDatabase	EID

	Source	Abbreviation
17.	Earth in Upheaval by Immanuel Velikovsky, Dell Publishing Co., Inc., 1 Dag Hammarskjold Plaza, New York NY 10017	E in U
18.	Encyclopedia Britannica 1958, Published by William Benton, Chicago London Toronto	En Br
19.	Engineering Mechanics, Dynamics, by Meriam & Kraige, John Wiley & Sons Inc, New York, London	Eng Mech
20.	Field Notes from a Catastrophe, By Elizabeth Kolbert 2007, Bloomsbury, USA	Notes
21.	1st International Conference on Creationism Vol II	Conf 1
22.	Funk & Wagnalls New Encyclopedia, Edited by Robert S. Phillips, Funk and Wagnalls	F & W
23.	Grand Canyon, The Story Behind the Scenery, By Merrill D. Beal, KC Publications Inc., P.O. Box 14883, Las Vegas, Nevada 89114	Grand
24.	Handbook of Chemistry and Physics, 52nd Edition, The Chemical Rubber Publishing Company, 18901 Cranwood Parkway, Cleveland Ohio 44128	Handbook
25.	Historical Geology, By Carl Owen Dunbar, John Wiley & Sons Inc., New York London	Geology
26.	How It Works, The Magazine That feeds Minds, Image Publishing ltd. Bournemouth, Dorset, BH2 6EZ	HIW
27.	In the Minds of Men by Ian T. Taylor, TFE Publishing, Toronto	In the Minds
28.	Kronos Press, PO Box 313, Wynnewood PA 19096	Kronos
29.	MacLeans, 11th Floor, One Mount Pleasant Road, Toronto, ON M4Y 2Y5, Vol. 122, Number 24, June 29, 2009	Mac1
30.	Macleans, Aug. 14, 2000	Mac3
31.	Macleans, Aug. 24, 2009	Mac2
32.	Modern University Physics, By Richards, Sears, Wehr & Zemansky, Addison-Wesley Publishing Company Inc., Reading Mass., USA	Modern

	Source	Abbreviation
33.	National Geographic Society, 17th and M Streets, NW Washington DC 20036	Nat Geo
34.	Nature Alberta, By James Cavanagh, Lone Pine Publishing, Edmonton, Alberta	Nature
35.	Pensee, Student Academic Freedom Forum, P.O. Box 414, Portland Oregon 97207	Pensee
36.	Peoples of the Sea, by Immanuel Velikovsky, Doubleday & Company Inc., Garden City, New York	Peoples
37.	Petrified Forest, The Story Behind the Scenery, By Sidney R. Ash and David D. May, Petrified Forest Museum Association, Petrified Forest national Park, Holsbrook Arizona 86025	Pet For
38.	Physiology and Biophysics by Ruch and Patton, Nineteenth edition, W. B. Saunders Company, Philadelphia and Company	P&B
39.	Postcards from Mars, by Jim Bell, Penquin Group (USA), 375 Hudson Street, New York NY 10014	Postcards
40.	Principles of Microbiology Eighth Edition, By Alice Lorraine Smith, The C. V. Mosby Company, 11830 Westline Industrial Drive, Saint Lewis Missouri 63141	Principles
41.	Scientific American Inc., 415 Madison Ave., New York NY	Sci Am
42.	Silent Snow by Marla Cone, Grove Press, 841 Broadway, New York NY 10003	Silent
44.	The Beothucks or Red Indians, By James P. Howley, Prospero Canadian Collection	Beo
45.	The Big Splash, by Dr, Louis A. Frank, Avon Books,New York	Splash
46.	Brent Crater Trail, Ontario Parks	Brent
47.	The Concise Oxford Dictionary, Oxford University Press, Walton Street, Oxford, 0X2 6DP	Oxford
48.	The End of the World, by John Leslie, Routledge, London, USA, and Canada	The End
49.	The Genesis Flood, by Whitcomb & Morris, The Presbyterian and Reformed Publishing Company, Philadelphia, Pennsylvania, 29 West 35th Street, New York, NY 10001	The Flood

	Source	Abbreviation
50.	The Greatest Show on Earth, By Richard Dawkins, Free Press, New York	The Greatest
51.	The Living Cosmos by Chris Impey, Random House New York	Living
52.	The Moon Its Creation Form and Significance, By John C. Whitcomb/Donald B. DeYoung, BMH Books, Winona Lake, Indiana 46590	The Moon
53.	The New World of the Oceans, Men and Oceanography, by Daniel Behrman, Little Brown & Company	New
54.	The Oceans, By Sylvia Earle & Ellen Prager, McGraw-Hill Book Company, New York NY	Oceans
55.	The Rough Guide to the Universe by John Scalzi, Rough Guide Ltd., 80 Strand, London, WCR2 ORL	Rough
56.	The Scientific American Book of the Cosmos, Daniel H. Levy Editor, St. Martin's Press, New York, New York	Cosmos
57.	The Sea Around Us by Rachel Carson, Oxford University Press Inc. 2003, 198 Madison Ave., New York NY 10016	The Sea
58.	The Sun and Stars, By J C Brandt, 1966, McGraw Hill Book Company, New York NY	The Sun
59.	The Trouble with Physics by Lee Smolin, Houghton Mifflin Company, 215 Park Avenue South, New York, New York 10003	Trouble
60.	The Violent Face of Nature, by Kendrick Frazier, William Morrow and Company Inc., New York 1979	Violent
61.	Time Upside Down, By Erich A. Von Fange, 460 Pine Brae Drive, Ann Arbor MI 48105	Time
62.	Weather, A Visual Guide by Buckley, Hopkins and Whitaker, Firefly Books Ltd. 2008	Weather
63.	Worlds in Collision by Immanuel Velikovsky, Doubleday & Company, Garden City, New York	W in C
64.	In The Hills, published by MonoLog Communications Inc. R. R. 1, Orangeville, ON	Hills

Appendix 4 - References

Number	Source	Topic
1	Kronos X 3 p 25	95% of Martian craters on 1 side
2	EID	asteroid impacts spread over 10(9) yrs
3	Brent	Brent Crater features
4	Short p 190	Manson Crater Limestone
5	Short p 201	Chicuzulub 120m x 30m deep
6	Short p 217	Russian borehole
7	Short p 203	blinding flash
8	E in U p 330	360 day year
9	Can Geo M/J 95 p 35	> 1 million asteroids
10	Kronos X 3 p32	largest asteroids
11	Can Geo M/J 95 p 38	> 2000 > 1 km dia.
12	Can Geo S/O 99 p 74	> 100,000/yr added
13	Wiki 1685 Toro	Earth Trojan asteroid 8:5 w E. & 13:5 w V.
14	The Moon p 114	300 transients from lunar craters
15	The Moon p 106	transients looked on with distain
16	Short p 202, CM M/J 2007 p 3	Chicxulub under sed. 2-3 km
17	Nasa.gov/news	incoming material – millions of tons
18	Can Geo M/J 95 p 38	mass extinction
19	Climate p 14	small temp > disaster
20	CRSQ v 21 p 84	10 mile asteroid – hemisphere wipeout
21	Cosmos p 227	Mars below freezing
22	The End p 86	comet into Sun
23	F &W Vol 17 p 23	Mars orbital deviation
24	CRSQ V 32 p 70	5% orbit variation
25	Design p 249	Moon stabilizes axis of inclination
26	CSRQ v 21 p 82	10 mile asteroid – hemisphere wipeout
27	Can Geo Sept/Oct 1999 p 76	14 mass extinctions
28	Cosmic p 4	2x dia depth in 0.01 sec
29	Can Geo M/J 95 p 35	asteroid orbit shift due to Jupiter
30	Rough p 49	3.8 cm/year
31	Rough p	1100 hour day-night
32	Design p 78	1 billions years to contact
33	CRSQ V 30 p 74	Sun 75% of present, climate models crude
34	Climate p 19	surface temp below freezing
35	Comets p 79	fading Sun
36	CRSQ V 18 p 16	binary stars

Number	Source	Topic
37	Cosmos p 189	solar system formation
38	CRSQ V 26 p 50	solar collapse rate
39	CRSQ V 26 p 57	neutrinos from solar flares
40	An Ice Age p 13	over 60 ice age theories
41	Thelostsquadron.com	300 ft of blue ice
42	Cosmic Pursuit Spring 1999 p 27	orbit deviation a disaster
43	The Big Splash p 79	incoming comets
44	An Ice Age p 99	ice age glacier thickness
45	Violent p 189	Tambora dust
46	Violent p 188	chilled earth from Tambora dust
47	Violent p 207	Tambora & Krakatoa temp drop
48	An Ice Age p 18	Milankovitch, insufficient chill
49	CRSQ V 21 p 69	1 or 2% drop in insolation
50	Climate p 21	snowball Earth
51	Climate p 165	Ave sur temp +15C
52	Climate Wars p 2	methane and CO2 from bogs
53	An Ice Age p 45	drop in CO2 during Ice Age
54	An Ice Age p 45	CO2 absorbs into cold water
55	An Ice Age p 5	temp spiral to below freezing
56	CRSQ V 37 p 144	bare trenches
57	http://starchild.gsfc.nasa.gov	3000 tonnes per day
58	CRSQ V 33 p 85	axial tilt stability
59	The Big Splash p 36	loss of hydrogen from atmosphere
60	Smithsonian.com	giant snake and turtle
61	Can/Geo M/A 99 p 38	Harrington
62	E in U p 13	muck to 140 ft deep
63	Can/Geo M/A 99 p 44	small white horse
64	Nat Geo Apr 03 p 31	no explanation for extinction
65	Can Geo M/A 99 p 38	60 species at Whitehorse
66	An Ice Age p 193	drumlins are erodible
67	http://tcc.customer.centrex.ca	drumlins flow to northwest
68	The Living Cosmos p 291	life self-annihilation
69	En Br Vol 14 p 958	Mars eccentric orbit
70	CRSQ v 33 p 86	Mars angle of inclination is not stable
71	En Br Vol 15 p 780	Moon distance
72	What If? P 175	Neutrino detection
73	Nat Geo Dec 09 p 91	Ejection of small planet
74	The Big splash p 36	hydrogen leaving the Earth

Number	Source	Topic
75	Can Geo M/J 95 p 57	Chixulub killed the dinosaurs
76	Kronos X p 36	3000 craters on one side of Mars
77	Kronos X p 34	3 large craters on one side of Mars
78	Wiki Caloris Basin	Weird terrain on Mercury
79	Comets p 5	darkness lasting for months
80	E in U p 23	Erratic Boulders on At & Pa islands
81	E in U p 18	horses died with Woolly Mammoth
82	An Ice Age p 45	more CO2 in cold water
83	An Ice Age p 29	ocean near freezing
84	Climate Wars p 95	Russian bogs start melting in 2006
85	Wiki mass concentration	mascons on one side of Moon
86	The Moon p 145	lunar libration
87	An Ice Age p 5	water vapour stabilizes well below freezing
88	H it W Issue 59 p 85	asteroids hit the Moon
89	CRSQ V 32 p 76	5% tolerance
90	CRSQ V 29 p 190	insufficient light up north
91	CM M/A 2011 p 10	tidal effects on exo-planets
92	Nat Geo Dec 2009 p 91	queer universe
93	The Moon p 145	lunar libration
94	Sc Am April 1977 p 98	57% are multiples
95	Cosmos p 126	single stars are rare
96	www.nasa.gov/worldbook/star	75% are binary
97	Nat Geo Dec 2009 p 94	planet floater
98	CM J/F 2005 p 1	60 to 90% are multiples
99	Wiki multiples	list of multiples
100	Short p 247, Cosmic Pursuit Spring 1999	habitable zone 5% in 1% out
101	The End p 81	monster comet into the Sun
102	Short p 249	Moon stabilizes Earth's orbit
103	En Br V 15 p 778	Lunar orbit 5% offset
104	CRSQ v 32 p 76	5% closer, 1% further
105	Short p 248	molten interior necessary
106	The Toronto Star Apr 16, 1999	catastrophic effect of large planets
107	CRSQ V 37 p 185	1178 impact on Moon
108	The Moon p 97	asteroid shower on Moon
109	The Moon p 97	uneven cratering on Moon
110	The Moon p 86	200,000 craters on Moon
111	Nat Geo Atlas of the World 7[th]ed p 19	lunar maria
112	Cosmos p 227	low Martian temp

Number	Source	Topic
113	Cosmos p 191	stability produced by Jupiter
114	HIW Issue 60 p 13	Earth 2.0 found Keplar 186f
115	Cosmos p 319	pH of water, atmos. of Venus
116	E in U p 202	trees in coal
117	CRSQ v 20 p 218	amount of coal 50x biosphere
118	Field Notes p 43	CO2 380 ppm
119	CRSQ V 20 p 218	carbon in coal
120	www.icr.org	5% of stars hot enough
121	E in U p 22	Erratic Boulders
122	E in U p 23	Erratic Boulders
123	CRSQ v 20 p 215	Erratics in coal
124	CM M/A 02 p 1	magnetic field decreasing
125	www.universetoday.com	lunar far side bulge
126	Wiki, mass concentration	lunar mass concentration
127	Design p 78	lunar recession from Earth
128	Climate p 24	Earth crust starts to form 3.8 B yrs ago
129	CRSQ v 26 p 134	crust thickness
130	CM M/J 06	hot Moon
131	Kronos X p 59	Tharsus Bulge & volcanoes
132	Kronos X p 61	Tharsus Bulge & volcanoes
133	CRSQ v 31 p 153	Arizona Crater
134	EID	Ries Crater 24 km dia
135	C & A & C p 5	Erratic Boulders from Ries Crater
136	Wiki, Chicxulub	Chicxulub dimension
137	Can Geo M/J 95 p 38	Chixulub impactor
138	Kronos XI 1 p 58	diameter of Hellas
139	Wiki, Caloris Basin	diameter of Caloris
140	Wiki, Hellas	shape of Hellas
141	En Br V 14 p 959	mass of Mars
142	www.daviddarlington.info	Late Heavy Bombardment
143	ww.bbc.co.uk/science/earth	Jupiter & Saturn eject Neptune
144	Cosmos p 190	planet formation
145	The Big Bang Theory	electromagnetic universe
146	By Design p 238	co-rotation radius
147	www.icr.org	5% of stars are hot enough
148	Wiki Gliese 581c	extremophile forms of life
149	www.evolutionnews.org	prob of life on red dwarfs
150	HIW Issue 60 p 55	Kepler 70b close to Kepler 70

Number	Source	Topic
151	http://evolutionnews.com	Gliese 581g
152	Cosmos p 198	Caloris Basin
153	Cosmos p 199	rings of Caloris
154	Wiki, Hellas	Hellas formation
155	A Short p 296	life 3.5 billion years ago
156	http://www.lpi.usra.edu/education	Tycho Crater very young see 2.
157	www.windows2universe	Isidis Planitia
158	Wiki List of craters on Mars	hundreds of thousands
159	Kronos XI p 66	Mars Moons
160	www.space.com	spherical asteroid close to Earth
161	www.google.ca	photos of Phobos and Deimos
162	A Short p 297	development of oxygen
163	Nat Geo Dec 09 p 92	habitable zone
164	CM N/D 2010 p 3	heavy bombardment
165	Climate p 18	greenhouse gases
166	Climate p 19	positive feedback
167	An Ice Age p 109	ocean 4C
168	An Ice Age p 45	CO2 into cold water
169	Climate p 21	snowball Earth
170	F & W Vol 5 p 295	Carboniferous Period
171	CM M/A 2911 p 10	Ellesmere Is. trees
172	Maine.gov July 10, 2008	hurricane 4 ft of rain
173	CSRQ v 26 p 137	noctilucent after Krakatoa
174	Ashrae p 27.2	9% ultraviolet
175	En Br v 15 p 519	Earth albedo 0.44
176	Weather p 32	Hadley & Ferrel cells
177	F & W v 3 p 519	warm stratosphere
178	CRSQ v 15 p 31	fossil dragonflies
179	HIW Issue 21 p 051	larger insects impossible
180	By Design p 76	large flocculi in teradactyl
181	Wiki Pterosaur	large wings on teradactyl
182	In the Minds p 319	airborne capability of pteradactyl
183	By Design p 78	lift improved by 30%
184	Modern p 934	neutrons slow nuclear reactions
185	CRSQ v 29 p 170	carbon14 production
186	Cosmos p 294	Goldilock's orbit
187	s8int.com, Out of place artefacts	artefacts in coal
188	CRSQ v 14 p 101	coal has C14 count

Number	Source	Topic
189	CRSQ v 7 p 35	wrong C14 count in coal
190	CRSQ v 29 p 189	trees on Axel Heiberg
191	Nat Post Dec 18,1998	warm Arctic
192	Can Geo Nov/Dec 1999 p 43	trees in Arctic
193	E in U p 19	forest on New Siberian Islands
194	E in U p 16	Elephants & Rhinos in Siberia
195	An Ice Age p 28	palm trees in Alaska
196	CRSQ v 28 p 31	Antarctic coal
197	E in U p 52	coal in Antarctica
198	CRSQ v 32 p 47	Spitzbergen coal and dinosaur prints
199	Ice Age p 45	higher CO2 during warm Earth
200	The Sea p 82	Submarine Canyons
201	En Br v 7 p 391	size of dinosaurs
202	Wiki, Baluchitherium	18 ft high
203	CRSQ v 26 p 137	Megatherium
204	E in U p 72	giant swine
205	Short p 347	giant camels
206	Wiki, Pterosaur	35' wingspan
207	Nat Geo news June 25, 2007	giant penquins, warm climate
208	Tor Star Sept 16, 1980 A 16	Teratorn
209	Det. 3:11 I Sam. 17:4	giant man
210	In The Minds p 319	giant snakes
211	CSRQ v 10 p 40	Hypothalamus Dis. Theory
212	CSRQ v 17 p 108	higher atmos. Pressure ok
213	CSRQ v 15 p 3	higher pressure speeds healing
214	CSRQ v 28 p 60	healing rate 25% greater
215	Short p 338	higher O2 in ancient shells
216	CSRQ v 29 p 190	Axel Heiberg trees
217	Rough p 169	Axel Heiberg trees broken off
218	Can Geo N/D 1999 p 45	forest up north
219	CSRQ v 26 p 137	Noctilucent Clouds
220	CSRQ v 26 p 15	clouds 50 miles up
221	CSRQ v 26 p 137	Krakatoa 1883
222	Splash p 103	moisture trapped up high
223	College p 489	looming
224	CSRQ v 17 p 65	too dark up north
225	CSRQ v 17 p 67	watery heaven
226	CSRQ v 17 p 65	tin – metal of heaven

Number	Source	Topic
227	W in C p 89	sky collapsed
228	CSRQ v 24 p 133	Bible – expanse above firmament
229	E in U p 94	Lake Triton in Sahara
230	E in U p 97	Arabia became barren
231	E in U p 98	Civilization in Gobi Desert
232	Weather p 32	Hadley cells
233	Climate p 52	Hadley cells – hot air to surface
234	E in U p 97	warm climate – paradise on Earth
235	Violent p 189	1816 famine
236	CSRQ v 15 p 32	no ozone with vapour canopy
237	The Flood p 375	warm atmosphere – more water vapour
238	CSRQ v 28 p 126	Atmos. Expansion with canopy collapse
239	Sc Am April 1977 p 60	14" of rain in 1 ½ hours
240	Silent p 128	Faroe Islands – narrow temp range
241	An Ice Age p 27	limited moisture in atmosphere
242	An Ice Age p 71-73	slow drop in ocean temp
243	Westjet Magazine June 2015 p 54/57	Whitehorse illumination
244	CRSQ v 18, p 86	gas cloud expansion
245	CESQ v 7, p 11	gas cloud expansion
246	Design p 16, CRSQ v 38, p 43	Jean's length
247	Nat Geo Atlas of the World 7th ed. P 5	ice age over 2 million years
248	En Br v 10 p 379	lower sea level during ice age
249	CRSQ v 7 p 13	> 60 ice age theories
250	Cosmic Pursuit spring 1999	ocean depth 2.5 miles land elevation 400 ft
251	CRSQ v 8 p 24	dying magnetic field
252	Time p 25	dead magnetic field by 3991
253	CSRQ v 28 p 29	boulder in coal
254	CSRQ v 20 p 216	type of trees in coal
255	CSRQ v 20 p 215	12 ft of peat = one foot of coal
256	E in U p 51	coal seam 30' deep
257	En Br v 4 p 24	Carboniferous period 60 million years
258	CSRQ v 12 p 24	no meteorites found in coal
259	Can Geo M/J 95 p 32	meteors continue to hit the Earth
260	CSRQ v 7 p 56	C14 count in coal
261	CSRQ v 14 p 101	coal formed recently
262	CSRQ v 14 p 103	coal formed recently
263	In the Minds p 433	polonium after radon222
264	CSRQ v 14 p 105	instant formation of rocks

Number	Source	Topic
265	CM M/J 2007 p 10	mangrove trees in coal
266	E in U p 202	sea shells and rock layers in coal
267	E in U p 204	numerous layers with sea shells
268	Pensee fall 72 p 20	split coal seams
269	CSRQ v 20 p 215	50x carbon in coal as biosphere
270	Notes p 201	atmospheric CO2 > 380 ppm
271	CSRQ v 29 p 173	½ life of carbon14
272	CSRQ v 29 p 173	production rate of carbon14
273	CSRQ v 5 p 78	decay rate of C14 < production rate
274	In the Minds p 318	decay rate of C14 < production rate
275	CSRQ v 7 p 62	coal C14 count about 50.000 yrs
276	CSRQ v 20 p 171	Libby knew rates were different
277	Wiki Caloris Basin	Caloris impact time
278	Wiki, Impact craters	asteroid impact list
279	Critique p 39	recently molten rock 22 million yrs old
280	Nat Geo Atlas of the World p 6	Europe separating from North America
281	Nat Geo June 1973 p 13	Europe from NA 2" per year
282	www.sciencenews.org v 123 J8,1983 p 20,21	no widening of Atlantic
283	E in U p 118,119	no wandering continents
284	CRSQ v 29, p 13	plate tectonics needs magnetic data
285	News p 109	moving ocean floor had to stop
286	Int Geo Review v 10, p 275	magnetic data unreliable
287	Oil & Gas Journal v 84 p 115	magnetic directions confusing
288	E in U p 130	magnetism too strong
289	CRSQ v 26 p 133	magnetic reversals happen
290	CRSQ v 26 p 134	ocean crust 3 km thick
291	CRSQ v 37 p 144	ocean trenches rock bare
292	CRSQ v 38 p 93	slumping back
293	Nat Geo Atlas of the World p 6	spreading rate very small
294	CRSQ v 38 p 97	hotspots
295	files.usgwarchives.net	plates move over hotspots
296	www.soest.hawii.edu	plates move over hotspots
297	CM J/A 2002 p 1	plate tectonics invalid
298	CRSQ v 29 p 17	sulphide problem
299	On Pet v 9 p 264	magnetic stripes not valid
300	CRSQ v 9 p 49	lightning can magnetize rocks
301	CRSQ v 21 p 174	magnetostriction
302	CRSQ v 9 p 50	magnetostriction happened

Number	Source	Topic
303	Climate p 19	Greenhouse Effect components
304	Weather, a Visual Guide p 177	corpuscular rays
305	CSRQ v 7 p 11	gravity of space dust too low to form sun
306	Climate p 150	Earth heating up
307	Nat Geo Dec 2009 p 80	photo of an exo-planet
308	Nat Geo Oct 2003 p 91	mass extinction included dinosaurs
309	CRSQ v 44 p 68	Burkle Crater
310	CM N/D 2007 p 7	Igneous Provinces
311	www,igsb.uiowa.edc	Manson Crater
312	Wiki, Manson Crater	Manson Crater
313	Cosmos p 184	Chicxulub mass extinction
314	The End p 83	mass extinction
315	CRSQ v 21 p 82	single impact devastate life in hemisphere
316	The End p 83	75% to 96% death
317	Nat Geo Oct 2003 p 91	mass extinction killed dinosaurs
318	Critique p 48	mag. field indicates 20,000 yrs max
319	Critique p 46	vapour canopy meant less C14
320	Nat Geo Dec 2009 p 93	many extinctions, millions of species
321	www.icr.org	continental erosion
322	F & W v 3 p 42	water vapour 190 ppm to 42,000 ppm
323	Design p 113	gravitational collapse, 20,000,000 yrs.
324	Design p 121	Sun pulsating and shrinking
325	Wiki Gliese 581C	CO2 & water freeze on cold side
326	Cosmos p 190	nebular theory dead
327	evolutionnews.org	Gliese 581d & 581g nonexistent
328	Design & Origins p 123	turning on a star is difficult
329	Design & Origins p 123	1 million K not enough
330	CM M/A 2011 p 10	no life around Red Dwarfs
331	In the Minds p 318	C14 production 30% more than decay
332	Cosmic Pursuit spring 1985	0.83 to 1.2 solar masses
333	Wiki, Gliese 581c	ice one side, hot other side
334	Nat Geo July 2014 p 43	extra-solar planet size
335	HIW Issue 36 p 57	Russian borehole 180C
336	HIW Issue 39 p 54	Lunar temperatures
337	Nat Geo At p 19	Lunar temperatures
338	Kronos II p 30	Lunar temperatures
339	CRSQ vol 28 p 76	Eagle Lander pads 2.5 ft. Dia.
340	Sc Am June 16, 2005	Asteroids once liquid

Number	Source	Topic
341	Field Notes p 50	Greenland ice> 100,000 yrs.
342	The Lost Squadron.com	Greenland Ice
343	Field Notes p 187	Antarctic ice> 400,000 yrs.
344	WIKI, Mass Concentrations	Mascons on Mars
345	Tor Star Aug. 16, 1999 A3	First exo-planets
346	Field Notes p 52	Greenland glacier shrinking
347	HIW Issue 26 p 9	Apollo lunar landing site
348	Pensee I p 19	impact marks on molten surface
349	HIW issue 39 p 55	volcanic origin of maria
350	www.lpi.usra.edu/education	Tycho crater very young
351	http://adsabs.harvard.edu/full	asteroids from lava
352	Design & Order p 78	rapid tides = temp increase
353	Oceans p 105-106	Krakatoa waves
354	En Br Vol 15 p 780	lunar diameter
355	Maine.gov July 10, 2008	4 ft of rain from hurricane
356	CSRQ V 26 p 137	noctilucent clouds
357	CSRQ v 15 p 153	albedo of Earth
358	Weather p 32	Hadley Cell
359	CSRQ V 15 p 31	dragonflies 2' wingspan
360	By Design p 76	massive floculi
361	Wikipedia, pterosaur	large wings
362	In the minds p 319	airborne pterosaur
363	By Design p 78	70% increase in lift
364	F & W v 3 p 42	Atmos. Temp
365	Modern p 934	reaction slowdown
366	CSRQ v 29 p 172	½ life of C14
367	s8int.com out of place artefacts	out of place artefacts
368	CSRQ v 7 p 56	coal 50,000 yrs old
369	CSRQ v 29 p 171	nitrogen into C14
370	E in U p 19	trees up north
371	Nat Post Vol 1 No 46	Champsosaurs up north
372	E in U p 51	coal on Spitzbergen
373	An Ice Age p 28	palm trees in Alaska
374	An Ice Age p 27	slow drop in ocean temp
375	Silent Snow p 128	Faroe Islands
376	Maine.gov July 10, 2008	4' of rain
377	CRSQ v 28 p 126	cooling by expansion
378	CSRQ v 17 p 65	canopy lit up

Number	Source	Topic
379	The Flood p 375	vapour canopy collapse
380	CSRQ v 15 p 32	no ozone with canopy
381	Violent p 189	cold summer 1816
382	E in U p 97	paradise on Earth
383	Climate p 52	Hadley Cells
384	Weather p 32	30 d Hadley Cells
385	E in U p 98	deserts cultivated
386	E in U p 97	barren land resulted
387	E in U p 95	Lake Triton, Sahara
388	CSRQ v 24 p 133	expanse separates waters
389	W in C p 89	sky collapsed
390	CSRQ v 17 p 65	metal of heaven
391	CSRQ v 17 p 67	god Canopus
392	Splash p 103	comets into atmosphere
393	CRSQ v 26 p 137	Krakatoa
394	CSRQ v 26 p 15	Noctilucent Clouds
395	Rough p 169	London lit up at night
396	CSRQ v 29 p 190	dark Arctic
397	En Br v 7 p 391	dinosaur size
398	Wikipedia, Baluchitherium	Baluchitherium
399	Tor Star Sept 16, 1980	giant Teratorn
400	I Sam 17:4	giant man, Goliath 6 cubits & a span
401	Wikipedia, Pterosaur	Pterosaur
402	Nat Geo April 2003 p 18	Megatherium
403	E in U p 72	giant swine
404	Nat Geo News June 25, 2007	giant penquins
405	CSRQ v 19 p 40	hypothalimus
406	CSRQ v 17 p 108	At pressure
407	CSRQ v 15 p 31	positive effect on healing
408	CSRQ v 28 p 60	positive effect on healing
409	Short p 339	higher O2 in the past
410	CSRQ v 29 p 189, 190	Axel Heiberg trees
411	Apocalypse When p 10	Tunguska
412	Can Geo Nov/Dec 1999 p 43	Axel Heiberg trees
413	Can Geo Mar/Apr 1999 p 44	giant animals
414	The Greatest p 268	giant wombats
415	E in U p 205	Erratic Boulders in coal
416	www.biblebelievers.org	steel/wood hammer in coal

Number	Source	Topic
417	Climate p 194	Earth surface temp rising
418	EID	Sudbury Crater
419	E in U p 131	heat necessary for ice age
420	E in U p 130	oceans steamed
421	Climate p 2	ocean 2 ½ miles deep
422	En Br v 14 p 958	Martian day
423	Violent p 205	Katmai dust around the world
424	Exodus 40:36	darkness after exodus
425	An Ice Age p 37	Nuclear Winter
426	Violent p 189	volcanic dust
427	Violent p 188	Starvation in England
428	Violent p 204	Mount Katmai darkness
429	Violent p 207	Krakatoa
430	Violent p 206	Krakatoa darkness
431	E in U p 143	Bandai eruption
432	Violent p 198	Mount Pelee, Saint Pierre
433	The End p 82	cooling from Chicxulub
434	Cosmic p 3	darkness and cooling from Chicxulub
435	Can Geo M/J 95 p 37	Chicxulub extinction
436	The End p 83	75 to 96% extinction
437	CRSQ v 41 p 212	Chesapeake Bay
438	Cosmic p 5	Ries Crater, Germany
439	Can Geo S/O 99 p 25	1.5 km rock would kill millions
440	Nat Geo April 2013 p 44	Mammoth tusks from Russia
441	Can Geo M/A 99 p 40	land bridges
442	An Ice Age p 42	surface cooling from cloudiness
443	Short p 222	Mount St. Helens
444	Cosmos p 126	angular momentum problem
445	http://neo.jpl.nasa.gov/images	asteroid Ida and Dactyl
446	Can Geo s/o 1999 p 74	asteroid Ida
447	Cosmos p 216	Phobos
448	Cosmic p 5	tektite fields
449	Can Geo m/j 95 p 38	asteroid every 100 million years
450	Cosmic p 5	great dying
451	Nat Post Dec 18, 1998 p A2	champsosaurs up north
452	CRSQ v 11 p 213	temp drop since chalk
453	Comets p 3	impact cloud bobbed like cork
454	Sc Am April 1977 p 60	darkness before sunset

Number	Source	Topic
455	CRSQ v 41 p212	temp drop due to aerosols
456	Violent p 207	Krakatoa temp drop
457	Comets p 5	ice age triggered
458	Violent p 172	Buffalo snowstorm
459	CRSQ v 21 p 69	no ice age from Milankovich
460	Wikipedia Ice Age	orbital change not enough
461	Wikipedia, ice age	no ice age from Milankovitch
462	An Ice Age p 16	orbit change long time
463	An Ice Age p 116	ice age termination 10,000 years
464	E in U p 158	large water flow
465	CM S/O 1995 p 3	Greenland ice accretion
466	By Design p 149	Origin of Life prize
467	E in U p 128	heat needed for ice age
468	CRSQ v 11 p 215	ice ages a riddle
469	CRSQ v 11 p 216	ice ages a riddle
470	E in U p 131	heat and cold needed
471	An Ice Age p 69	volcanic activity during ice age
472	Short p 273	Antarctic ice = 200 ft of ocean
473	En Br v 10 p 379	sea level 300 ft lower
474	Wiki submarine canyons	sea level 125m lower
475	Nat Post Dec 18, 1998 A2	warm Earth
476	An Ice Age p 71	warm oceans
477	CM N/D 2007 p 7	Igneous Provinces
478	An Ice Age p 69	volcanism during ice age
479	Cosmic p 3	temp drop after impact
480	In The Minds p 97	coal on Antarctica
481	F & W v 5 p 294	CO_2 into water
482	En Br v 4 p 838	CO_2 into water
483	CRSQ v 21 p 70	albedo of snow
484	CRSQ v 21 p 73	cold over snow
485	En Br v 14 p 83	little uv on cloudy days
486	En Br v 10 p 375	ice cover during ice age
487	An Ice Age p 41	sea level higher
488	En Br World Atlas plate 21	land below 1000 ft
489	College p 207	vapour pressure
490	An Ice Age p 99	glacier thickness
491	By Design p 147	definition for Theory of Evolution
492	CM s/o 2005 p 2	ice cores 100,000 years

Number	Source	Topic
493	The Lost Squadron.com	planes in Greenland ice
494	Violent p 171	warm lake, cool land = snow
495	Beo p 150	3 ft snow one night
496	An Ice Age p 98	high volcanism = more ice
497	E in U p 158	Hudson River flow
498	An Ice Age p 109	ocean 10C at ice age max
499	CRSQ v 9 p 213	no ice northern Greenland
500	E in U p 50	no ice up north
501	An Ice Age p 83	mild temp during ice age
502	Weather p 250	warm ocean = more rain
503	An Ice Age p 204	present ocean temp 4C
504	CRSQ v 11 p 213	ocean cooled since chalk
505	Cosmos p 189	protosun formed & blew away gas
506	CRSQ v 7 p 9	upsetting physical laws
507	CRSQ v 18 p 86	collapse forbidden
508	CRSQ v 7 p 11	100K outward pressure too much
509	Rough p 31	Sun 99% matter, 1% angular momentum
510	Win C p68	satellites not in 2 directions
511	Design p 100	Neptune's Triton retrograde orbit
512	Nat Geo At p 5	Planetary rotation times
513	W in C p 8	no planetary globes formed
514	W in C p 5	different orbits non-circular
515	W in C p 7	satellites in different directions
516	www.NASA	doubles and triples
517	Cosmos p 136	old star & new star together
518	www.space.com	small spherical asteroid
519	CRSQ v 14 p 105	instant formation of rock
520	CRSQ v 20 p 212	coal reserves
521	www.cbc.ca	Southern ON. snowstorm 5' snow
522	Rough p 167	Asteroid Eros
523	Violent p 196	volcanic bombs 100 tons
524	By Design p 150	start of life too difficult to explain
525	Cosmos p 185	1 km asteroid no mass extinction
526	CM N/D 2007 p 7	Igneous Provinces
527	The Big Bang p 240	electromagnetic forces in galaxy
528	The Big Bang p 256	no black holes
529	CRSQ v 24 p 54	floating continents
530	The Big Bang p 7	Creation 'myth' similar to the Big Bang

Number	Source	Topic
531	F&W v 5 p 295	Carboniferous Period
532	Petrified Forest p 4	millions of years needed
533	CSRQ v 45 p 71	protoavis before dinosaurs
534	http://tycho/usno.navy.mil/leapsec	1 sec lost/500 days
535	www.scientificamerican.com	molten Vesta
536	asteroid name not found	splash mark on asteroid
537	W in C p 16	heavy metals in crust
538	Short p 342	99.99% extinct
539	Can Geo N/D 1999 p 45	Earth warm 15 to 20C
540	Wiki, axial tilt	axial tilt cycle 41,000 years
541	Nat Geo July 2013 p 45	The Nice Model
542	C M M/A 2011 p 10	for 2nd Earth need 2nd Sun
543	Tor Star Feb 23, 2017	7 planets around Red Dwarf
544	Cosmos p 144	remaining asteroid mass = 0.05% of Earth mass
545	Comets p 213	mass of remaining asteroids
546	Cosmic p 4	asteroid speed very high
547	HIW Issue 36 p 57	Kola Borehole
548	Kronos X 3 p 32	asteroid 90% of crater
549	The Greatest p 249, 256	Evolution very improbable
550	The Sea p 187	Bay of Fundy tides
551	By Design p 17	30 vol of En Br needed 3 or 4 times for DNA
552	Greatest p 81	millions of centuries needed for evolution
553	Rough p 49	day 6 hours long
554	CM N/D 2012 p 1	humans evolved from fish
555	Can Geo M/A 99 p 38	ice age animals
556	E in U p 13	Animals died after Ice Age
557	Nat Geo April 03 p 51	Ice Age animals vanished
558	Can Geo M/A 99 p 44	rich grass under snow
559	An Ice Age p 125	Ice Age mammals

Appendix 5 - Energy Table

Planet	Speed	Dis. To Sun	Orbit Length	Orbit Time	Mass	Total Energy
	X 10(3)	x 10(6)	x 10(8)		x 10(23)	x 10(30)
Earth1	29.6 m/s	148.64 km	9.336 km	365 ¼ days	60 kg	-2628.48 joules
Earth2	29.72 m/s	147.04 km	9.239 km	360 days	60 kg	-2651.26 joules
Venus	34.35 m/s	108.0 km	6.78 km	227 days		
Mars	24.8 m/s	228.0 km	14.32 km	687 days		
Jupiter	13.0 m/s	778.3 km	48.88 km	4,332 days		
Asteroid	18.0 m/s	550.0 km	34.5 km	2,225 days	0.312.0 kg	22.508 joules
Earth2 – Earth1 -Energy to raise Earth from 360 day/yr orbit to 365 1/4 day/yr orbit						22.48 joules
Mercury1	47.6 m/s	57.6 km	3.62 km	88 days	2.22 kg	-251.0 joules
Mercury2	53.5 m/s	45.6 km	2.86 km	62 days	2.22 kg	-317.7 joules
Mercury2 – Mercury1 Energy to raise Mercury from 62 day orbit to 88 day orbit						66.7 joules
Asteroid kinetic energy at Mercury's				62 day orbit	0.233 kg	66.7 joules

Appendix 6 - List of Illustrations and Photos

Number	Location	Name
1.	2.12.2	The Water Vapour Layer
2.	2.12.5	Atmospheric Pressure Profile
3.	2.12.8	Solar Energy without a Vapour Canopy
4.	2.12.8	Solar Energy with a Vapour Canopy
5.	2.12.8	Vapour Canopy Temperature Increase
6.	2.12.9	Atmospheric Temperature Profile
7.	3.1.1	The Angle of Inclination (Axial Tilt)
8.	3.1.2	Tidal Effect on the Moon
9.	3.3.8	Time Constant Curve and Half-Life Curve
10.	3.3.8	Leaky Bucket
11.	3.3.8	Carbon14 Buildup
12.	4.2.6	Okotoks Erratic Boulder
13.	4.2.6	Adirondack Erratic Boulder
14.	7.3.5	Vapour Capacity of Air
15.	7.6.1	Atmospheric Circulation
16.	7.6.4	Greenland Gripe2 Ice Core
17.	7.7.2	Genesis Human Age Data
18.	8.4.3	The Axial Tilt 23 ½ degrees

Appendix 7 - Index

www.ingramcontent.com/pod-product-compliance
Lightning Source LLC
Chambersburg PA
CBHW051333200326
41519CB00026B/7406